NEW DEVELOPMENTS IN FLUE GAS
DESULFURIZATION TECHNOLOGY

NEW DEVELOPMENTS IN FLUE GAS DESULFURIZATION TECHNOLOGY

Edited by M. Satriana

NOYES DATA CORPORATION
Park Ridge, New Jersey, U.S.A.
1981

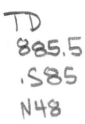

TD
885.5
.S85
N48

Published in the United States of America by
Noyes Data Corporation
Noyes Building, Park Ridge, New Jersey 07656

Library of Congress Cataloging in Publication Data
Main entry under title:

New developments in flue gas desulfurization
    technology.

    (Pollution technology review ; 82)
    Bibliography: p.
    Includes index.
    1. Flue gases--Desulphurization.
I. Satriana, M. J. II. Series.
TD885.5.S85N48   628.5'32   81-11045
ISBN 0-8155-0863-8   AACR2

FOREWORD

This book covers the latest developments in flue gas desulfurization (FGD) technology, both nationally and internationally. Advanced systems are surveyed, with emphasis placed on those processes which currently seem to offer the best prospects for efficient removal of sulfur oxides from flue gases. Conventional technology is also reviewed.

With the constantly changing state of availability of oil and natural gas, increasing consideration is being given to the use of coal-based energy systems. As coal utilization increases, the potential impact on air quality of SO_2 emissions from coal-fired units will become more significant. FGD is considered to be one of the most commercially developed means of continuous control of SO_2 emissions from coal-fired boilers.

The majority of commercial FGD systems are based on lime/limestone scrubbing. Adipic acid modifications, sodium carbonate, magnesium oxide, and double alkali systems are also used. Emerging systems include copper oxide adsorption, dry adsorption, dry alkali scrubbing and citrate processes. Stated differently, FGD processes may be said to fall into three broad categories—throwaway processes, gypsum processes, and regenerative processes—in which eventual products such as gypsum are disposed of entirely, or primary reactants are regenerated.

Included in the discussion of the various FGD methods are process descriptions, operating parameters, development status, problem areas, and case studies of operating systems. A chapter on comparative costs has also been included.

Because the information in this book is taken from multiple sources, it is possible that certain portions of this book may disagree or conflict with other parts of the book. These different points of view are included, however, in order to make the book more valuable to the reader. Cost figures provided are those given in the report cited, the date of which is given.

Advanced composition and production methods developed by Noyes Data are employed to bring these durably bound books to you in a minimum of time. Special techniques are used to close the gap between "manuscript" and "completed book." Industrial technology is progressing so rapidly that time-honored, conventional typesetting, binding and shipping methods are no longer suitable. We have bypassed the delays in the conventional book publishing cycle

1298430

and provide the user with an effective and convenient means of reviewing up-to-date information in depth.

The expanded table of contents is organized in such a way as to serve as a subject index and provides easy access to the information contained in this book which is based on studies produced by and for various governmental agencies. These sources are listed at the end of the book in the Sources Utilized.

Some of the illustrations in this book may be less clear than could be desired; however, they are reproduced from the best material available to us.

CONTENTS AND SUBJECT INDEX

INTRODUCTION

The information in the following three sections is based on *Flue Gas Desulfurization Pilot Study—Phase I—Survey of Major Installations—Summary of Survey Reports on Flue Gas Desulfurization Processes,* January 1979, NTIS PB-295 001, prepared by the Flue Gas Desulfurization Study Group of the NATO Committee on the Challenges of Modern Society (NATO-CCMS Study), F. Princiotta of the U.S. Environmental Protection Agency, Chairman.

BACKGROUND

The first commercial application of flue gas desulfurization (FGD) to power plant sulfur oxide control was in the United Kingdom in the early 1930s. The Battersea A Power Plant (228 MW) of the London Power Company, London, UK, began flue-gas washing in 1933. The process utilized wet scrubbing with Thames River water providing most of the alkaline absorbent. The spent absorbent was discharged back into the Thames after settling and oxidation. The FGD system operated successfully at up to 95% SO_2 removal efficiency until the Battersea A Power Plant closed down in 1975. A similar FGD system operated on the Battersea B Power Plant (245 MW) between 1949 and 1969, when FGD operation was temporarily suspended because of adverse effects on the Thames water quality.

The ICI Howden process, also developed in England, was developed to avoid discharging scrubber effluent into the Thames. A solid sludge was produced and barged to sea for dumping. This process was applied to the Swansea Power Plant in 1935 and the Fulham Power Plant in 1937. These systems operated successfully until early World War II when they were shut down.

The next FGD unit was installed at the Electrolytic Zinc Company in Tasmania in 1949. Tidal water was used there as the absorbent for SO_2 from smelter gas.

1

In 1952 the first unit of the new oil-fired Bankside Power Station in London, UK, was commissioned. This FGD system is an improved version of the Battersea system, using water from the Thames. This system is still operating at up to 98% removal efficiency and with a present capacity of 240 MW.

The 1950s and 1960s were a time of laboratory and pilot plant investigations of new processes. During the 1950s the Tennessee Valley Authority (USA) investigated lime/limestone systems, both dry and wet, and dilute acid processes; in Germany the first major carbon adsorption processes were developed. During the 1960s the magnesium oxide, copper oxide, and sulfite scrubbing processes were investigated among others.

Lime/limestone processes were installed in 1964 on an iron ore sintering plant in Russia and on a large sulfuric acid plant in Japan in 1966.

In 1966 Combustion Engineering developed a dry limestone injection process, which was installed at five boilers in the United States by 1972. Because of major problems associated with dry limestone injection including plugging (especially of the boiler tubes), low sulfur dioxide removal, and reduced particulate collection in the electrostatic precipitators, these systems proved inadequate. The five installations are now either closed down or converted to other control systems.

Japan has at present the largest installed capacity of FGD systems. Most of these systems were built between 1975 and 1977. Systems based on lime/limestone predominate. As of December 1977, the United States had 27 operating utility units treating about 35×10^6 Nm^3/h (normal conditions are 0°C and 1 bar), 34 units under construction to treat 50×10^6 Nm^3/h, and 20 units planned to treat 34×10^6 Nm^3/h; in addition 25 industrial boilers were operational or under construction with a capacity of about 5×10^6 Nm^3/h. In the Federal Republic of Germany, approximately 775,000 Nm^3/h of flue gas are being treated by lime and carbon adsorption processes, and in Norway 155,000 Nm^3/h are treated by seawater scrubbers on oil-fired boilers.

PROCESS CATEGORIES

FGD processes can most conveniently and usefully be categorized by the manner in which the sulfur compounds removed from the flue gases are eventually produced for disposal. In this way three main categories result:

(1) Throwaway processes, in which the eventual product is disposed of entirely as waste. Disposal can include landfill, ponding, discharge to water course or ocean, or discharge to a worked-out mine.

The processes in this category involve wet scrubbing of the flue gases for absorption, followed by various methods for neutralizing the acidity, separating the sulfur compounds from the scrubbing liquor, and usually recycling at least part of the scrubbing liquor.

(2) Gypsum processes, which are designed to produce gypsum of sufficient quality either for use as an alternative to natural gypsum or as a well-defined waste product with good disposal characteristics.

As with the throwaway processes, this category involves wet scrubbing for absorption followed by various methods of neutralizing the lime or limestone and recovering the sulfur compound. An oxidation step is included to insure recovery of the sulfur compounds in the form of gypsum.

(3)   Regenerative processes, which are designed specifically to regenerate the primary reactants and concentrate the sulfur dioxide that has been removed from the flue gases. Further chemical processing can then convert the concentrated SO_2 into sulfuric acid or elemental sulfur, or physical processing into liquefied sulfur dioxide. The surveyed processes in this catagory contain both wet scrubbing and dry adsorption processes.

Tables 1.1, 1.2, and 1.3 present a summarized description of processes in each of these 3 categories.

STATUS OF OPERATING FGD SYSTEMS

There are now 144 known FGD systems operating on fossil-fueled combustion sources. Of these, 74 are operating on power utility boilers, representing about 70×10^6 Nm^3/h of flue gas capacity from about 20 GW generation. The remainder are operating on industrial combustion sources, principally boiler plants, but also on iron ore sinter plants and petroleum refinery plants.

In the NATO-CCMS Process Status Reports, 35 FGD systems were surveyed. Although it was originally desired to include only the larger, commercially available installations with adequate operating experience, it was necessary to include data on some smaller scale operations and some that had been installed originally for demonstration purposes in order to provide sufficient comparability between the various processes.

Table 1.4 summarizes basic technical data on the FGD systems surveyed in the NATO-CCMS study and presents information on fuel burned and system operation.

Throwaway Processes: This category includes the three processes that produce a calcium sulfite/sulfate sludge and also the seawater scrubbing process. The sludge processes are becoming the most widely used FGD systems; 30 systems are in operation and 53 are planned or under construction. They have been used successfully on both coal- and oil-fired plants with a wide range of fuel sulfur contents and are reported to have high SO_2 removal efficiencies, whereas plant availabilities are variable.

The double alkali process is a further development designed to overcome the scaling, plugging, and erosion problems that have been generally associated with lime or limestone systems. Double alkali systems are presently used mainly with small to medium sized boilers where the extra process equipment can be offset by lower maintenance costs.

Gypsum Processes: These processes are designed to produce a quality of gypsum that may be used in place of natural gypsum in such markets as plaster or plaster wallboard, or as a setting retarder in cement manufacture. It is expected that if sufficient markets are not available for gypsum, it will be disposed of as a solid waste. Gypsum has better settling characteristics than sludges containing calcium sulfite from the throwaway processes.

Table 1.1: Summary Description of FGD Processes—Throwaway Category

FGD Process (year of first use)	Primary Reagent	Product for Disposal	Operating Principles	Present Status
Limestone/sludge (1933)	$CaCO_3$	$CaSO_3/CaSO_4$ wet sludge	Flue gases scrubbed with limestone slurry. SO_2 reacts to form insoluble calcium compounds, which are separated and partly dewatered for disposal.	Commercial operation (coal and oil), principally in U.S.
Lime/sludge (1972)	CaO	$CaSO_3/CaSO_4$ wet sludge	Flue gases scrubbed with slurry of hydrated lime. SO_2 reacts to form insoluble calcium compounds, which are separated and partly dewatered for disposal.	Commercial operation (coal and oil), principally in U.S.
Double alkali/sludge (1974)	NaOH or Na_2CO_3 and CaO or $CaCO_3$	$CaSO_3/CaSO_4$ wet sludge	Flue gases scrubbed with alkaline sodium based compounds in solution. Used solution regenerated with lime or limestone in separate reactor to form insoluble calcium compounds, which are separated and partly dewatered for disposal.	Recent commercial operation on industrial sized coal-fired boilers in U.S.
Seawater scrubbing (1975)	Seawater	Dilute sulfate solution	Flue gases scrubbed with seawater on a once-through system, diluted, oxidized and neutralized, and discharged back to sea.	Commercial operation on small oil-fired boilers in Norway.

Source: PB-295 001

Table 1.2: Summary Description of FGD Processes—Gypsum Category

FGD Process (year of first use)	Primary Reagent	Product for Disposal or Sale	Operating Principles	Present Status
Limestone/gypsum (1973)	$CaCO_3$	Gypsum	Flue gases scrubbed with limestone slurry. Resulting calcium-sulfur compounds are air-oxidized to gypsum, separated and dewatered.	Commercial operation (coal and oil), principally in Japan.
Lime/gypsum (1964)	CaO	Gypsum	Flue gases scrubbed with slurry of hydrated lime. Resulting calcium-sulfur compounds are air-oxidized to gypsum, separated and dewatered.	Commercial operation (coal and oil), principally in Japan; also recently in West Germany.
Double alkali/gypsum (1973)	$NaOH$ or Na_2CO_3 and CaO or $CaCO_3$	Gypsum	Flue gases scrubbed with alkaline sodium compounds in solution. Used solution regenerated with lime or limestone in separate reactor and the insoluble calcium-sulfur compounds are air-oxidized to gypsum, separated and dewatered.	Commercial operation on oil-fired boilers in Japan.
Dilute sulfuric acid/ gypsum (Chiyoda) (1972)	H_2SO_4 and $CaCO_3$	Gypsum	Flue gases are scrubbed with dilute sulfuric acid containing ferric ion catalyst. Air oxidation produces more sulfuric acid, which is bled off and neutralized with limestone. The gypsum is separated and dewatered.	Commercial operation on oil-fired boilers in Japan.

Source: PB-295 001

Table 1.3: Summary Description of FGD Processes—Regenerative Category

FGD Process (year of first use)	Primary Reagent	Primary Product	Operating Principles	Present Status
Sodium sulfite (Wellman-Lord) (1971)	Na_2CO_3 NaOH	SO_2	Flue gases are scrubbed with sodium sulfite solution and the resulting sodium bisulfite is thermally regenerated to sodium sulfite and SO_2. The SO_2 is then further treated to yield marketable products such as sulfuric acid or sulfur.	Commercial operation on oil-fired boilers in Japan. Recent commercial application on coal-fired boilers in U.S. Also on chemical/oil refinery plants.
Magnesium oxide (1971)	MgO	SO_2	Flue gases scrubbed with slurry of hydrated magnesium oxide. Insoluble magnesium sulfite is separated and calcined for recycle of MgO and generation of SO_2 for further treatment.	Commercial operation on oil-fired boilers in Japan. Demonstration plant on coal-fired boiler in U.S.
Carbon adsorption (1974)	Carbon	SO_2	SO_2 from flue gases is adsorbed on activated carbon, which is then regenerated thermally, releasing SO_2 for recovery.	Demonstration
Copper oxide (Shell) (1973)	CuO	SO_2	SO_2 from flue gases is adsorbed on copper oxide coated alumina. This is then treated chemically to regenerate the adsorbent and produce SO_2 for recovery.	Demonstration

Source: PB-295 001

Table 1.4: Summarized Data on FGD Systems Installed on Boiler Plants

FGD Process	Number of Systems Surveyed	MWe Range[a]	Predominant Fuel Type	Predominant Fuel Sulfur, %	Reported SO$_2$ Removal Efficiency Range[b], %	Reported Plant Availability Range[b], %	Process Capacity Range, Nm^3/s[c]
Limestone/sludge	3	115-820	Coal	0.5-5.4	57-92	88-94	82-108
Lime/sludge	6	65-835	Coal	0.8-4.7	70-92	83-100	28-139
Double alkali/sludge	2	20-32	Coal	0.9-5.0	90-99	80	25-35
Seawater scrubbing	1	(50)[d]	Oil	3.0	90-97[e]	96-99	17
Limestone/gypsum	6	250-850	Coal	0.6-2.2	90-96	97-100	103-250
Lime/gypsum	2[f]	37-750	g	2.9	80-92	96-98	35-333
Double alkali/gypsum	4	26-450	Oil	2.0-2.9	90-95	95-98	23-355
Dilute sulfuric acid/gypsum	2[f]	20-350	g	3.6[h]	96	100	270
Sodium sulfite	2	115-220	g	2.5-3.5	90-92	97-99	99-172
Magnesium oxide	4	95-155	g	2.0-3.8	90-95	i	102
Carbon adsorption	2	20-49	Coal	0.9-5.0	70-80	NA	10-42
Copper oxide	1	36	Oil	2.5	80-90	NA	30

[a] Range of MWe units served by FGD system
[b] Whole numbers
[c] Range of individual scrubber (or equivalent absorber) capacity
[d] Equivalent MWe
[e] Efficiency dependent on bypass (reheat) flow
[f] Includes one coal-fired demonstration unit
[g] Equal number of coal and oil firing systems were surveyed
[h] Calculated value
[i] Insufficient data base
NA: Not available

Source: PB-295 001

By late 1977, 92 systems were operating, of which 91 were in Japan and 1 in West Germany. They have been used successfully on both coal- and oil-fired plants and are reported to have high SO_2 removal efficiencies and high availabilities.

The gypsum processes have been applied particularly in Japan. Production of FGD-gypsum and phosphate fertilizer-gypsum has led to an annual oversupply of about 2×10^6 t/y. This gypsum oversupply and the reportedly lower cost of low sulfur fuel (desulfurized fuel oil and low sulfur crudes) are mentioned as reasons why no new gypsum producing FGD plants will be built on oil-fired boilers.

As with the throwaway process, the double alkali and the dilute sulfuric acid processes were developed to overcome some of the operating problems originally present with the lime or limestone scrubbing processes.

Regenerative Processes: Interest in the two surveyed regenerative processes that use wet scrubbing of the flue gases, i.e., the sodium sulfite and magnesium oxide processes, is largely due to the regeneration of the absorbent and production of a saleable sulfur product. No additional land area for solid waste disposal is required (beyond that for ash disposal), and raw material consumption is greatly reduced.

These processes have been successfully operated on oil-fired plants and are now being applied to coal-fired boilers. High SO_2 removal efficiencies are reported. With one of the processes (magnesium oxide), there is insufficient data to report on availability, but high availability is reported for the two Wellman-Lord systems surveyed for short periods.

The dry adsorption regenerative processes (carbon adsorption and copper oxide) are being investigated because of their potential for recovery of a sulfur by-product without the disadvantages of wet scrubbing systems, and with less energy requirement for flue gas reheat. These processes have not yet been applied on a commercial scale.

All of the regenerative processes require additional processing plants for converting the concentrated SO_2 into saleable form, e.g., by purification and liquefaction, by conversion into sulfuric acid, or by chemical reduction to elemental sulfur. The processes may be more suitable for operation near industrial areas where the sulfur products may be marketed. The wet scrubbing processes also offer the possibility of utilizing a single, centralized regeneration facility to service a number of FGD systems.

APPLICATION AND PERFORMANCE OF FGD TECHNOLOGY IN THE UNITED STATES

The information in this section is based on "Status of Flue Gas Desulfurization in the United States," prepared by B.A. Laseke and T.W. Devitt of PEDCo Environmental, Inc., in *Proceedings: Symposium on Flue Gas Desulfurization, Las Vegas, Nevada, March 1979,* Vol. I, EPA-60017-79-167a, compiled by F.A. Ayer of Research Triangle Institute for the U.S. Environmental Protection Agency. These proceedings will be hereinafter referred to as Symposium—Vol 1.

Significant development of FGD technology in the United States dates from about the 1950s when bench-scale and limited pilot plant programs were initiated. Major pilot plant investigations first started in 1961, and between 1961 and 1978 more than 60 systems representing generating capacity of approximately 75 MW were investigated at the pilot plant level in the utility sector. Concurrent with later pilot plant investigations, prototype, demonstration, and full-scale systems were installed. The first commercial application of an FGD system on a utility boiler occurred in 1968. Since then, 62 systems representing a generating capacity of approximately 17,600 MW have been operated at the prototype, demonstration, and full-scale levels. As of the end of November 1978, 46 systems representing a generating capacity of 16,054 MW were in service, and another 98 systems representing 46,453 MW were under construction or planned.

Since the early investigations considerable progress has been made in the development of FGD technology, and FGD is now considered to be the most commercially developed means of continuous control of sulfur dioxide emissions from coal-fired boilers.

The evolution of FGD technology from the limited pilot plant level to the full-scale commercial level can be attributed to several general process design and application considerations. Although it is difficult to quantify the impact that many of these factors have had on the development of the technology, an attempt is made to identify some of the general contributing factors in the balance of this section.

Process Design Strategy: Several general tendencies are evident in recent FGD design strategies. Generally, system designs incorporate an increased degree of flexibility and reliability. Specifically, trends are toward the paring of modules and ancillary components and the designing of less interdependent systems (i.e., systems in which major unit operations are not strongly affected by upstream component performance).

System Applications: Many recent FGD facilities are installed on large base-loaded units designed to fire coal from a specific source. This generally results in a flue gas with more constant and stable characteristics, which can improve system reliability because the system does not have to respond to as dramatic a variation in flue gas flow rate and composition. In many of the original FGD applications, the systems were required to operate on widely varying loads (cycling and peak) and coal types (low sulfur western, high sulfur eastern, and blends); such situations often demanded response to conditions beyond their process control capability. As a result, variations in the reagent feed rate, loss of chemical control, and the incidence of chemical and mechanical problems caused numerous forced outages and lower dependabilities.

System Supplier and Architectural-Engineering Experience: Later FGD system designs have benefited from experience gained in the operation of first-generation systems. Building on this experience, system suppliers and designers are providing better process design configurations and materials of construction. That many suppliers now offer broader quarantees covering sulfur dioxide removal, particulate loading, mist loading, waste stream quality/quantity, power consumption, water consumption, reagent consumption, reheat energy consumption, and availability is indicative of this trend.

Utility Experience: The utilities have also been gaining valuable operating and design experience. Many utilities have conducted or participated in FGD pilot plant programs and are thus better prepared to operate demonstration and full-scale systems. Operation of their first demonstration or full-scale system has also led to improved design and operation of subsequent systems.

Regulatory Agency Attitudes: As FGD technology has evolved from a research, development, and demonstration effort to a means of continuous compliance with applicable regulations, local, state, and Federal regulatory agencies have changed their attitudes toward enforcement, compelling utility companies to improve the reliability of FGD systems.

Process Chemistry: Although scale and corrosion are still encountered and are sometimes still severe, general knowledge concerning scale formation and the occurrence of corrosion has greatly improved. As a result, systems are being designed and operated so that problems experienced by the earlier units will not be encountered.

CONVENTIONAL PROCESSES

The information in this section is based on two appendices to the NATO-CCMS study, *Flue Gas Desulfurization Pilot Study—Phase I—Survey of Major Installations: Appendix 95-A—Limestone/Sludge Flue Gas Desulfurization Process,* January 1979, NTIS PB-295 002, prepared by F. Princiotta of the U.S. Environmental Protection Agency and R.W. Gerstle and E. Schindler of PEDCo Environmental and *Appendix 95-B—Lime/Sludge Flue Gas Desulfurization Process,* January 1979, NTIS PB-295 003, prepared by N. Haug of Umweltbundesamt and G. Oelert and G. Weisser of Battelle-Institut e.V.

Limestone/Sludge Process

General Process Description: The principles of all limestone scrubbing systems are essentially the same. When the limestone-water slurry comes in contact with flue gas containing SO_2, the SO_2 is absorbed into the slurry and reacts with the limestone to form an insoluble sludge. The by-products include gypsum ($CaSO_4 \cdot 2H_2O$) and calcium sulfite hemihydrate ($CaSO_3 \cdot \frac{1}{2}H_2O$). These sludge by-products are generally disposed of in a pond. Figure 1.1 is an example of a flow diagram of a 500-MW, coal-fired boiler with a limestone/sludge FGD system.

Process Chemistry: The overall reactions that take place in the absorber are:

(1) $$SO_2 + CaCO_3 \rightarrow CaSO_3 + CO_2$$

(2) $$SO_3 + CaCO_3 \rightarrow CaSO_4 + CO_2$$

Many intermediate steps also take place, however. The calcium ion is formed during slurry preparation.

(3) $$CaCO_3(s) \rightarrow CaCO_3(aq)$$

(4) $$CaCO_3 + H_2O \rightarrow Ca^{++} + HCO_3^- + OH^-$$

The SO_3 anion forms at the flue gas-slurry interface in the absorber.

(5) $SO_2(g) \rightarrow SO_2(aq)$

(6) $SO_2(aq) + H_2O \rightarrow H_2SO_3 \rightarrow HSO_3^- + H^+$

(7) $HSO_3^- \rightarrow H^+ + SO_3^=$

The sulfite ion (equation 7) then combines with the calcium ion (equation 4) to form the precipitate calcium sulfite hemihydrate (equation 8).

(8) $Ca^{++} + SO_3^= + \frac{1}{2}H_2O \rightarrow CaSO_3 \cdot \frac{1}{2}H_2O(s)\downarrow$

Gypsum, an additional precipitate, is formed as follows:

(9) $SO_3^= + \frac{1}{2}O_2 \rightarrow SO_4^=$

(10) $Ca^{++} + SO_4^= + 2H_2O \rightarrow CaSO_4 \cdot 2H_2O(s)\downarrow$

As reactions (8) and (10) proceed, the calcium cation is depleted from solution and additional $CaCO_3$ dissolves to react with the sulfite ion. In a limestone-sludge system, by-products occur from both reactions (8) and (10).

According to the molecular weights of limestone and SO_2, the theoretical requirement is 1 mol of limestone per mol of SO_2 removed. If a 20% excess stoichiometric amount and 95% purity of limestone are assumed, actual limestone required is 1.97 kg/kg of SO_2.

Dry sludge generated in the limestone process consists of calcium sulfite hemihydrate, carbonates, fly ash, and gypsum. Unused limestone and limestone impurities also combine with the sludge. The exact proportions of calcium sulfite hemihydrate and gypsum depend on system design; but if equal proportions are assumed, the sludge generated is 2.76 kg/kg of SO_2. When a venturi scrubber removes particulate matter, the particulates thus removed are also combined with the sludge. The final sludge to be disposed of contains at least 20% free water.

Description of Equipment Components: *Primary Particulate Removal* — The venturi scrubber is the first unit in most limestone FGD systems. This unit scrubs particulate from the gas; however, some SO_2 removal also occurs. The venturi scrubber has an advantage over an electrostatic precipitator (ESP) for particulate removal because the venturi cools and humidifies the flue gases before they enter the absorber section. A flue-gas cooler and humidifier must be used in connection with an ESP to cool the flue gases, generally to 50°C, prior to absorption. Moreover, if particulate is removed before the absorber, corrosion problems are reduced. Booster fans are sometimes installed in series with the venturi to provide the power necessary to force the gas through the scrubber system.

SO_2 Absorber — The absorber is the primary SO_2 removal unit in the system. Each of the many available designs employs a different method to contact the flue gas with the slurry. The most common unit designs include fixed packing, mobile-bed packing (hollow or solid spheres), and horizontal or vertical spray towers. Although each unit performs differently identical parameters have the same general effect on performance. Because of its simplicity, however, the spray tower is gaining popularity.

The scrubber must be constructed of materials that resist corrosion, erosion, and scaling. Scrubber bodies are fabricated of stainless steel or mild steel lined with an acid-resistant coating such as fiber glass-reinforced polyester (FRP), rubber or glass flake. Scrubber internals are made of a variety of materials such as stainless steel, which has a tendency to pit; high nickel alloys, which are expensive; or FRP, which is fragile. No one material seems to stand above the others.

Figure 1.1: Coal-Fired Boiler and Limestone Sludge FGD System

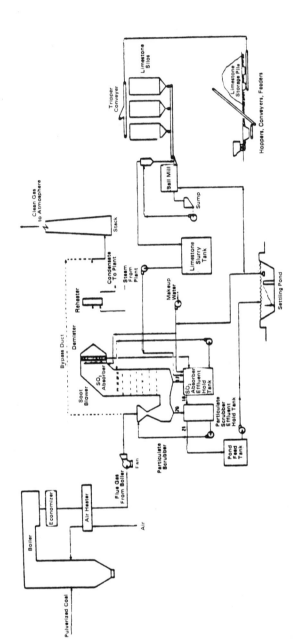

Based on Tennessee Valley Authority report, "Detailed Cost Estimate for Advanced Effluent Desulfurization Process."

Source: PB-295 002

The size and number of modules in a scrubber system are directly related to boiler size, turndown (reduction in boiler output) requirements, system availability, and gas-liquid distribution. Boiler system loads fluctuate, and the scrubber system must change to maintain optimum scrubber performance. One method of adjusting to turndown is to shut down scrubber modules as the load decreases. The more modules in the system, the smoother the transition. Scrubber modules not being used can be scheduled for cleaning and maintenance during periods of low system load, thereby reducing overall scrubber downtime. The use of multiple modules also has the advantage of permitting the modules to be smaller. Smaller cross-sectional areas in the scrubber module promote uniform gas-liquid distribution and improve efficiency. Scrubber module sizes range from about 25 to 200 MW.

Demister — A demister is necessary to remove entrained droplets from the scrubber outlet gas to reduce downstream equipment corrosion and scaling and to reduce reheat requirements. Most of the droplets are large enough to be removed with a simple change in flue gas direction; this is provided by baffles. Two banks of demisters are usually sufficient, but more can be added for additional demisting capability. Demisters are also subject to scaling and corrosion. Soot blowers or spray washes are installed to reduce this tendency, and materials of construction must be carefully selected.

Chevron and cyclonic-vane-type demisters are both effective. Although either can be operated in a horizontal or vertical position, the horizontal position predominates in the United States. There is much discussion pertaining to the advantages of each position, and a final selection depends on site-specific considerations.

Reheater — Reheating of stack gas is generally necessary to increase its bouyancy and reduce downstream corrosion. Thus, reheat not only helps meet ambient air standards; it also protects downstream equipment and prevents formation of acid mist. Reheating can be accomplished by installing a gas or low sulfur oil burner that exhausts directly into the stack, installing steam coils to heat the flue gas, injecting preheated air into the stack, or by-passing some hot flue gas around the FGD system directly into the stack (increasing emissions of SO_2). In-line heat exchangers are the most popular because of their low initial capital cost, but they tend to corrode and scale. Soot blowers, better demisters, and better materials of construction reduce these problems.

Slurry Makeup — Limestone can be received in a crushed and milled state or can be crushed and milled on site. In the latter case, the limestone is ground (wet or dry) in a ball mill to a size not larger than 200-mesh and often finer than 325-mesh. Finer grinding reduces the amount of limestone that remains unreacted and would otherwise be disposed of in the sludge. Water is added until the solids content reaches 15 to 25%. The slurry is sent first to a feed tank, then to the absorber holding tank, where it is mixed with absorber effluent. The slurry from the absorber holding tank is pumped to the absorber, where it reacts with SO_2 in the flue gas and is then returned to the holding tank. Slurry from the absorber holding tank is pumped to the venturi holding tank and from there to the venturi to scrub out fly ash. The slurry containing the fly ash returns to the venturi holding tank, from which it is pumped to the sludge disposal area for final treatment. Other designs have only one holding tank for the venturi and absorber.

Sludge Disposal — Sludge disposal can require 200,000 m^2 at a small plant and as much as 4,000,000 m^2 at a large plant. Disposal practices are very site specific. A power plant in an arid location might pump the sludge into an unlined pond, allowing the water to evaporate or seep into the ground. In an area where surface runoff or leaching could be a problem, the sludge sometimes is dewatered before being pumped into a lined pond. The water is returned to the system or purged after treatment to reduce chloride ions in the slurry. Plants located in areas where land is relatively expensive dewater the sludge to at most 20% liquid level by using clarifiers, thickeners, and vacuum filters and then fix the sludge by adding a proprietary compound. Fixation is necessary to allow the sludge pond to be used as a construction site at a later date. A totally closed system where no water is purged into a nearby stream or river has been achieved at the Sherburne County Station, Becker, Minnesota. Closed-loop systems are more subject to chloride ion concentration buildup in the slurry, especially when coal containing chloride is burned.

Advantages and Disadvantages: Limestone/sludge systems are the least expensive of all flue gas desulfurization systems currently used in the United States. Limestone is cheap and generally locally available, as is the land necessary for sludge disposal. In other countries, limestone appears to be available in sufficient quantities to make it inexpensive, but space for sludge disposal is generally not as available as in the United States.

The process is well developed chemically, but mechanical problems are still encountered in certain facilities. These problems include: fan vibration; pump and pipe erosion; scale buildup in the scrubber, demister and reheat sections; potential pollution in open-water systems; and corrosion and erosion. Essentially all these problems have been solved at particular installations. The similar limestone/gypsum systems in Japan offer the best examples of well-designed and well-operated limestone systems. Rapid response to changing boiler rates and flue gas conditions is essential in all limestone systems, and only highly trained chemical technicians or well-designed automatic controls can make the changes fast enough. The Japanese have succeeded in developing these automatic process controls on their limestone/gypsum systems.

The system operates well on large boilers fueled with either coal or oil. On small systems with low operating factors, labor and capital charges can be a limiting factor. Strict solid waste and water regulations either in force or imminent could necessitate more careful consideration of sludge disposal approaches. It may be necessary to incorporate an oxidation step and/or employ sludge stabilization/fixation to produce acceptable materials for landfill disposal.

The advantages of the limestone/sludge systems are as follows:

(1) The basic process is fairly simple and has few process steps.

(2) Capital and operating costs are relatively low, and reserves of limestone are fairly abundant.

(3) SO_2 removal efficiencies can be as high as 95%.

(4) The two-stage treatment of flue gases permits removal of SO_2 and particulates.

(5) Many years of operating experience have led to a greater understanding of the basic principles of this process.

(6) The process has demonstrated successful performance, especially in the removal of SO_2 from coal-fired systems.

(7) Fly ash does not adversely affect the system.

The disadvantages/problems of the limestone/sludge systems are as follows:

(1)   Large quantities of waste must be disposed of in an acceptable manner.

(2)   If not designed carefully or operated attentively, limestone systems have a tendency toward chemical scaling, plugging, and erosion which can frequently halt its operation.

(3)   The scrubber requires high liquid-to-gas (L/G) ratios necessitating large pumps with attendant electrical requirements.

(4)   The sludge may have poor settling properties when it has high sulfite content. Forced oxidation or soluble Mg in the slurry have been shown to lower sulfite content.

Lime/Sludge Process

General Process Description: The principles of all lime scrubbing systems are essentially the same. When the lime-water slurry or solution comes in contact with flue gas containing SO_2, the SO_2 is absorbed into the slurry and reacts with the lime to form an insoluble sludge. The by-products include gypsum ($CaSO_4 \cdot 2H_2O$), calcium sulfite hemihydrate ($CaSO_3 \cdot \frac{1}{2}H_2O$), and calcium carbonate ($CaCO_3$). These sludge by-products are generally disposed of in a pond. The process is schematically shown in Figure 1.2.

Process Chemistry: *Chemical Reactions —*

(1) $$CaO + H_2O \rightarrow Ca(OH)_2$$

(2) $$Ca(OH)_2 + SO_2 \rightarrow CaSO_3 + H_2O$$

(3) $$CaSO_3 + SO_2 + H_2O \rightarrow Ca(HSO_3)_2$$

(4) $$Ca(HSO_3)_2 + Ca(OH)_2 \rightarrow 2CaSO_3 + 2H_2O$$

Sulfate formation (detrimental), scaling:

(5) $$2CaSO_3 + O_2 \rightarrow 2CaSO_4$$

Control Theory — Scrubbing liquor is a slurried mixture of calcium hydroxide and calcium sulfite in water. The pH of slurry entering the scrubber is 8 to 10; low pH can cause gypsum scaling; high pH can cause formation of carbonates. Reaction tank (if used) pH is 10 (addition of lime); addition of crystal seed causes homogeneous crystallization. Presence of MgO in the lime allows a subsaturated mode of operation and improves the SO_2-removal efficiency.

Reaction Kinetics — Reaction with SO_2 in the flue gas takes place in the liquid phase; dissolution of calcium sulfite is the rate controlling step for SO_2-absorption. In other cases the mass transfer through the interface between gas and liquid is the rate controlling step.

Stoichiometric Requirement — The equipment components are similar to those described for the limestone/sludge process.

Advantages and Disadvantages: Considering the application of lime/sludge flue gas desulfurization (FGD) systems, the space necessary for the sludge disposal is among others an important factor. Therefore, the lime/sludge (and the limestone/sludge) technology is economically advantageous in the U.S. due to the availability of inexpensive land, whereas in Europe and in Japan the availability of sufficient space is often prohibitive.

Generally inexpensive lime can be provided to the FGD plants and, as far as available, carbide sludge from chemical industry or alkaline fly ash can be utilized as scrubbing agent. The lime/sludge technology is well developed.

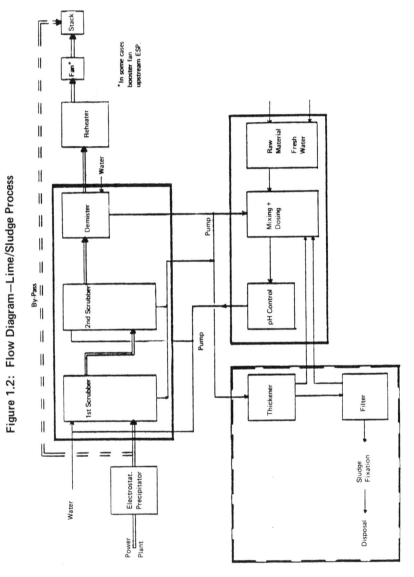

Figure 1.2: Flow Diagram—Lime/Sludge Process

Source: PB-295 003

Through operation of large scale plants in different countries considerable experience was gathered and reliability of the systems was proven.

Current R&D efforts aim at the following chemical, mechanical and design areas:

- Precipitation of calcium sulfate (gypsum) may cause scaling, which is particularly unwanted in mist eliminators.

- Dissolved salts in the scrubbing agent and chloride built-up in the recycle water can cause corrosion, which is possibly aggravated by the erosive nature of the slurry.

- Pumps, fans and agitators allow mechanical improvements as to their use in this technology.

- Interrelated mechanical and chemical factors may influence the lifetime of expansion joints and piping.

- Finally the optimization of the design parameters like gas flow and slurry distribution, liquid-to-gas ratio, control instrumentation and accessibility for maintenance has to be mentioned.

The advantages of the lime/sludge system are similar to those listed for the limestone/sludge process.

The disadvantages and problems are also similar to those listed for the limestone/sludge process. In addition, although fly ash does not adversely affect the process in general it can adversely affect the process by intensification of mechanical wear and erosion in the washing cycle and by increased load of the thickener.

The limestone/gypsum and lime/gypsum processes described in Table 1.2 may also be considered conventional processes.

RECENT ADVANCES
IN CONVENTIONAL TECHNOLOGY

SURVEY OF IMPROVEMENTS IN CONVENTIONAL TECHNOLOGY

The information in this section is based on "Status of Flue Gas Desulfurization in the United States" prepared by B.A. Laseke and T.W. Devitt of PEDCo Environmental, Inc. in Symposium–Vol. 1.

Conventional processes include all direct lime and limestone systems. Because these systems are the most widely applied, they are the ones with the most operating experience. Furthermore, they will have the greatest utilization in the very near future. For these reasons, lime and limestone systems have been subjected to extensive investigations, the results of which are summarized, along with general conclusions concerning process design.

Process Chemistry–Scaling, Sulfur Dioxide Removal, and Reagent Utilization

The use of reagents with low reactivity, such as limestone or lime, for sulfur dioxide removal from gas streams that can vary widely in flow and composition over short periods of time has resulted in several obvious and sometimes severe chemical limitations, including scale formation, low sulfur dioxide removal, and poor reagent utilization. A number of important findings regarding process chemistry and its effects on system performance have been made in recent years, and many of the process design and operating features that are now being incorporated into commercial systems are using these findings to improve performance. A brief summary of these features follows.

Two basic modes of lime/limestone FGD system operation will enable scale-free operation: coprecipitation and desupersaturation.

Coprecipitation involves the removal of calcium sulfate from the system as part of a calcium sulfite/sulfate solid solution. If a system is operated so that the maximum oxidation in the slurry circuit is about 16%, the scrubbing liquor remains subsaturated with respect to calcium sulfate (gypsum), and no hard scale occurs. If the degree of oxidation exceeds this level, more calcium sulfate is formed in the slurry circuit than can leave the system in a coprecipitated form.

This causes the system to operate supersaturated with respect to gypsum. If relative saturation levels approaching the critical value of 1.4 is reached, the formation of hard scale can occur within the system.

Desupersaturation involves the removal of calcium sulfate from the system through the use of calcium sulfate (gypsum) seed crystals, which provide nucleation sites for the precipitation of calcium sulfate as well as through the calcium sulfite/sulfate solid solution. The seed crystals control sulfate scaling in a closed-loop system operating in a supersaturated mode. Crystal growth occurs on the seed crystals, the sulfate is removed from the system as gypsum, and relative saturation levels are kept below the critical 1.4 value.

Another approach to improving the chemistry of lime/limestone slurry systems so as to reduce scaling, increase sulfur dioxide removal, and improve reagent utilization is the use of additives in the slurry.

Use of Magnesium Additives: Magnesium additives have proved to be effective. Increasing the magnesium ion concentration increases the liquid-phase alkalinity of the scrubbing slurry and increases the amount of sulfite and sulfate the scrubbing slurry can hold without exceeding solubility limits. The overall effect is a subsaturated operation that produces higher sulfur dioxide removal efficiencies and higher utilization. Experimental operating experience with magnesium additives has been accumulated at the Shawnee TVA/EPA Alkali Scrubbing Test Facility and Paddys Run of Louisville Gas and Electric. Experimental and full-scale operating experience has been obtained at Phillips and Elrama of Duquesne Light, Bruce Mansfield of Pennsylvania Power, and Conesville of Columbus and Southern Ohio Electric.

Use of Other Additives: The use of organic acids as additives is also under consideration. The use of carboxylic acids in lime/limestone scrubbing has been tested by the Tennessee Valley Authority (TVA) and TVA/EPA. This research has concentrated on the use of benzoic acid and, more recently, adipic acid at Shawnee. These acids are generally stronger than carbonic acid, but weaker than sulfurous acid. The addition of these acids has two effects: first, it aids mass transfer by buffering the pH of the liquid film at the gas/liquid interface; and second, because sulfurous acid is a stronger acid, the benzoate or adipate ion acts as a base in the sulfur dioxide absorption step. Thus, the addition of organic acids increases the total liquid phase alkalinity of the scrubbing liquor in much the same fashion as an increase in alkalinity because of magnesium ion. Intensive testing with adipic acid was recently performed at the TVA/EPA Shawnee alkali scrubbing test facility. In July 1978 initial test runs were performed without adipic acid to establish base lines for both lime and limestone scrubbing. These runs were followed by adipic acid testing, which continued throughout the balance of the year.

The preliminary results of this test program indicate agreement with initial expectations of higher sulfur dioxide removal efficiencies, and higher reagent utilizations. It has also been shown to be effective when used in conjunction with forced oxidation and when chlorides are present—conditions which adversely affect magnesium additives. One negative result, however, has been the unexpectedly high deterioration or decomposition of adipic acid that takes place in the scrubber. Actual feed rates of adipic acid were two to three times higher than could be accounted for in the system discharge sludge.

Design Changes: A number of process design innovations also have been developed to eliminate scaling, increase sulfur dioxide removal efficiency, and improve reagent utilization. Forced oxidation is one technique that has been successfully piloted and used in commercial installations. On a number of systems the use of forced oxidation has contributed to scale-free operation, to meeting or exceeding design sulfur dioxide removals, and to achieving high reagent utilization levels and, most importantly, in improving the quality of the waste sludge. This has been especially true for limestone systems treating flue gas with low sulfur dioxide loadings. For these applications, the low sulfur dioxide levels coupled with the long liquid retention times result in high "natural" oxidation levels. This makes forced oxidation a particularly attractive method to improve sludge quality and minimize scaling, increase sulfur dioxide removal and improve reagent utilization in these systems.

Operating Changes: Several operating parameters have an important impact on scaling, sulfur dioxide removal, and reagent utilization.

Control of pH — Excluding all other factors, differences in optimum operating pH can affect the performance of lime/limestone slurry systems in two ways: operation at low pH generally promotes the formation of hard calcium sulfate scale (gypsum), and operation at high pH generally promotes the formation of softer calcium sulfite scale. Operating experience indicates that optimum pH levels are generally maintained between 8.0 to 8.5 for lime and 5.5 to 6.0 for limestone.

Solids Level — If all other variables are held constant, increases in slurry solids levels will increase the amount of seed crystal area available for homogeneous crystallization. This is especially true for systems that control sulfate levels by desupersaturation. For these systems, an optimum amount of slurry seed crystals is maintained in the system by maintaining an optimum level of slurry solids. Thus, when the solids level drops, the seed crystal level drops correspondingly and causes the impairment or loss of homogeneous crystallization, the onset of heterogeneous crystallization, and subsequent scale development. Some of these systems aid desupersaturation by forcibly oxidizing all the sulfite to sulfate and then precipitate the calcium sulfate with the aid of gypsum seed crystals. Minor episodes of sulfate scaling have also occurred in these systems, in every case as a result of dilution of the slurry solids level after the mist eliminator wash water rate was increased to improve cleanliness. The increased levels of makeup water in the system decreased the slurry solids level and the seed crystal level, which impaired desupersaturation and resulted in scale formation. Reestablishment of slurry solids levels prevented further episodes of scaling.

Liquid-to-Gas Ratio (L/G) — If all other variables are held constant, increasing the L/G reduces the sulfur dioxide pickup per volume of scrubbing liquor. Thus, the relative saturation in the circulating slurry can be reduced by increasing L/G, assuming that desupersaturation takes place in the hold tank.

Process Chemistry—Corrosion

In simple terms, corrosion is the dissolving of metal surfaces. The incidence of corrosion in lime/limestone slurry FGD systems and design and operating measures taken to minimize or eliminate this problem are discussed briefly in the following paragraphs.

Two corrosive agents are present in the process: (1) sulfurous and sulfuric acids, and (2) chlorides. These two agents contribute to several specific types of corrosion: general corrosion, pitting, crevice corrosion, intergranular corrosion, stress-corrosion cracking, and erosion-corrosion.

A number of successful design, construction, fabrication, and operation measures have been developed to minimize the rate of corrosion or prevent it altogether. These measures are summarized briefly.

A selective process design approach developed by the major system suppliers allows highly corrosive environments to be isolated in discrete areas of the FGD system. This approach involves the separation of the scrubbing loop into separate multiple loops so that a different set of chemical conditions is maintained for quenching or prescrubbing, sulfur dioxide absorption, and mist eliminator washing (wash trays). In such designs, the quencher or prescrubber bears the first full brunt of the incoming hot flue gas. It encounters the total chloride content from the fuel fired without dilution and is the area in which low-pH, chloride-laden, return water is used freely in lime/limestone slurry systems. Isolation of such a corrosive environment to the quencher or prescrubber is advantageous because this area is small and discrete enough that it can be constructed of chloride-resistant materials without drastically increasing system cost. Alloys that have been tested and specified for full-scale systems are listed in ascending order of molybdenum content, pitting resistance, and cost: 317L stainless steel, Incoloy Alloy 825, Hastelloy G, Inconel Alloy 625, and Hastelloy C-276.

In many of the initial lime/limestone slurry systems, the incoming hot gas contacted the reactive absorbent suspension, which resulted in the accumulation of solids at the wet/dry interface. These deposits in some cases provided convenient sites for the accumulation of chloride at concentrations approaching 50,000 ppm. The result was severe episodes of pitting, stress corrosion, crevice corrosion, stress-corrosion cracking, and erosion-corrosion. This problem has been largely overcome by better control of process chemistry, use of self-cleaning devices, selective use of superior construction materials, and the use of multiple-loop designs. Better control of process chemistry eliminates the formation of scale in the system and thus prevents appearance of convenient chloride ion host sites. A number of systems are equipped with soot blowers in the approach ducts to the scrubber modules, which allows the inevitable buildup of solids at the wet/dry interface to be cleaned automatically and periodically.

Emission Control Strategy

In recent years the trend in design of particulate and sulfur dioxide emission control systems has been toward combined electrostatic precipitator (ESP)/FGD or fabric filter (FF)/FGD strategies over simultaneous or two-stage wet scrubbing strategies. This preference is due to the high reliability afforded by ESPs and FFs, which enables selective bypass of scrubber modules without reduction of load or shutdown of the unit. Other benefits include the following:

- The potential for corrosion at wet/dry interfaces and erosion-corrosion in the FGD systems is minimized.
- Exotic construction materials can be used more selectively and in smaller amounts.
- Balanced-draft and booster fans can precede rather than follow the FGD system.
- Sludge blending and stabilization processes that use dry fly ash as an additive are permitted.

Equipment Design Improvements

Specific design and operating improvements for FGD-related equipment are as follows:

In addition to the placement of balanced-draft or booster fans upstream of the FGD system, another development is the use of variable-pitch, axial flow fans. The main advantage of this design is its consistently higher efficiency (versus centrifugal fans) over the entire boiler operating range, which results in a substantial power savings. Other advantages are superior flow control, arrangement flexibility, easy access and maintenance, less severe construction requirements, and increased design reliability.

For scrubber modules, most suppliers now prefer 316 or 316L stainless steel because this material has demonstrated superior resistance to corrosion, erosion, and scale development compared with carbon steel, 304 stainless steel, and 304L stainless steel. This preference for 316 and 316L stainless steel is based primarily on the smooth mating surfaces and molybdenum content of these steels. The former attribute minimizes the presence of crevices that provide convenient sites for buildup of soluble chloride. The molybdenum content (2.50 to 2.75% minimum) of stainless steel increases corrosion resistance to localized attack such as pitting and crevice corrosion.

For mist eliminators, which have been very susceptible to corrosion, most suppliers now recommend the use of fiber glass-reinforced plastics, polypropylene, and corrugated plastics over stainless steels and other alloys because these materials are relatively lightweight, inexpensive, and do not corrode. Reheaters have been especially susceptible to corrosion, and the trend is toward indirect hot-air reheat because in-line reheat systems have been subject to corrosion and tube plugging.

Process Design

Several advances in the process design of lime/limestone slurry FGD systems have improved system dependability and sulfur dioxide removal. The following subsections describe these advances and some methods developed to reduce problems with major FGD equipment.

Dampers: Bypass and isolation dampers are used to regulate the flow of flue gas into and around the FGD system. The primary purpose of isolation is continuance of unit operation while the scrubber modules are under maintenance. Efficient and reliable dampers allow maintenance crews to service the modules in an efficient and timely manner. Common designs include slide-gate (guillotine), single-blade butterfly, and multiblade parallel (louver) dampers. Corrosion and erosion of the various types of dampers and damper seals have been common. In some cases, dampers have failed or been so inefficient that the modules could not be maintained during bypass situations. The current trend is toward two-stage louver dampers having a pressurized seal-air system that maintains a positive pressure between them. Pressurized seal-air increases the energy demand of the system because of increased fan power requirements, but it contributes significantly to successful damper operation.

Scrubbers: Several recent design innovations have increased dependability and removal efficiency of scrubbers. Cooling the gas to its adiabatic saturation temperature prior to contact with the scrubbing slurry increases sulfur dioxide removal capability and minimizes the potential for scaling and corrosion at the slurry/gas interface area. Presaturators or quenchers were not incorporated into the design of many of the initial FGD systems. In such systems, the incoming, hot, pollutant-laden flue gas contacted the suspended reactive absorbent and resulted in solids accumulation and subsequent corrosion. For this reason, presaturators or quenchers that use clear liquor or spent slurry for the absorbing stage (multiple slurry loops) are now used (or plans call for their use) in systems that include dry-phase particulate precollection.

The trend in the design of lime/limestone slurry FGD systems is away from venturis and packed beds to spray towers and combination towers. The venturi design was abandoned largely because the short liquid/gas contact time results in relatively low sulfur dioxide absorption. Scaling, plugging, and corrosion of internals occurred in some of the packed-bed designs (fixed and mobile) and tray towers. Spray towers, on the other hand, have few internal components in the gas/liquid contact zone and therefore offer the potential for greater dependability because there are fewer sites for deposition of solids in the form of scale, collected fly ash, and unused reagent. To date, spray tower operation has been very successful. Although high dependability and sulfur dioxide removal have been reported for almost all the FGD systems incorporating spray tower designs, several limitations also have been encountered. Mass transfer limitations tend to restrict spray tower design applications so that only low- and medium-sulfur coal can be used in conventional lime/limestone slurry systems. When high-sulfur coal is burned, spray towers in service and scheduled for operation use or plan to use special reagents (carbide lime, magnesium-promoted lime, and limestone) to compensate chemically for mass transfer limitations. In addition, the greater tendency for slurry carry-over in spray towers requires either increased tower height or special mist eliminator designs (wash trays, bulk entrainment separators), both of which increase capital and annual cost requirements.

These limitations have given rise to the development of combination towers. These towers combine the features of venturi, packed, tray, and spray towers into one module. Examples of combination tower designs now offered by some of the major system suppliers include spray/packed towers, venturi/spray towers, and tray/packed towers. These designs offer greater flexibility because extreme operating conditions can be segregated into discrete areas of the scrubber, allowing separate chemical and physical conditions to be maintained. This permits the use of the two-loop slurry concept, in which low pH liquor contacts the entering flue gas in an initial scrubbing loop, where some sulfur dioxide removal takes place. High pH liquor is contacted with the gas in the second scrubbing loop, where the bulk of the sulfur dioxide removal takes place. Spent slurry from this loop is discharged to the first loop where the unused reagent is consumed. Fresh makeup reagent is added only in the second loop. This type of design takes advantage of the concept of contacting the flue gas containing the highest sulfur dioxide concentration with the lowest liquor alkalinity and the highest liquor alkalinity with the lowest sulfur dioxide concentration. Performance has verified the potentially high removals and utilization afforded by such designs.

Reaction Tanks: Coinciding with gas-side staging is liquid-side staging, in which hold tanks are arranged in series to simulate plug flow reactor designs. (A plug flow design is one that allows the reacting liquor to flow through the reactor without backmixing. A plug flow situation can be approximated by arranging agitated tanks in series.) This design concept was originally piloted at IERL-RTP and further tested at Shawnee. A number of full-scale systems that incorporate liquid-side staging have resulted; all are low-sulfur, limestone-slurry systems. Sulfur dioxide removals and dependabilities greater than 90% have been reported.

Mist Elimination: Chevron and baffle-type mist eliminators continue to be the only designs used in the U.S. utility FGD systems. Several different designs have been tested (including wire-mesh, tube-tank, gull-wing, ESP, and radial vane), but the performance and economics associated with these and other design alternatives indicate that the exclusive use of chevron and baffle types will probably prevail. The popularity of these separators is due primarily to design simplicity and flexibility, adequate collection efficiency for medium to large size drops, relatively low pressure drop, open construction, easy access for maintenance, and relatively low cost.

Within these two preferred types of mist eliminators, a number of specific design, construction, and operation improvements have been implemented, with the following results:

(1) Chevron designs of continuous-vane construction now predominate over noncontinuous-vane construction because of their greater strength and lower cost.

(2) Multiple-stage designs predominate over single-stage designs on limestone systems. This tendency is process-sensitive in that limestone systems, which outnumber lime systems in the United States, generally require two or three stages for effective mist entrainment separation.

(3) Single-stage designs are successful for lime systems because of the superior reactivity of lime and the correspondingly higher utilization.

(4) The number of passes per stage also tends to be process-sensitive. Four-pass designs are generally used for lime and three-pass designs for limestone. More passes are required for lime systems because the single-stage design is used, whereas fewer passes are required for limestone systems because the multiple-stage design is used.

(5) Fiber glass-reinforced plastics, polypropylene, and corrugated plastics are now used in almost every operational system and specified for use in nearly every planned system. These materials are preferred because they are relatively lightweight, inexpensive, and superior in resistance to corrosion. Potential problems associated with high temperature excursions have been minimized by specifying materials that can withstand exposure up to 400°F.

(6) Vane spacings of 1.5 to 3.0 inches are generally used in single or first stages and 0.9 to 1.0 inch for second stages. Multiple staging permits the use of finer spacings, which provide increased mist-separation capability for smaller particles.

(7) The horizontal configuration (vertical gas flow) is still widely used because of its adequate performance (to date), its operational and design simplicity, and its lower capital cost. The vertical configuration offers a number of advantages over the horizontal configuration. For example, reentrainment due to the gas flow opposing the path of drainage is eliminated, and limitations on wash water quality and quantity (as well as wash direction) are eliminated. Two systems recently started operations using vertical configurations. Initial results indicate adequate operation performance and no major operating problems.

(8) Special features, such as hooks and pockets on the vanes, are desirable for prevention of reentrainment.

(9) Bulk separation devices, impingement plates, single baffle deflectors, and gas direction changes are becoming integral parts of mist eliminators because they increase removal efficiency and design flexibility.

(10) Wash and knock-out trays have been incorporated into a number of mist eliminators to conserve fresh water and increase (or extend) the quantity of water available for washing.

(11) Wash systems that use blended water consisting of pond return water or thickener overflow and fresh water are used over other strategies (total return or total makeup). Intermittent, high-pressure, high-velocity wash systems are preferred to continuous wash systems because they have less impact on water balance and chemistry.

(12) Optimum distances between stages are generally 4 to 5 ft, and freeboard distances are 4 to 5 ft. The former is the minimum distance permitting easy access for maintenance. The latter is the distance at which carry-over can be minimized without drastically increasing tower height and pressure drop.

(13) Superior overall operation is obtained when fly ash is collected prior to the scrubbing system. This is because scrubbing systems in which fly ash is not used usually have a low slurry solids content. The lower the slurry solids content, the less likely the tendency for mist eliminator fouling.

Reheaters: A pronounced preference for stack gas reheat versus no reheat (wet stack) is still evident for those systems in service and committed for future operation. The use of wet stacks at a number of systems has been abandoned in favor of reheat because of problems encountered with corrosion, plume dispersion, and plume visibility. Such problems are more pronounced where high-sulfur coal is burned because sulfur dioxide loadings are higher. A number of developments concerning reheater design and construction are noted:

(1) Six methods are available to increase the temperature of the gas from a scrubber prior to discharge to the stack: in-line reheat, direct combustion reheat, indirect hot air reheat, gas bypass reheat, exit gas recirculation reheat, and waste heat recovery reheat.

(2) Of the six methods now available only four are applied in commercial operations in the United States: in-line reheat, direct combustion reheat, indirect hot air reheat, and gas bypass reheat.

(3) Among the systems that have operated or are currently in service, in-line reheat has proved to be the most popular strategy.

(4) The trend in reheat strategies, as evidenced by FGD systems scheduled for immediate and future operation, is away from in-line and direct combustion methods and toward indirect hot air reheat. This is largely due to the problems encountered with in-line reheaters and the need for oil or natural gas with direct combustion reheaters. In-line reheat systems have been

subject to corrosion and plugging in the tubes. The corrosion in many cases has been so severe that even the heartier alloys have been unsatisfactory under many operating conditions. Many of these problems have been attributed to upstream mist eliminator inefficiency and inadequate self-cleaning techniques (soot blowers).

(5) A number of the major system suppliers still recommend in-line reheaters, especially when minimization of the energy demand is desired. It has been determined that corrosion of high-alloy materials is attributed to stress corrosion caused by chloride, whereas carbon steel is more susceptible to acid corrosion caused by sulfur dioxide. Therefore, if low sulfur/low chloride, low sulfur/high chloride, or high sulfur/low chloride environments can be accurately predicted, in such applications in-line reheaters may be used successfully.

(6) Indirect hot air reheat has the undesirable effect of increasing the energy demand of the FGD system and increasing overall system cost.

(7) Bypass reheat may be used for low-sulfur coal FGD applications when the required degree of reheat is not seriously constrained by emission standards.

(8) The use of efficient mist eliminators reduces the load on the reheat system by removing water droplets from the flue gas stream.

(9) ΔTs of 14° to 28°C (25° to 50°F) adequately prevent downstream water condensation.

(10) Waste heat recovery reheat (regenerative) is now being specified for two full-scale systems planned for future operation. In this system, the sensible heat of the incoming flue gas is recovered in an in-line heat exchanger placed upstream of the air preheater. This heat is then used to reheat the scrubbed gas stream. The in-line heat exchanger can be a direct gas-gas heat exchanger or a gas-liquid heat exchanger that uses a fluid of high heat capacity. Experience with these systems has been reported for experimental, small-scale, pilot plant (1 MW) tests.

Solids Separation (Sludge Dewatering): The major development in this area is the increased emphasis placed on clarification, centrifugation, and vacuum filtration, and the corresponding decreased emphasis on interim ponding. Formerly, an interim pond was relied on to fulfill three functions: clarification, dewatering, and temporary or final sludge storage. The realization that a single pond cannot perform all three functions adequately encouraged the development of the other techniques. Furthermore, the increased emphasis placed on off-site disposal for landfill and structural fills and on attaining closed water-loop operations also stimulated the use of clarifiers, centrifuges, and vacuum filters. In addition to using these techniques, several installations are using forced-oxidation strategies to enhance solids settling and filtration properties, to improve process chemistry, to improve waste sludge quality, and to decrease land requirements for sludge disposal.

Process Control and Instrumentation

Because of the complex nature of lime/limestone scrubbing chemistry, which has been the primary source of operating problems in full-scale systems, process control is considered a crucial item. The following is a brief review of some of the essential findings and innovations in the development of process control technology:

(1) Virtually all of the operating full-scale systems regulate reagent feed rate by controlling slurry pH. A pH sensor provides a signal for modulating the flow of reagent to the FGD system in a feedback control mode. The pH signal regulates the position of control valves for controlling the rate of reagent feed.

(2) The major problems encountered with pH control systems are sensor plugging, calibration drift, breakage, false indication, and erosion/corrosion damage.

(3) Sufficient operating experience has been obtained so that most of the reagent feed control problems have been identified. Once identified, these problems have been resolved for the most part through design modifications and/or new operating and maintenance procedures.

(4) Concerning selection of hardware for pH control, it has been noted that dip-type sensors are more successful than in-line sensors because they are easier to clean and calibrate. In-line, flow-through sensors are generally subject to more wear and abrasion and generally require more frequent maintenance.

(5) Other reagent feed control systems have been or are being evaluated on full-scale systems. One type involves feed-forward reagent control on the inlet flue gas flow rate and sulfur dioxide concentration, with trim provided by slurry pH. Another type involves control of reagent feed rate by using the outlet sulfur dioxide as the control variable. Limited success has been reported on both of these systems, primarily because of the difficulty in obtaining accurate and consistent readings from sulfur dioxide gas analyzers. This has been especially difficult on high-sulfur coal applications.

Construction Materials

Analysis of FGD experience in the United States to date, when examined at all levels of development, makes one point clear: generalization is difficult because of the many factors and apparent contradictions in FGD operation. This is especially evident in the area of construction materials. Many examples can be cited in which seemingly inferior construction materials have been adequate, whereas apparently adequate materials have failed. Although construction materials have been discussed throughout this section with respect to process and equipment design improvements, a brief review of the trend in the construction of critical elements in FGD systems is provided as follows:

(1) The use of 316 and 316L stainless steel is generally preferred as the construction material in critical areas of the FGD system.

(2) Some designers avoid 316 and 316L stainless steels whenever possible and instead use carbon steel with a surface lining or coating that physically shields the bare metal from the corrosive environment. These linings are usually resins applied in liquid or semiliquid form, by spray or trowel, and allowed to cure. The use of lined or coated carbon steel offers the potential benefits of being able to withstand low pH/high chloride environments much better than the 316s and being considerably less expensive. Actual operating experience, however, has shown that these materials are very susceptible to high temperature excursions, and often require an additional capital investment for an auxiliary power system. Corrosion resistance is also a limiting factor for many of the materials used. In addition, application and reapplication of linings have been suspect, especially reapplications, where proper preparation of the metal surface becomes more difficult.

(3) The use of chemically resistant masonry materials (e.g., silicone carbide) and ceramic liners in the throat areas of venturi scrubbers (usually applied to 316 or 316L stainless steel at the converging section of the venturi, where the gas velocity and erosive nature of the fly ash are highest), has been quite successful.

(4) The use of natural rubber, neoprene, polyvinyl chloride (PVC), fiber-reinforced plastic (FRP), and flaked-glass polyester generally predominates in the liquid and thickener circuits of the FGD system (tanks, pumps, agitators, piping, and thickeners) where the metal parts are 316 stainless steel or Alloy 20. For filtration systems, neoprene, polypropylene, FRP, and Alloy 20 are generally employed.

(5) Many systems are also being constructed of more resistant alloys in trouble spots (e.g., wet/dry, high-temperature, and high-chloride environments such as a prescrubber or presaturator). Alloys such as Hastelloy C-276, Hastelloy G, Alloy 20, Inconel 625, Incoloy 825, 317 low-carbon stainless steel, 904 low-carbon stainless steel, Jessop JS-700, and E-Brite 26-1 are being selected in minimum amounts.

(6) The use of stack lining or coating materials has been troublesome, especially where high-sulfur eastern coal is burned. Lining failures have also been reported in systems that included an apparently adequate degree of reheat. Recent developments in this area indicate that some progress has been made. A proprietary, spray-on, elastomer has been applied on three of the four flues at a station burning high-sulfur eastern coal (without reheat) with apparent success. The use of acid brick also appears to be successful in a similar application. At installations burning low-sulfur coal, adequate linings have not been as great a problem; many installations can get by with unprotected carbon steel flues.

ADIPIC ACID AS SCRUBBER ADDITIVE

The information in this section is based on "Recent Results from EPA's Lime/Limestone Scrubbing Programs—Adipic Acid as Scrubber Additive" by H.N. Head, S.C. Wang and D.T. Rabb of Bechtel National, Inc. and R.H. Borgwardt, J.E. Williams and M.A. Maxwell of Industrial Environmental Research Laboratory, U.S. Environmental Protection Agency in Symposium—Vol. 1.

Adipic acid is a commercially available organic acid, [$HOOC(CH_2)_4COOH$], that buffers the pH in SO_2 absorbers when present at low concentrations in the scrubbing liquor. Adipic acid has two buffering points. In the absence of chloride these are at pH 5.5 and 4.5. Chloride concentrations in the range of 5,000 to 7,000 ppm depress these buffering points to about pH 5.0 and 4.0. The theoretical basis for adipic acid's effect on the performance of limestone and lime scrubbers was developed in detail by Rochelle (1): The buffering action limits the drop in pH that normally occurs at the gas/liquid interface during SO_2 absorption, and the higher concentration of SO_2 in the surface film resulting from this buffering action accelerates the liquid-phase mass transfer. The capacity of the bulk liquor for reaction with SO_2 is also increased by the presence of calcium adipate in solution. SO_2 absorption is therefore less dependent upon the dissolution of limestone in the absorber. The overall result of these effects is improved SO_2 removal in limestone or lime scrubbers of a given type operating at a given L/G. In the case of limestone, it follows that a given SO_2 removal efficiency can be achieved at a lower stoichiometric ratio.

Further analysis by Rochelle (2) indicated that the use of additives would be most attractive economically when used in scrubbers employing forced oxidation. If no decomposition or volatilization of the additive occurs, the makeup requirements for the additive would be minimized by the tightly closed loop resulting from the better dewatering properties of oxidized sludge. For this reason, the scrubber tests reported here emphasized the evaluation of adipic acid in combination with forced oxidation.

Adipic acid has several potential advantages over other additives, such as MgO, which are also known to increase SO_2 removal, but do so by means of reactions involving the sulfite/bisulfite equilibrium. Since adipic acid does not depend on this mechanism for its buffering activity, it is not adversely affected by oxidation of the sulfite in the scrubbing liquor. It should therefore be especially useful in single loop scrubbers that employ forced oxidation. Furthermore, the buffering mechanism by which adipic acid enhances SO_2 absorption is not affected by the presence of chloride. The lack of interference by chloride means that adipic acid should be fully effective in the most tightly closed loop systems.

Preliminary economic evaluations have shown that adipic acid can reduce the operating cost of limestone systems while simultaneously increasing performance. Adipic acid is a major ingredient in the manufacture of nylon and is sufficiently available that widespread use in commercial FGD systems would have little effect on the market.

Beginning in 1977, initial studies conducted by EPA with the 0.1 MW pilot plant at the Industrial Environmental Research Laboratory located at Research Triangle Park, North Carolina (IERL-RTP) demonstrated as predicted by Rochelle that adipic acid was indeed an attractive additive. Results of these tests were first reported at the EPA Industry Briefing at Research Triangle Park in August 1978 (3).

Based on the findings at the IERL-RTP pilot plant, a program was set up at the EPA sponsored Test Facility located at the TVA Shawnee Steam Plant near Paducah, Kentucky, to develop commercially usable design data for adipic acid as a chemical additive. Testing with adipic acid was initiated in July 1978 on the 10 MW EPA prototype scrubbers and has continued since as a major part of the Shawnee Advanced Test Program.

Configurations successfully demonstrated during this period were:

- Adipic acid enhanced limestone scrubbing in a venturi/ spray tower system with two scrubbing loops and forced oxidation in the first loop.

- Adipic acid enhanced limestone scrubbing in a Turbulent Contact Absorber (TCA) system with no forced oxidation.

In addition, preliminary tests have been conducted in the same configurations using adipic acid enhanced lime slurry.

Test Facility and Program

There are two scrubber systems operating at the EPA sponsored Shawnee Test Facility, each with its own independent slurry handling facilities. Both systems were tested with adipic acid additive. The systems have the following scrubbers:

- A venturi followed by a spray tower (V/ST) (35,000 acfm capacity @ 300°F)

- A Turbulent Contact Absorber (TCA) (30,000 acfm capacity @ 300°F)

The scrubbers receive flue gas from TVA Shawnee coal-fired boiler No. 10. The boiler normally burns a high-to-medium sulfur bituminous coal producing SO_2 concentrations of 1,500 to 4,500 ppm. Flue gas can be taken from either upstream or downstream of the boiler No. 10 particulate removal equipment, allowing testing with high fly ash loadings (3 to 6 grains/scf dry) or low loadings (0.04 to 0.2 grain/scf dry). All tests in the adipic acid series were made with high fly ash loadings.

The test program was conducted with the scrubbing loop fully closed. Chlorides from the flue gas were concentrated in the scrubber slurry liquor over a range of 1,000 to 10,000 ppm depending on the tightness of the scrubber water balance and the chloride concentration in the coal burned.

Evaluation of Adipic Acid Enhanced Scrubbing at the IERL-RTP Pilot Plant

The initial testing of adipic acid as a scrubber additive was carried out in the EPA pilot plant at Research Triangle Park. A single-loop limestone scrubber (4) was used for this purpose, operated with forced oxidation in the scrubbing loop. This configuration would be expected to provide a sensitive response to any effect of adipic acid on oxidizer performance, because it operated at a higher pH than the two-loop system at Shawnee. In addition to effects on SO_2 removal and oxidation efficiencies, these tests also sought to determine whether adipic acid caused any change in the properties of the oxidized sludge. So that these properties could be clearly seen, the system was operated without fly ash. Chloride was added as HCl and controlled at the high levels expected for tightly closed-loop systems.

The results of the tests showed adipic acid to be very effective in improving SO_2 removal efficiency, even when operating at chloride levels as high as 17,000 ppm. A TCA scrubber, which removed 82% of the inlet SO_2 without the additive, yielded 89% SO_2 removal with 700 ppm adipic acid, 91% removal with 1,000 ppm, and 93% removal with 2,000 ppm adipic acid. The limestone utilization was concurrently increased from 77% without the additive, to 91% with 1,600 ppm adipic acid. The observed effects thus confirmed the theoretical expectations in all respects. In addition, the tests showed no serious interference by adipic acid on the performance of the oxidizer, operating at pH 6.1.

The quality of the oxidized sludge was similar to that obtained when operating without adipic acid, although small differences were detected. For example, the filtered sludge averaged 80% solids (for 13 one-week tests) vs 84% solids for 11 tests without the additive, when operating at 97 to 99% oxidation in both cases. The settling rate of the slurry (fly ash free at 50°C) averaged 2.3 cm/min during the adipic acid tests and 3.4 cm/min without adipic acid; bulk settled densities averaged 1.0 and 1.2 g solids/cm^3 slurry, respectively. It was concluded from these results that the large improvements in sludge quality that can be achieved by forced oxidation are unaffected by the use of adipic acid as a scrubber additive.

Tests without forced oxidation also demonstrated the efficacy of adipic acid. Operating a TCA scrubber with 2,000 ppm adipic acid and 6 inches H_2O pressure drop, 92% SO_2 removal was obtained at a limestone utilization level of 88%. By comparison, only 75% SO_2 removal would be expected in the pilot plant at these test conditions without the additive. At this adipic acid level the unoxidized sludge filtered to 49% solids; at lower adipic acid levels (1,500 ppm or less) the filterability of the slurry was the same as that obtained without additives: 55% solids.

During all tests with adipic acid, the scrubbing liquor had a noticeable odor even though the additive feed did not. The odor has been identified as valeric acid $[CH_3(CH_2)_3COOH]$, which is an intermediate product formed by side reactions that degrade adipic acid at scrubber operating conditions. Tests conducted by Radian Corporation (5) at IERL-RTP showed 40 to 50 ppm valeric acid in the scrubbing liquor, and about 1 ppm in the effluent flue gas when operating without forced oxidation and a 2,000 ppm adipic acid level. Although laboratory tests show that adipic acid ultimately degrades to lower molecular weight (C_1 to C_4) paraffinic hydrocarbons, no degradation products other than valeric acid have been detected (detection threshold = 10 ppb) in the pilot plant effluents. Efforts by Radian to identify the chemical mechanism responsible for degradation are continuing.

Material balances were carried out with and without forced oxidation to compare the adipic acid makeup requirements. The results showed that the degradation was greater when operating with forced oxidation, in which case 64% of the feed was unaccounted for. Without forced oxidation, the estimated loss was 24%. In spite of the greater rate of degradation with forced oxidation, the pilot plant material balance indicated that the adipic acid makeup should still be minimal for this mode: the reduction in liquor loss resulting from improved slurry dewatering properties more than compensates for the additional degradation.

Adipic Acid Enhanced Scrubbing in the Shawnee Venturi/Spray Tower System

Since January 1977, the venturi/spray tower system has operated with two scrubber loops in series and with forced oxidation accomplished by sparging air into the first scrubber loop (venturi) hold tank. Successful demonstration of this mode of scrubbing has already been reported with three alkali types: limestone, lime, and limestone with added magnesium (6)(7).

Since July 1978, tests have been conducted with added adipic acid in both lime and limestone systems. With adipic acid, removals as high as 98% have been achieved with both alkalis while concurrently oxidizing the product to gypsum. Results of these adipic acid enhanced, forced-oxidation tests are reported below.

System Description: The venturi/spray tower system was modified for two-loop scrubber operation with forced oxidation as shown in Figure 2.1. To separate the venturi and spray tower scrubber loops, a catch funnel was installed beneath the bottom spray header of the spray tower. To minimize slurry entrainment through the catch funnel, the bottom spray header was turned upward.

The hold tank in the first scrubber (venturi) recirculation loop was used as the oxidation tank. The arrangement of this tank is shown in Figure 2.2. The tank was 8 ft in diameter and could be operated at 10, 14, or 18-ft slurry levels. In early tests the tank contained an air sparger ring made of straight 3-inch 316L SS pipe pieces welded into an octagon approximately 4 ft in diameter. It was located 6 inches from the bottom of the tank. Sparger rings had either 130 $\frac{1}{8}$-inch diameter holes or 40 $\frac{1}{4}$-inch diameter holes pointing downward. The sparger ring was fed with compressed air to which sufficient water was added to assure humidification. (Dry air can evaporate water at the sparger orifice and cause scaling.) In more recent tests the sparger ring was replaced by a 3-inch diameter pipe with an open elbow discharging air downward at the center of the tank about 3 inches from the tank bottom. In all tests with adipic acid enhancement, the 3-inch pipe was used for air discharge.

The oxidation tank had an agitator with two axial flow turbines, both pumping downward. Each turbine was 52 inches in diameter and contained 4 blades. The bottom turbine was 10 inches above the air sparger. The agitator rotated at 56 rpm and was rated at 17 brake hp.

A 10-ft diameter desupersaturation tank, operating at a 5-ft slurry level, followed the oxidation tank to provide time for gypsum precipitation and to provide air-free pump suction.

Provision was made to add alkali to either loop. Adipic acid was added as a dry powder to the spray tower effluent hold tank. The dewatering system consisted of a clarifier followed by a rotary drum vacuum filter. Clarified liquor from the dewatering system can be returned to either scrubber loop or to the mist eliminator wash circuit.

Two-Loop Forced-Oxidation Test Results Using Limestone Slurry with Added Adipic Acid: Beginning in July 1978, a series of 9 limestone tests with forced oxidation on the two-loop venturi/spray tower system were conducted to demonstrate adipic acid as an additive to enhance SO_2 removal. Results of these tests are summarized in Table 2.1. Controlled run conditions common to all of the tests were:

(1) Fly ash loading: High (3 to 6 grains/scf dry)

(2) Venturi slurry rate: 600 gpm (21 gal/Mcf)

(3) Venturi slurry solids concentration: 15%

Figure 2.1: Flow Diagram for Adipic Acid Enhanced Scrubbing in the Venturi/Spray Tower System

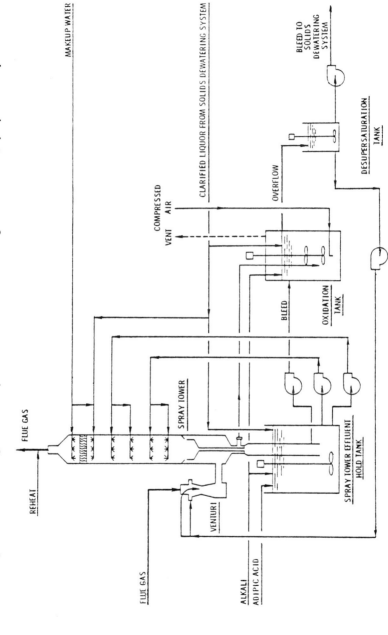

Source: Symposium—Vol. 1

Table 2.1: Venturi/Spray Tower Two-Loop Tests with Forced Oxidation (Adipic Acid Enhanced Limestone Slurry)

	901-1A	902-1A	902-1B	903-1A	904-1A	905-1A	906-1A	907-1A[7]	907-1B[7]
Major test conditions[1]									
Adipic acid concentration in venturi, ppm	0	(~2,000)	(~2,000)	3,500	3,500	(~2,000)	(~2,000)	(~2,000)	(~2,500)
Adipic acid concentration in ST (controlled), ppm	0	1,500	1,500	–	–	1,500[4]	1,500	1,500	1,500
Gas rate, acfm @ 300°F	35,000	35,000	35,000	35,000	18,000	35,000	35,000	18,000-35,000	20,000-35,000
ST slurry rate, gpm	1,600	1,600	1,600[2]	0[3]	0[3]	1,600	1,600	1,600	1,600
Residence times (min)/tank level (ft)									
Spray tower EHT	14.7/10	14.7/10	14.7/10	–	–	14.7/10	14.7/10	14.7/10	14.7/10
Oxidation tank	11.3/18	11.3/18	11.3/18	11.3/18	11.3/18	11.3/18	11.3/18	11.3/18	11.3/18
Desupersaturation tank	4.7/4.8	4.7/4.8	4.7/4.8	4.7/4.8	4.7/4.8	4.7/4.8	4.7/4.8	4.7/4.8	4.7/4.8
Venturi inlet, pH (estimate, not controlled)	(~4.5)	(~4.5)	(~4.5)	(~5.5)	(~5.5)	4.5[5]	4.5[5]	4.5[5]	4.5[5]
Venturi limestone stoichiometric ratio (controlled)	–	–	–	1.3	1.3	–	–	–	–
ST limestone stoichiometric ratio (controlled)	1.4	1.4	1.4	–	–	1.45	1.70	1.70	1.70
Air rate to oxidizer, scfm	210	260	260	210	130	260	260	160	260
Run average results									
Start-of-run date	7/12/78	8/3/78	8/22/78	8/28/78	9/5/78	9/8/78	9/27/78	10/8/78	11/13/78
On-stream hours	187	373	65	183	72	439	153	719	1,666
SO₂ removal, %	57	91	89	53	67	86	93	97.5	97
Inlet SO₂ concentration, ppm	2,800	2,450	2,400	2,600	2,350	2,200	2,700	2,350	2,500
Spray tower solids recirculated, wt %	5.9	4.9	4.8	–	–	3.2	6.8[6]	6.1[6]	5.9[6]
Venturi inlet, pH	4.50	4.45	4.00	4.7	4.75	4.55	4.5	4.65	4.65

(continued)

Table 2.1: (continued)

	901-1A	902-1A	902-1B	903-1A	904-1A	905-1A	906-1A	907-1A[7]	907-1B[7]
Spray tower inlet, pH	5.45	5.25	5.05	—	—	4.75	5.15	5.45	5.35
Spray tower limestone stoichiometric ratio	1.36	1.65	1.38	—	—	1.53	1.49	1.77	1.70
Spray tower inlet liquor gypsum saturation, %	95	110	115	—	—	130	100	105	110
Spray tower sulfite oxidation, %	28	40	27	—	—	63	28	29	25
Overall sulfite oxidation, %	98	98	99	98	98	99	98.5	98.5	98
Overall limestone utilization, %	97	94	98	74	76	91	88	88	92
Venturi inlet liquor gypsum saturation, %	95	100	90	100	100	100	110	110	105
Venturi inlet liquor $SO_3^=$ concentration, ppm	80	55	45	65	60	60	70	65	75
Adipic acid concentration in venturi, ppm	0	2,045	2,355	3,510	3,690	2,315	2,495	2,360	2,180
Adipic acid concentration in ST, ppm	0	1,445	1,615	—	—	1,410	1,485	1,560	1,510
Air stoichiometry, atoms O/mol SO_2 absorbed	2.30	2.05	2.15	2.65	2.80	2.40	1.80	2.0-3.85	1.9-3.3
Filter cake solids, wt %	85	86	88	83	80	85	84	87	85
Mist eliminator restriction, %	0.2	0.2	—	—	0	0	0	0	—

[1] All runs were made with 9 in H_2O venturi pressure drop except Runs 907-1B in which pressure drop was 9 in H_2O at 35,000 acfm.

[2] Intermittent spray tower operation. Spray slurry flow turned off 30 minutes every 8 hours. SO_2 removal averaged ~30% by venturi only operation.

[3] Venturi alone operation. The spray tower slurry recirculation pumps were turned off for the entire run.

[4] During the initial portion of the run (unsteady state operation) adipic acid was allowed to deplete to observe SO_2 removal/adipic acid relationship.

[5] Venturi inlet pH was controlled by separate limestone addition.

[6] Clarified liquor in excess of mist eliminator wash was returned to spray tower EHT, except for Runs 906-1A, 907-1A, and 907-1B where a fraction of this stream was diverted to the oxidation tank to control spray tower slurry solids at 6%.

[7] Long-term reliability test.

Source: Symposium—Vol. 1

Figure 2.2: Arrangement of the Venturi/Spray Tower Oxidation Tank

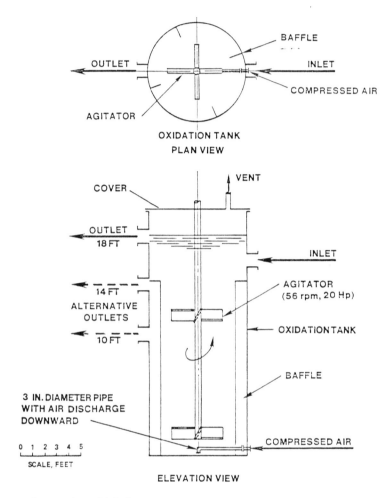

Source: Symposium—Vol. 1

After operating variables were explored in runs lasting about a week each, SO$_2$ removals above 95% (2,500 ppm inlet SO$_2$ concentration) were achieved in long-term tests at limestone utilizations of about 90% and with near complete oxidation.

Effect of Adipic Acid Concentration – The effect of adipic acid on enhancing SO$_2$ removal can be seen by comparing Run 901-1A (no adipic acid) with Run 902-1A (nominally 1,500 ppm adipic acid in the spray tower slurry). Both runs were purposely made at high limestone utilization (97 and 94% in Runs 901-1A and 902-1A, respectively) to demonstrate the effect of adipic acid. The addition of adipic acid increased SO$_2$ removal from 57% with no acid to 91% at nominally 1,500 ppm. These runs were made before conditions for operating with adipic acid were optimized. After optimization, SO_2 removals in excess of 95% were routinely achieved at 1,500 ppm adipic acid.

Actual adipic acid concentrations were 1,445 ppm in the spray tower loop and 2,045 ppm in the venturi loop. Dissolved solids are concentrated in the venturi loop because water is evaporated in humidifying the hot flue gas entering the scrubber.

Results at the IERL-RTP pilot plant indicated that most of the enhancement effect with adipic acid may be achieved at concentrations as low as 700 ppm in the slurry liquor.

SO_2 Removal in the Venturi — Three runs were made to determine SO_2 removal in the venturi alone. In Run 902-1B the spray tower recirculation pumps were turned off for 30 minutes every 8 hours. In Runs 903-1A and 904-1A, only the venturi was operated. Results were:

V/ST Run	Flue Gas Rate (acfm)	Adipic Acid in Venturi (ppm)	Venturi Inlet pH	SO_2 Removal in Venturi (%)
902-1B	35,000	2,355	4.0	30
903-1A	35,000	3,510	4.7	53
904-1A	18,000	3,690	4.8	67

In Run 902-1B, SO_2 removal was no greater than usually achieved without adipic acid. This is not unexpected because the operating pH was below the 4.0 to 5.0 buffering range of adipic acid.

In Run 903-1A, the pH was increased to 4.7 (limestone stoichiometric ratio of 1.35 mols Ca/mol SO_2 removed) and adipic acid concentration was increased to nominally 3,500 ppm with a resultant increase in SO_2 removal to 53%. Finally in Run 904-1A, the flue gas flow rate was cut in half to double the liquid-to-gas ratio, resulting in a further increase in SO_2 removal to 67%. These tests showed that high removal efficiencies cannot be achieved with reasonably low limestone stoichiometry in the venturi alone. Although not yet demonstrated, it may be possible to increase pH and get high removal efficiency using lime with adipic acid in the venturi alone.

System Control — In previous limestone runs without adipic acid, the limestone stoichiometry in the spray tower loop was controlled by adding limestone to the spray tower hold tank. By using a spray tower limestone stoichiometric ratio of about 1.4, there was sufficient residual limestone in the bleed from the spray tower loop to the venturi loop to maintain the venturi inlet slurry liquor pH above 5.0. This was not the case with adipic acid addition. With adipic acid, the venturi inlet pH fluctuated between 4.0 and 4.5 which was too low to get good removal.

Beginning with Run 905-1A, limestone was added to both the spray tower and venturi loops in an attempt to raise the venturi inlet pH. The pH could not be raised above 4.9 even with a venturi limestone stoichiometry greater than 2.0. For reasons not yet clear, adipic acid has a depressing effect on the pH in the venturi loop.

In subsequent runs, the spray tower limestone stoichiometric ratio was increased to 1.7 to maintain a venturi inlet pH of 4.5. This pH was established as the highest pH in which an overall limestone utilization of about 90% could be achieved. Occasionally, limestone is still added to the venturi loop if the venturi inlet pH drops much below 4.5.

Effect of Slurry Solids Concentration — In unenhanced limestone systems, slurry liquor pH and, consequently, SO_2 removal are sensitive to slurry solids concentration (i.e., limestone surface area available for dissolution) below 8%. Because of increased liquor buffer capacity with adipic acid enhancement, the system should be less sensitive to solids concentration. However, in early runs with adipic acid, the slurry solids concentration was allowed to drop to sensitive levels. In these runs, clarified liquor from the thickener was returned to the spray tower loop which resulted in low solids concentrations in the spray tower slurry in the range of 3 to 5%.

Beginning with Run 906-1A, the spray tower solids concentration was controlled at nominally 6% by splitting the clarified liquor return between the spray tower and the venturi loops. The spray tower pH and, consequently, SO_2 removal were increased as can be seen by the following comparison:

V/ST Run	Spray Tower Solids (%)	Spray Tower Inlet pH	Spray Tower Limestone Stoichiometry	Inlet SO_2 Concentration (ppm)	SO_2 Removal (%)
905-1A	3.2	4.75	1.5	2,200	86
906-1A	6.8	5.15	1.5	2,700	93

Based on an examination of day-to-day operations, with 1,500 ppm adipic acid in limestone slurry, it appears that SO_2 removal increases with spray tower slurry solids concentration up to about 6 wt %.

Forced Oxidation — Forced oxidation of the scrubber slurry has been shown to occur equally well with or without adipic acid. In fact, adipic acid buffers the slurry liquor at a low pH favorable for forced oxidation. The adipic acid enhanced runs on the two-loop venturi/spray tower system were all operated at air stoichiometric ratios in the range of 1.8 to 2.4 atoms oxygen/mol SO_2 absorbed. Near complete oxidation was achieved in all runs and the filter cake solids concentration of the oxidized slurry was consistently high, in the range of 85% or better.

Venturi/Spray Tower Demonstration Runs 907-1A and 907-1B — Run 907-1A was a month long adipic acid enhanced limestone run with forced oxidation designed to demonstrate operational reliability with respect to scaling and plugging and to demonstrate the removal enhancement capability of the adipic acid additive.

This run was controlled at a nominal 1.7 limestone stoichiometric ratio (up from 1.4 in previous runs) and 1,500 ppm adipic acid in the spray tower. Spray tower slurry solids concentration was controlled at 6% by splitting the clarified liquor return between the spray tower and venturi loops. Venturi inlet slurry liquor pH was nominally 4.5. Occasionally, limestone addition to the venturi loop was required to maintain this pH.

Flue gas flow rate was varied from 18,000 acfm to a maximum of 35,000 acfm (spray tower gas velocity between 4.8 and 9.4 ft/sec) to follow the daily boiler load cycle which normally fluctuated between 100 and 150 MW. The adjustable venturi plug was fixed in a position such that the pressure drop across the venturi was 9 inches H_2O at 35,000 acfm maximum gas rate. Actual pressure drop ranged from 3.0 to 9.6 inches H_2O.

The slurry recirculation rates to the venturi and spray tower were fixed at 600 gpm (L/G = 21 to 42 gal/Mcf) and 1,600 gpm (L/G = 57 to 111 gal/Mcf), respectively.

The oxidation tank level was 18 ft and the air flow rate was held constant at 160 scfm.

The run began on October 8, 1978 and terminated November 13, 1978. It ran for 719 on-stream hours (30 days) with no unscheduled outages. The scrubber was down once for a scheduled 3-hour inspection and again when the boiler came down for 135 hours to install a new station power transformer.

Average SO_2 removal for the run was 97.5% at 2,350 ppm average inlet SO_2 concentration. The SO_2 removal stayed within a narrow range of 96 to 99% throughout almost the entire run. On October 19 and on October 27, SO_2 removal dropped briefly to less than 90% when the pump which supplies the slurry to the top two spray headers was brought off stream for repacking and the spray tower slurry flow rate was cut in half to 800 gpm. At the reduced slurry recirculation rate, SO_2 removal was 82 to 87%.

Venturi and spray tower inlet pH averaged 4.65 and 5.45, respectively. Overall limestone utilization was 88% and the spray tower limestone utilization was 56%, demonstrating the advantage of good limestone utilization in a two-scrubber-loop operation.

Average adipic acid concentrations were 2,360 ppm in the venturi loop and 1,560 ppm in the spray tower loop.

Sulfite oxidation in the system bleed slurry averaged 98.5% with the air stoichiometric ratio varying between 2.0 and 3.85 atoms oxygen/mol SO_2 absorbed. The filter cake solids content was 87%.

The mist eliminator was clean during the entire run. Inspections after 207 hours, 603 hours, and 719 hours (at the end of the run) showed that the mist eliminator was completely free of solids deposits. The system was free of plugging and scaling. There was no increase in solids or scale deposits on the scrubber internals during Run 907-1A.

Following Run 907-1A, a second adipic acid enhanced limestone run with forced oxidation was made during which flue gas monitoring procedures were evaluated by EPA. This run, 907-1B, was made under the same conditions as Run 907-1A except that a "typical" daily boiler load cycle was established for the gas flow rate to follow rather than the Unit No. 10 Boiler load. The gas rate was changed as follows to simulate a "typical" boiler load cycle:

Time, hours	Gas Rate, acfm @ 300°F
100	20,000
500	30,000
700	35,000
1,100	30,000
1,700	35,000
2,300	30,000

Run 907-1B began on November 13, 1978 and terminated January 29, 1979. It ran for 1,666 on-stream hours (69 days) with only 27 hours of scrubber related outages. The scrubber was also out of service 146 hours when Unit 10 came down for replacement of a broken turbine thrust bearing. Scrubber related outages were:

	Hours
Plugged slurry line	11
Rebuild bleed pump	10
Miscellaneous mechanical	4
Leak in water supply line	2
Total	27

Excluding boiler outages and scheduled inspections, the combined Runs 907-1A and 907-1B operated for a period of over 3 months with an on-stream factor of 98.9%. No deposits whatsoever were observed in the mist eliminator for the entire 3-month test period. On only one occasion did solids accumulation cause an outage. The cross-over line carrying slurry effluent from the venturi to the oxidation tank plugged with soft solids and had to be cleaned out. Because of problems associated with converting the Shawnee venturi/spray tower system to two-scrubber-loop operation, this cross-over line follows a tortuous path. A properly designed new system would not have this problem.

Results of Run 907-1B were equally as good in every respect as those of Run 907-1A. Average SO_2 removal was 97% at 2,500 ppm average inlet SO_2. With few exceptions, SO_2 removal remained within a narrow band of 95 to 99%. SO_2 removal dropped briefly (typically 30 minutes) below 90% five times when one of the two spray tower recirculation pumps was taken out of service for maintenance, effectively cutting the slurry recirculation rate in half.

Overall limestone utilization during this run was 92%. Sulfite oxidation averaged 98% and the waste sludge filter cake quality was excellent, having a solids content of 85%.

SO_2 emissions for Run 907-1A and 907-1B were calculated based on an assumed coal heating value of 10,500 Btu/lb, on 100% sulfur overhead (none in bottom ash), and on an assumed excess air of 30%. This excess air corresponds to about 700 ppm inlet SO_2 per 1.0 wt % sulfur in coal for the above conditions.

The average SO_2 emission for the entire 3-month operating period was only 0.20 lb/MM Btu. The highest 24-hour average SO_2 emission during Run 907-1A was 0.37 lb/MM Btu (October 18) and during Run 907-1B was 0.41 lb/MM Btu (January 20). These values compare with the federal new source performance standard of 1.2 lb/MM Btu.

Adipic Acid Consumption — From the onset of testing, adipic acid consumption was higher than anticipated. A material balance calculation for the adipic acid consumption was made for the entire Run 907-1B for a period of 76 days from November 13, 1978 to January 29, 1979. During this period, a breakdown of the average adipic acid addition rate was:

	. . . Adipic Acid Consumption . . .	
	lb/hr	lb/ton limestone feed
Discharged with filter cake	0.8	1.8
Unaccounted	2.9	6.5
Total	3.7	8.3

These losses were higher than experienced in the IERL-RTP pilot plant. The reasons for the differences are not yet clear.

The unaccounted adipic acid loss in this run with forced oxidation was somewhat higher than that recorded in TCA Run 932-2A without forced oxidation (2.2 lb/hr). Based on this comparison, it appears that forced oxidation in the scrubber system may increase the unaccounted adipic acid losses.

As already mentioned Radian Corporation is investigating the mechanism of this unaccounted loss. Preliminary results indicate that adipic acid decomposes to valeric acid [$CH_3(CH_2)_3COOH$] and other components. Valeric acid creates an odor even in the small concentrations present in the scrubber slurry. An unpleasant odor was apparent immediately above the effluent hold tanks and in the filter and centrifuge building where dewatering takes place. About 600 cubic yards of the gypsum/fly ash mixture from the 3-month combined venturi/ spray tower Runs 907-1A and 907-1B were saved in a pile near the test facility. During this winter no odor was apparent even when the pile was worked by a bulldozer.

Two-Loop Forced Oxidation Test Results Using Lime Slurry with Added Adipic Acid: The emphasis at Shawnee has been on adipic acid enhancement with limestone because this combination may prove to be the most economical route for achieving high removal in throwaway systems. However, two one-week runs made in July 1978 demonstrated that adipic acid is also effective in enhancing SO_2 removal in lime systems.

Effect of Adipic Acid Concentration – Two runs were conducted with lime on the venturi/spray tower two-loop system with forced oxidation. Run 951-1A was a base case without adipic acid and Run 952-1A was under identical conditions with nominally 1,500 ppm adipic acid in the spray tower slurry liquor. The addition of adipic acid increased SO_2 removal from 66% with no acid to 98% at 1,500 ppm (2,600 to 2,750 ppm average inlet SO_2 concentration).

Effect of Slurry Solids Concentration – In these lime runs, all clarified liquor from the dewatering system was returned to the spray tower loop resulting in relatively low spray tower solids concentrations. Apparently, a low solids concentration of 4.5% in the adipic acid enhanced lime run was not detrimental because 98% SO_2 removal was achieved. This may be attributed to the high pH inherent with lime systems, at which adipic acid becomes fully effective. However, in the lime run without adipic acid, the low solids concentration was detrimental. Run 951-1A averaged 66% SO_2 removal at 5.7% spray tower slurry solids concentration. This run can be compared with a previous one-month lime demonstration run under essentially the same conditions which averaged 88% SO_2 removal with 10.4% average spray tower slurry solids concentration.

Adipic Acid Enhanced Scrubbing in the Shawnee TCA System

Beginning in July and continuing to November 1978, a series of tests were conducted on the TCA system with adipic acid enhanced lime and limestone slurries. These tests were conducted without forced oxidation. SO_2 removals ranging up to 95% have been achieved with adipic acid concentrations of up to 1,500 ppm in the recirculating slurry liquor. Results of these adipic acid enhanced tests without forced oxidation are reported below.

System Description: The TCA (turbulent contact absorber) was operated in a single-loop scrubbing configuration as shown in Figure 2.3. The TCA contained three beds of $1\frac{7}{8}$ inch diameter, 11.5 gram nitrile foam spheres retained between bar grids. Each bed contained 5 inches static height of spheres.

The effluent hold tank was 7 feet in diameter and operated at a 17 ft slurry level giving a 4.1 minute residence time at the slurry recirculation rate of 1,200 gpm. Alkali and adipic acid were added directly to the effluent hold tank.

The dewatering system consisted of a clarifier followed by a solid bowl centrifuge. All clarified liquor from the dewatering system was returned to the scrubber loop either via the effluent hold tank or the mist eliminator underside sprays.

Test Results on the TCA with Adipic Acid Enhanced Limestone Scrubbing: Beginning in July 1978, a series of 7 limestone tests were conducted on the TCA system to demonstrate the effects of adipic acid on enhancing SO_2 removal in a system without forced oxidation. Results of these tests are summarized in Table 2.2. Unless otherwise specified, controlled run conditions common to all of the tests were:

> Fly ash loading: High (3 to 6 grains/scf dry)
> Slurry flow rate: 1,200 gpm (50 gal/Mcf)
> Slurry solids concentration: 15%

Figure 2.3: Flow Diagram for Adipic Acid Enhanced Scrubbing in the TCA System

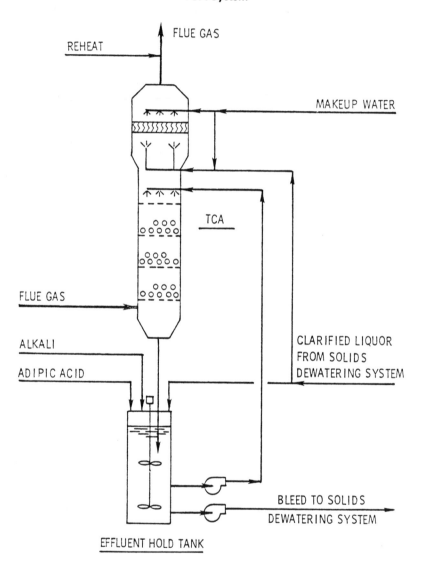

Source: Symposium—Vol. 1

Table 2.2: TCA Single-Loop Tests Without Forced Oxidation
(Adipic Acid Enhanced Limestone Slurry)

	926-2A	927-2A	928-2A	929-2A	930-2A	931-2A	932-2A[3]
Major test conditions[1]							
Adipic acid concentration, ppm	0	300	1,500	750	750	750	1,500
Gas rate, acfm @ 300°F	30,000	30,000	30,000	30,000	30,000	30,000	20,000-30,000
LS stoichiometric ratio controlled at	1.2	1.2	1.2	1.2	1.35	1.05	1.2
EHT residence time, min	4.1	4.1	4.1	4.1	4.1	4.1	4.1
EHT level, ft	17	17	17	17	17	17	17
Limestone addition point[2]	EHT	EHT	EHT	EHT	EHT	EHT	EHT
Run average results							
Start-of-run date	7/4/78	8/4/78	8/22/78	9/5/78	9/13/78	9/20/78	9/26/78
On-stream hours	192	374	252	186	162	137	833
SO_2 removal, wt %	71	75	93	92	93	77	96
Inlet SO_2 concentration, ppm	2,750	2,300	2,600	2,300	2,550	2,300	2,450
SO_2 make-per-pass, mmol/liter	10.1	9.1	12.8	11.2	12.6	9.4	4-18
TCA inlet liquor gypsum saturation, %	90	110	100	110	80	110	110
Sulfite oxidation, %	13	19	14	10	13	24	21
Limestone utilization, %	80	84	85	80	75	93	82
TCA inlet, pH	5.65	5.30	5.40	5.50	5.60	4.95	5.30
Adipic acid concentration, ppm	0	350	1,600	885	700	840	1,620
Centrifuge solids, wt %	37[4]	62	59	59	58	62	61
Mist eliminator restriction, %	1	3	0	0	0	0	0

[1] All runs were made with 3 beds and 5 inches per bed of 1⅞ inch diameter, 11.5 gram nitrile foam spheres.

[2] EHT = in the effluent hold tank.

[3] Long-term reliability test.

[4] Clarifier only.

Source: Symposium—Vol. 1

Effect of Adipic Acid Concentration — TCA limestone Runs 926-2A through 929-2A were all made under identical conditions except for adipic acid concentration. In these runs limestone stoichiometry was controlled at 1.2 mols Ca/mol SO_2 absorbed. The effect of adipic acid on SO_2 removal was:

TCA Run	Actual Adipic Acid Concentration (ppm)	SO_2 Removal (%)
926-2A	0	71
927-2A	350	75
929-2A	885	92
928-2A	1,600	93

SO_2 removal increased from 71% with no adipic acid to 92% with 885 ppm in the slurry liquor. Increasing adipic acid further to 1,600 ppm increased SO_2 removal only slightly to 93%. Thus, the majority of the adipic acid enhancement was achieved in the TCA at a concentration somewhere between 350 and 885 ppm.

Effect of Limestone Stoichiometry and pH — Limestone stoichiometry was explored at a nominal adipic acid concentration of 750 ppm and a liquid-to-gas ratio of 50 gal/Mcf with the following results:

TCA Run	Limestone Stoichiometry (mols Ca/mol SO_2 absorbed)	TCA Inlet pH	SO_2 Removal (%)
931-2A	1.05	4.9	77
929-2A	1.20	5.5	92
930-2A	1.35	5.6	93

Thus, it is apparent that the system required a limestone stoichiometry of only about 1.2 to maintain sufficiently high pH to achieve the full SO_2 removal enhancement with adipic acid. Additional limestone did not significantly increase SO_2 removal.

TCA Demonstration Run 932-2A — Run 932-2A, a month long demonstration run, was made with adipic acid enhanced limestone slurry to demonstrate both operational reliability with respect to scaling and plugging and the removal enhancement capability of the adipic acid additive.

The run began on September 26, 1978 and terminated on November 2, 1978 for a total of 833 on-stream hours (35 days). During the run, the scrubber was out of service for 48 hours due to a boiler outage caused by a tube leak, 5 hours for a scheduled inspection, and 8 hours for unscheduled outages to clean and repair the scrubber I.D. fan damper.

Excluding boiler outages and scheduled inspections, Run 932-2A operated with an on-stream factor of 99.0%.

The run was controlled at a nominal 1.2 limestone stoichiometric ratio and 1,500 ppm adipic acid concentration in the slurry liquor. Slurry solids concentration was controlled at 15%. The flue gas flow rate was varied between 20,000 and 30,000 acfm (8.4 to 12.5 ft/sec superficial gas velocity) as the boiler load fluctuated between 100 and 150 MW. The slurry recirculation rate was fixed at 1,200 gpm (L/G = 50 to 75 gal/Mcf). As with all runs during this test block, the effluent hold tank residence time was only 4.1 minutes.

SO_2 removal during the run averaged 96% at an average inlet SO_2 concentration of 2,450 ppm. Excluding the first few days of unsteady-state operation, SO_2 removal stayed within the narrow range of 94 to 98% as the inlet SO_2 concentration varied widely between 1,400 and 3,500 ppm.

SO$_2$ emissions were calculated for Run 932-2A on the same basis as for the venturi/spray tower Run 907-1A. During the first seven days (September 26 through October 3), SO$_2$ emissions were relatively high and widely varying. The highest daily average emissions were 1.1 lb SO$_2$/MMBtu on September 27 and 0.9 lb SO$_2$/MMBtu on both September 30 and October 3. It should be noted, however, that the new source performance standard of 1.2 lb/MMBtu was never exceeded on a daily average basis.

The relatively high SO$_2$ emissions resulted from frequent excursions to low pH in the scrubber slurry liquor as the test personnel were trying to control the limestone stoichiometric ratio at 1.2 with widely varying inlet SO$_2$ concentrations (1,000 to 3,500 ppm). Beginning on October 6 after the boiler outage, a scrubber inlet pH underride of 5.1 was implemented in addition to the limestone stoichiometric ratio control value of 1.2. This combined stoichiometry/pH control produced the improved results shown for the remainder of the run.

SO$_2$ emissions for the 27-day period from October 6 through the end of the run on November 2 averaged only 0.26 lb/MMBtu. The highest 24-hour average SO$_2$ emission during this period was only 0.44 lb/MMBtu.

The mist eliminator was completely clean at the end of the run and the entire scrubber system was free of scaling and plugging.

Limestone utilization during the run averaged 82%. Discharge solids from the centrifuge averaged about 61% which is typical of unoxidized limestone sludge.

In summary, the objectives of this run were met. High removal was consistently achieved at a good limestone utilization and no fouling, scaling, or plugging occurred.

Adipic Acid Consumption — As with the forced-oxidation runs on the venturi/spray tower system, adipic acid consumption was greater than anticipated. An adipic acid material balance calculation was made during Run 932-2A between October 10 and October 30, 1978, a total of 21 days:

	. . .Adipic Acid Consumption. . .	
	lb/hr	lb/ton limestone feed
Discharged with centrifuge cake	1.9	4.2
Unaccounted	2.2	5.0
Total	4.1	9.2

As already discussed, the unaccounted loss in this run without forced oxidation of 2.2 lb/hr compares with 2.9 lb/hr for venturi/spray tower Run 907-1B with forced oxidation. Thus, it appears that air sparging for forced oxidation may increase adipic acid losses.

Test Results on the TCA with Adipic Acid Enhanced Lime Scrubbing: Although the majority of the adipic acid enhanced tests on the TCA were with limestone slurry, a single week-long test was conducted with lime slurry.

Effect of Adipic Acid Concentration — The runs were conducted with lime on the TCA system, two without adipic acid and one with. Run 976-2A was a lime run at a scrubber inlet pH of 7 with 15% slurry solids concentration; SO$_2$ removal averaged 70%. In Run 976-2B the slurry solids concentration was dropped to 8% and SO$_2$ removal dropped slightly to 67%. Finally, Run 977-2A was made with 8% slurry solids and nominally 300 ppm (actual average 420 ppm) adipic acid in the slurry liquor. SO$_2$ removal increased to 80%, an enhancement of 10 to 13 percentage points over the base cases with only a small addition of adipic acid.

It should be pointed out, however, that SO_2 removals in the mid 80s can be achieved in the Shawnee TCA system with lime alone by raising the scrubber inlet pH to 8 at 15% slurry solids.

Dewatering Characteristics of Adipic Acid Enhanced Limestone Slurry at Shawnee

Settling and dewatering characteristics of slurry solids are routinely monitored in the Shawnee laboratory by cylinder settling tests and vacuum funnel filtration tests. A comparison of the results of these monitoring tests for limestone slurry with and without adipic acid addition is presented in this section and summarized in Table 2.3.

Table 2.3: Comparison of Shawnee Waste Slurry Dewatering Characteristics With and Without Adipic Acid Addition

Alkali	Fly Ash Loading	Oxidation	Adipic Acid	Initial Settling Rate (cm/min) Average	Range	Ultimate Settling Solids (wt %) Average	Range	Funnel Test Cake Solids (wt %) Average	Range
LS	High	Yes	No	1.0	0.6-1.5	72	62-86	72	65-88
LS	High	Yes	Yes	0.5	0.3-0.9	72	59-83	69	59-77
LS	High	No	No	0.2	0.1-0.5	54	41-67	57	48-66
LS	High	No	Yes	0.2	0.1-0.3	51	42-69	56	49-73

Source: Symposium—Vol. 1

Cylinder settling tests are performed with slurries containing 15% solids at room temperature in a 1,000 ml cylinder containing a rake which rotates at 0.16 rpm. The initial settling rate and ultimate settled solids concentration are recorded as indices of dewatering characteristics. The initial settling rate is only a qualitative index of the solids settling properties. Design rates for sizing clarifiers must take into consideration the hindered settling rate as the solids concentrate. The ultimate settled solids from the cylinder tests represent the highest solids concentration achievable in a settling pond.

Funnel filtration tests are performed in a Buchner funnel with a Whatman 2 filter paper under a vacuum of 25 in Hg. The funnel tests correlate well with the Shawnee rotary drum vacuum filter when not blinded but the funnel test cakes tend to have lower solids concentration.

Table 2.3 lists the effects of adipic acid on both oxidized and unoxidized limestone slurries. The data reported are for a range of adipic acid concentration of 1,500 to 3,000 ppm. All samples were with high fly ash loadings in which about 40% of the slurry solids was fly ash. All tests were conducted with samples containing 15% slurry solids.

As reported previously, settling and filtration characteristics of oxidized slurry are much superior to the characteristics of unoxidized slurry. The same trend exists with the slurry samples containing adipic acid.

The average initial settling rate for oxidized limestone slurry decreased from 1.0 cm/min with no adipic acid to 0.5 cm/min with adipic acid. However, this rate was still considerably higher than settling rates without forced oxidation. Without forced oxidation, the initial settling rate averaged 0.2 cm/min with or without adipic acid.

Adipic acid had little or no effect on ultimate settled solids or funnel test cake solids. The data indicated a slight decline in solids quality with adipic acid but the decline was small compared with the difference between limestone slurry with forced oxidation and without.

In addition to the funnel filtration tests, the filter cake solids from the rotary drum vacuum filter were monitored. During forced-oxidation operation with adipic acid enhanced limestone, the rotary drum filter cake solids concentration ranged from 80 to 87 wt % solids (see Table 2.1). This range is no different than that obtained when operating with oxidized sludge in the absence of adipic acid.

In summary, the only significant effect of adipic acid addition on dewatering characteristics was a decline in initial settling rate with oxidized limestone slurry. These observations agree generally with those at the IERL-RTP pilot plant.

Summary of Characteristics of Adipic Acid as a Scrubber Additive

Based on testing at the IERL-RTP pilot plant and at the Shawnee Test Facility and on the preliminary economic evaluations conducted by TVA, the characteristics of adipic acid as a scrubber additive can be summarized as follows:

Beneficial Effects

- Significantly enhances SO_2 removal efficiency.

- Increases alkali utilization, hence decreases waste solids disposal requirements.

- When used with limestone, has projected lower capital and operating costs than unenhanced limestone or limestone/ MgO.

- Can be used with both lime and limestone in either conventional or forced-oxidation modes for both new and existing installations.

- Is not adversely affected by chlorides as is the limestone/ MgO process.

- Does not significantly affect solids quality (filterability/ settling rate) as can occur with high magnesium ion concentrations.

- Should promote use of less expensive and less energy intensive limestone rather than lime.

- With proper pH control, steady outlet SO_2 concentrations can be maintained even with wide fluctuations of inlet SO_2 concentrations.

Negative Aspects

- Has unpleasant odor associated with adipic acid decomposition product.

- Adipic acid decomposition requires adding up to 5 times that theoretically required. (However, consumption over the ranges anticipated has negligible economic impact.)

- Other possible secondary environmental effects have not yet been determined. Separate studies are underway to determine if any such problems might exist.

References

(1) Rochelle, G.T., "The Effect of Additives on Mass Transfer in CaCO$_3$ and CaO Slurry Scrubbing of SO$_2$ from Waste Gases," *Ind. Eng. Chem.*, pp 67-75, 1977.

(2) Rochelle, G.T., "Process Alternatives for Stack Gas Desulfurization by Throwaway Scrubbing," Proceedings of 2nd Pacific Chemical Engineering Congress, Vol 1, p 264, August 1977.

(3) Borgwardt, R.H., Significant EPA/IERL-RTP Pilot Plant Results, EPA Industry Briefing, Research Triangle Park, NC, August 29, 1978.

(4) Borgwardt, R.H., *Effect of Forced Oxidation on Limestone/SO$_x$ Scrubber Performance*, in Proceedings: Symposium on Flue Gas Desulfurization --Hollywood, FL, November 1977, EPA-600/7-78-058a, (NTIS PB 282090), March 1978 (pp 205-228).

(5) Meserole, F.B., *Adipic Acid Degradation in FGD Systems*, progress report for EPA Contract 68-02-2608, Task 58, Radian Corporation, Austin, TX, December 1978.

(6) Head, H.N. et al, *Results of Lime and Limestone Testing with Forced Oxidation at the EPA Alkali Scrubbing Test Facility*, in Proceedings: Symposium on Flue Gas Desulfurization - Hollywood, FL, November 1977, EPA-600/7-78-058a, (NTIS PB 282090), March 1978 (pp 170-204).

(7) Head, H.N., *Results of Lime and Limestone Testing with Forced Oxidation at the EPA Alkali Scrubbing Test Facility --Second Report*, EPA Industry Briefing, Research Triangle Park, NC, August 29, 1978.

ADIPIC ACID DEGRADATION MECHANISM

The information in this section is based on *Adipic Acid Degradation Mechanism in Aqueous FGD Systems*, September 1979, EPA-600/7-79-224, by F.B. Meserole, D.L. Lewis, A.W. Nichols and G. Rochelle of Radian Corporation for the U.S. Environmental Protection Agency.

Long-term scrubber tests at Industrial Environmental Research Laboratory-Research Triangle Park (IERL-RTP) and EPA's test facility at the TVA Shawnee Power Plant have experienced substantial losses of adipic acid. These losses were in excess of that expected from the scrubber solution removed with the filter cake solids.

Tests have also shown that the adipic acid loss rate is considerably higher during forced oxidation runs than during natural oxidation conditions. Unaccounted losses of as much as 80% of the adipic acid makeup rate were observed in the Shawnee venturi/spray tower scrubber system.

Several possibilities could explain this disappearance including:

- inaccurate measurement of adipic acid in the scrubber solution,
- precipitation of insoluble adipate salts,
- chemical complexation and/or decomposition of adipic acid,
- biodegradation of adipic acid,
- physical adsorption on, or coprecipitation with the scrubber solids, and
- volatilization of adipic acid into the flue gas.

Radian was contracted by the EPA to select methods to accurately measure levels of adipic acid in FGD scrubber solutions and conduct material balance measurements at the IERL-RTP pilot plant and at the Shawnee Test Facility. The primary objective was to verify adipic acid loss and to determine its cause. Other objectives were to identify by-products if degradation occurs, determine mechanism, and suggest possible courses for limiting the loss.

Between October, 1978 and February, 1979 Radian personnel conducted three sampling trips to RTP and one sampling trip to the Shawnee Test Facility to collect samples at the various slurry and gas streams around the system for adipic and valeric acid determination. The data obtained were used to perform material balance calculations.

A series of laboratory studies aimed at simulating scrubber conditions was conducted. These tests were designed such that operating conditions were variable and degradation could be monitored by the collection and analysis of decomposition products. The results are presented in this report.

Summary of Results

The samples collected during this program from the RTP and Shawnee SO_2 scrubbers during the adipic acid addition tests were analyzed to determine the concentrations of adipic acid and any degradation products. Gaseous, liquid and solid samples were taken using a variety of sampling techniques and were then characterized using several analytical procedures including: ion chromatography, gas chromatography, gas chromatography-mass spectrometry, infrared spectroscopy and total organic carbon determination.

The major degradation products of adipic acid identified in the samples collected from the RTP and Shawnee test units were:

- valeric acid – $CH_3(CH_2)_3COOH$, and
- glutaric acid – $HOOC(CH_2)_3COOH$.

Trace quantities of the following were also found:

- butyric acid – $CH_3(CH_2)_2COOH$, and
- succinic acid – $HOOC(CH_2)_2COOH$.

All four of the above acids were found in the liquid phase samples, but only valeric acid was found in the scrubbed flue gases. No measurable amounts of organics could be detected in the washed solids using infrared analysis. The detection limit using this technique is about 0.1% on a weight basis.

The products measured in the samples collected during the laboratory degradation studies were the following: valeric acid, carbon dioxide, methane, and butane.

In some cases, these degradation products were found to account for most of the adipic acid lost whether in the field or laboratory tests. However, in general, only 25 to 50% of the adipic acid added could be accounted for. The results of the field sampling efforts associated with this program are summarized in Table 2.4. The material balance calculations were performed using analytical data measured by Radian personnel and flow rates that were supplied by on-site personnel.

In addition to the field measurements, laboratory programs were carried out to establish the conditions necessary for adipic acid degradation and to identify the degradation products. Conclusions drawn from these tests include:

- Adipic acid degradation can occur by a chemical process as opposed to a microbial process.

- The similarity in the degradation rates between the laboratory and field tests indicates that the catalytic and microbial effects are small, if present.

- The oxidation of sulfite in the presence of adipic acid results in degradation, supposedly through a free-radical mechanism.

Table 2.4: Summary of Field Sampling Results

Facility	Date	Scrubber Type	Oxidation Mode	Adipic Acid Decomposition (%)	Fraction of Products Measured (%)
RTP	10/25/78	TCA	Natural	28	84-97
RTP	11/15/78	TCA	Forced	69	42-45
RTP	12/18/78	TCA	Forced	54	53
Shawnee	2/2/79	Venturi/ Spray Tower	Forced	91	17

Source: EPA-600/7-79-224

During the laboratory phase, a series of material balance tests was made using various techniques to trap low molecular weight hydrocarbon products. The results of these tests are presented in Table 2.5.

Table 2.5: Material Balance Results of Dynamic Laboratory Degradation Studies

Hydrocarbon Collection Techniques	Initial Adipic Acid	Final Adipic Acid	Valeric Acid	Methane	Ethane	Propane	Butane	CO_2	Total
	. . . . (mmol C)								
Poropak Q Trap	18.7	11.4	4.4	1.6	0.4	0.2	0.9	1.4	20.3
Recirculation of Product Gases	19.0	11.5	2.4	—	—	—	—	3.2	17.1
	9.5	4.1	—	—	—	—	—	1.1	5.2
Silica Gel Trap	19.2	12.6	1.5	—	—	—	0.02	1.6	15.7

Source: EPA-600/7-79-224

The degradation rate of adipic acid was found to depend upon the degree of sulfite oxidation and adipic acid concentration in solution. Field and laboratory data were compared using this correlation. Although there is significant scatter a general trend appears to prevail. Much of this variability may be a result of the different scrubber systems used to collect the data and different analytical techniques used to measure adipic acid.

Electron micrographs of a limited number of solid samples suggest that the average particle size decreases as the adipic acid increases. However, the effect was not large enough to yield a detectable difference at Shawnee when filter-cake solids were compared with and without the use of adipic acid in the scrubber.

DOUBLE-LOOP SYSTEM AT MARTIN LAKE STEAM ELECTRIC STATION

The information in this section is based on "Limestone FGD Operation at Martin Lake Steam Electric Station" by M. Richman of Research-Cottrell in Symposium—Vol. 1.

Design Criteria

Martin Lake Steam Electric Station consists of four 750,000 kilowatt lignite-fired units. The station, located in Rusk County, Texas, is surrounded on three sides by Martin Lake, which serves as a source of makeup and cooling water. The station is also less than ten miles from the source of its lignite fuel.

The Texas lignite fuel for the Martin Lake S.E.S. has the following characteristics:

	Average
Heating value, Btu/lb	7,380
Ash content, %	8.0
Sulfur content, %	0.9
Moisture content, %	33.0

To produce 750,000 kilowatts of power with this fuel, each unit requires approximately 1,000,000 lb/hr of lignite fired. To meet this demand, the utility operates a drag line with a 94 cubic yard bucket at the mine source.

The particulate emissions code for Martin Lake is 0.1 lb/MMBtu. A 99.4% efficiency Research-Cottrell Double-Deck Electrostatic Precipitator is provided to achieve this requirement. The SO_2 emissions code for Martin Lake is 1.2 lb/MMBtu. This computes to about 71% SO_2 removal for the worst case (i.e., maximum sulfur) fuel. Other requirements for the FGD System included: totally closed loop operation; a minimum of 25°F reheat at the stack entry; and the production of a truckable, dumpable disposal product that can be transported back to the mine site.

FGD System Design

The process provided by Research-Cottrell for SO_2 removal at Martin Lake is the Double-Loop Limestone System. A schematic of the basic process design is shown in Figure 2.4. The Double-Loop process differs from simple single-loop processes in that two separate sets of chemical conditions are maintained. In a single loop process the limestone slurry contacting the flue gas is kept at a fixed set of chemical conditions (i.e., solids composition and pH) for the required level of SO_2 removal. Since high levels of SO_2 removal efficiency can only be obtained with sufficiently high inventories of limestone reagent present, and since the solids blowdown to the solids handling/dewatering system is at the same chemical conditions as the scrubbing slurry, it is difficult to achieve both high SO_2 removal efficiencies and high reagent utilization with this type of system. As a result, typical first generation systems of this type achieve reagent utilization levels no better than 70 to 80%.

In a two-loop process, Figure 2.5, the first slurry loop (i.e., the Absorber Loop) operates identically as the single loop process operates to insure high SO_2 removal efficiencies. The second slurry loop (i.e., the Quencher Loop) receives the slurry discharge from the first loop and reuses it to obtain reagent utilization levels exceeding 90%, even with absorber tower SO_2 removal efficiencies exceeding 95%.

Figure 2.4: Double-Loop Process

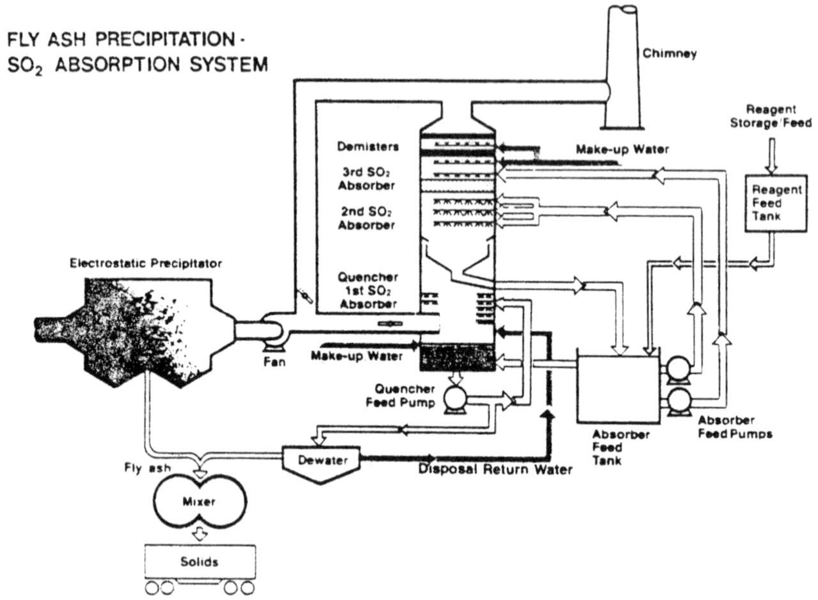

FLY ASH PRECIPITATION -
SO₂ ABSORPTION SYSTEM

Source: Symposium—Vol. 1

Figure 2.5: Two-Loop System

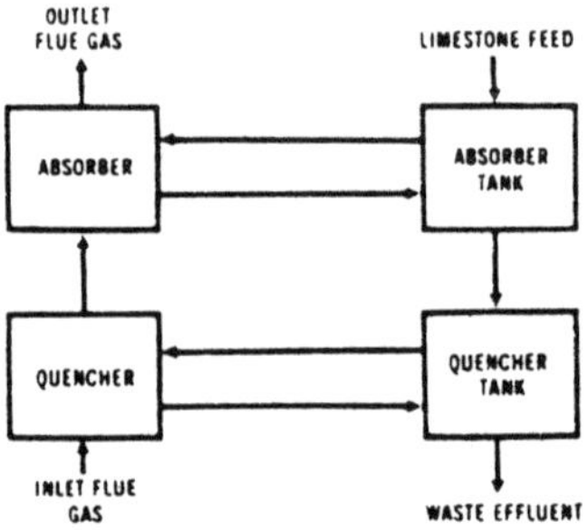

Source: Symposium—Vol. 1

Using the Double-Loop approach, six absorber towers each were provided for Martin Lake #1 and for Martin Lake #2. A schematic of the arrangement of the absorber area is shown in Figure 2.6. Each absorber tower is designed to treat 12.5% of the maximum boiler flue gas output. Therefore the entire FGD System can treat up to 75% of the maximum boiler output with all six towers in operation. In order to achieve the 71% overall SO₂ removal requirement for the worst fuel case, this means that each tower must be capable of achieving 95% SO₂ removal efficiency.

Figure 2.6: Absorber Area Schematic

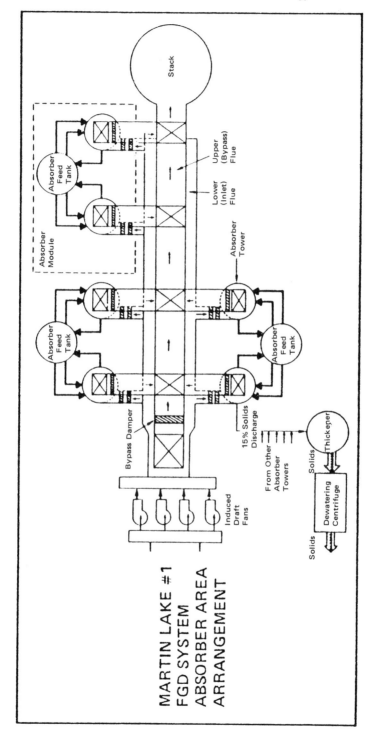

MARTIN LAKE #1
FGD SYSTEM
ABSORBER AREA
ARRANGEMENT

Source: Symposium—Vol. 1

The Martin Lake absorber tower design is presented in Figure 2.7. Each absorber tower, 28 feet in diameter by 100 feet tall, features three stages of SO_2 removal units followed by a two-stage mist elimination system. Boiler flue gas at 335°F enters the tower tangentially into the cyclonic quencher, a cocurrent spray chamber feeding quencher (i.e., second slurry loop) slurry. Here the gas is quenched to saturation, most of the fly ash not removed in the ESP is removed, and 25 to 30% of the SO_2 is removed. After passing through a liquid-gas separator which recycles absorber slurry back to the Absorber Feed Tank, the gas passes through three levels of absorber sprays and two feet of the fixed bed wetted film contactor. These two SO_2 removal sections remove virtually all of the remaining SO_2 by contacting the flue gas with absorber (i.e., first slurry loop) slurry. The cleaned gas then passes through two chevron mist eliminator stages and exits the absorber tower.

Figure 2.7: Absorber Tower

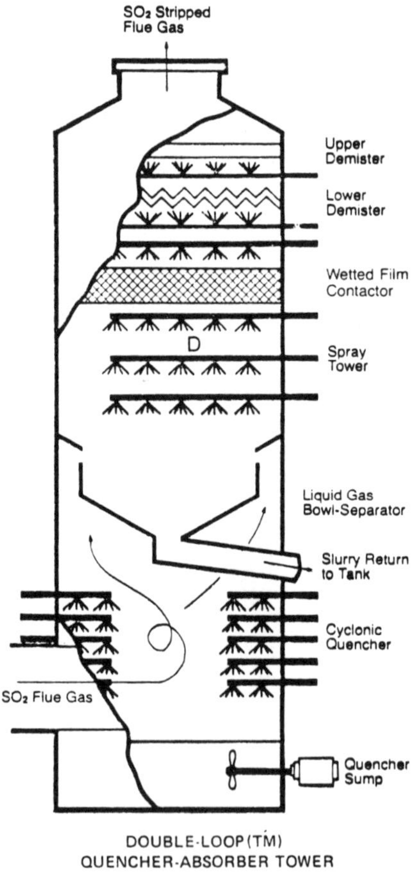

DOUBLE-LOOP(TM)
QUENCHER-ABSORBER TOWER

Source: Symposium—Vol. 1

Once the cleaned gases exit the absorber towers, they are recombined with the portion of the boiler flue gases that was not scrubbed to reheat the treated gases. Since no more than 75% of the gases is treated at any one time, a minimum of 48°F of reheat (well above the required 25°F) is obtained. Note that at lower sulfur levels, where less than 71% overall SO_2 removal is required, less gas is treated so that more gas is bypassed giving even greater levels of reheat.

Solids Handling System Design

Waste slurries from each absorber tower at 15% solids combine into a single stream to feed a 140 ft diameter by 12 ft sidewall gravity thickener. The thickener underflow at about 35% solids is fed to one of three centrifuges for additional dewatering. Cake from the centrifuges at 68 to 70% solids then discharges into a Muller-type blender. Here the cake combines with fly ash collected in the ESP to form a truckable, dumpable blend. Railcars receive the blend for ultimate disposal.

Thickener overflow and centrate streams are collected in wet wells for recycle back to the FGD System to insure fully closed loop operation. First generation FGD systems experienced process problems due to closed loop operation. As shown in Figure 2.8, these single loop systems do not isolate the return water to any part of the process, but instead expose the entire process to disposal return water.

The Double-Loop closed loop system works differently in that disposal return streams are concentrated only in the Quencher Loop. The Absorber Loop, with the wetted film contactor and demisters, runs as an open loop, even though the total process is closed loop. As a result, the Absorber Loop can operate unsaturated in calcium sulfate, minimizing the potential for harmful scale formation.

FGD System Performance

Several test programs were conducted by Research-Cottrell for TUGCo to evaluate the Martin Lake FGD System and to optimize operating setpoints. The following summarizes key results of the testing.

SO_2 **Removal:** Towers with four feet of wetted film contactor achieved greater than 99% SO_2 removal. Towers with no wetted film contactor achieved 80 to 85% removal at peak tower gas velocities. Based on this testing, two feet of WFC was installed in each tower and the maximum tower gas throughput was increased by 10%. Unit #1 acceptance tests in August of 1978 revealed absorber tower SO_2 removal to be greater than 98% with two feet of WFC at design velocities.

Reagent Utilization: Typically greater than 90%; average of 90 to 92% utilization. Tests on tower 1C-T200 on in situ forced oxidation showed that the tower can be operated with a reagent utilization in excess of 98% on a continuous basis.

Scale Control in Closed Loop Operation: Testing of the saturation levels of sulfate in each loop indicated an average Absorber Loop sulfate saturation level ~ 0.90 and an average Quencher Loop sulfate saturation level ~ 1.20. This indicates that, even during closed loop operation, the Absorber Loop operated unsaturated in SO_4, preventing sulfate scale formation in the absorber section of the towers. The Quencher Loop sulfate level shows, as expected, the sulfate level to be supersaturated but below the 1.3 times saturation critical level.

Figure 2.8: Single- and Double-Loop Closed-Loop Systems

SINGLE LOOP CLOSED LOOP SYSTEM

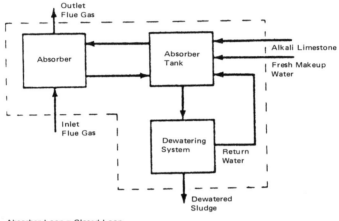

Absorber Loop = Closed Loop
Total Process = Closed Loop

DOUBLE-LOOP CLOSED LOOP SYSTEM

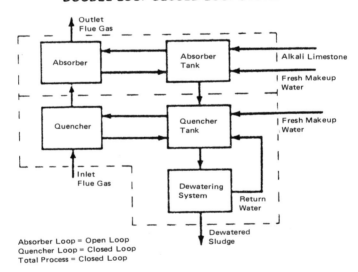

Absorber Loop = Open Loop
Quencher Loop = Closed Loop
Total Process = Closed Loop

Source: Symposium—Vol. 1

Disposal % Solids: Each system operated within 1 to 2% of design % solids with the thickener overflow $\sim$ 35% solids, the centrifuge cake $\sim$ 68% solids and the fly ash sludge blend at $\sim$ 82% solids.

Tower Pressure Drop: The absorber tower pressure drop at the design gas flow rate was within 0.1 IWC of the predicted value at 4.5 IWC. This does not include the two tower inlet louver dampers.

Degree of Oxidation: During normal operation, the quencher discharge has a sulfite:sulfate ratio of about 50:50 (or 1:1). A test program, using in situ forced oxidation on tower 1C-T200, produced 100 tons of solids with a sulfite: sulfate ratio of 1:99.

SURVEY OF ADVANCED SYSTEMS

EMERGING PROCESSES

The information in this section is based on "Status of Flue
Gas Desulfurization in the United States," prepared by
B.A. Laseke and T.W. Devitt of PEDCo Environmental,
Inc. in Symposium—Vol 1.

Considerable progress has been made in the development of conventional
and emerging or advanced FGD processes. Much of this information has been
acquired from the design and operation of first-generation FGD systems and
translated into more effective designs and improved operation of newer systems.
This section summarizes these emerging processes and presents a brief overview
of their current status.

For the purposes of this discussion, processes within the emerging or advanced category are defined as those that incorporate major design and operating
changes and thereby differ significantly from conventional direct lime/limestone
processes. Of the processes so categorized, several have been evaluated at pilot
and prototype development levels, and a few have progressed to the installation
and operation of demonstration or full-scale units. Table 3.1 provides a brief
summary of the emerging processes and highlights their current level of development and the extent of operating experience.

As is evident from Table 3.1, most of the previous operating experience
has been with wet-phase sodium- and magnesium-based processes that produce a
recoverable, marketable by-product or a high-quality filter cake. As might be
expected, this substantial experience has resulted in the commercial application
of dual alkali, Wellman-Lord, sodium carbonate, and magnesium oxide scrubbing
systems.

The most recent promising development in emerging processes involves dry
collection systems. Integrated processes designed to remove two or more pollutants simultaneously from the flue gas of a coal-fired boiler have commanded
considerable interest for economic operational reasons. The success of fabric
filters in removing particulates from the flue gases of coal-fired boilers prompted
investigations into the feasibility of using these filters to control both particulates and sulfur dioxide.

Table 3.1: Major Emerging FGD Processes

Process	Developer	Current Level of Development	Previous Operating Experience	Remarks
Aqueous carbonate	Rockwell International	100-MW system (planned)	Mohave pilot plant test program	Full-scale application not yet demonstrated.
Catalytic oxidation	Monsanto	100-MW system (terminated)	Wood River test program	No further process development.
Chiyoda Thorough-bred 101	Chiyoda International	20-MW system (terminated)	Scholz prototype test program	Development of the process has ceased in favor of a new design concept (Thorough-bred 121 which employs limestone reagent in a jet bubbler reactor).
Copper oxide absorption	Shell/Universal Oil Products	pilot plant	Big Bend test program	Process available for prototype or demonstra-tion application. No systems planned at present.
Dual alkali	A.D. Little/Combustion Equipment Associates	277-MW system (construction)	Scholz prototype test program	Full-scale application not yet demonstrated.
	FMC	250-MW system (construction)	Industrial systems and utility pilot plants	Full-scale application not yet demonstrated.
	Buell/Envirotech	575-MW system (construction)	Gadsby pilot plant test program	Full-scale application not yet demonstrated
Magnesium oxide	Chemico	150-MW system (terminated) 95-MW system (terminated)	Mystic test program; Dickerson test program	Process demonstrated on full-scale oil- and coal-fired boilers. System now offered for commercial application.
	United Engineers	600-MW system (planned)	Eddystone test program	Eddystone 120-MW prototype test program still in progress. A full-scale 600-MW system planned for TVA's Johnsonville Station.
Sodium carbonate	A.D. Little/Combustion Equipment Associates	Three 115-MW systems (opera-tional)	Reid Gardner Station	Three full-scale sodium carbonate (trona) FGD systems have been in service on coal-fired boilers at Reid Gardner Station (Nevada Power). System performance has been good.
	Universal Oil Products	509-MW system (planned)	Jim Bridger pilot test program	Full-scale sodium carbonate (30% sodium car-bonate purge solution from soda ash plant) FGD system planned for Jim Bridger Station.

(continued)

Table 3.1: (continued)

Process	Developer	Current Level of Development	Previous Operating Experience	Remarks
Wellman Lord	Davy Powergas	115-MW system (operational) 375-MW system (operational) 340-MW system (operational)	Crane test program	115-MW NIPSCO/EPA test program in progress at Mitchell. Two full-scale systems have recently started operations at Public Service of New Mexico's San Juan Station.
Dry adsorption	Foster Wheeler-Bergbau Forschung	20-MW system (terminated)	Scholz prototype test program	No further development of system reported. Further evaluation of the sulfur reduction component (RESOX) in Germany.
Dry collection	Wheelabrator-Frye Rockwell International	410-MW system (planned)	Leland Olds and Bowen Engineering pilot plant test program	Successful testing has resulted in planning of full-scale spray dryer and fabric filter FGD system at the Coyote Station.
	Joy/Niro	455-MW system (planned)	Leland Olds pilot plant test program	Successful pilot plant testing has resulted in planning of full-scale system at Antelope Valley. This system will use lime slurry as the reagent in 2-stage atomizer/fabric filter design configuration.
	Babcock & Wilcox	550-MW system (planned)	Basin electric pilot plant test program	Successful pilot plant testing has resulted in planning full-scale system at Laramie River. This system will use lime slurry as reagent in a configuration which employs an ESP as the second stage collector.
	Carborundum/Delaval	pilot plant	Leland Olds pilot plant test program	Process similar in design to Wheelabrator-Frye/R.I. and Joy/Niro processes. No full-scale applications have been announced.

Source: Symposium-Vol 1

PROCESS DESCRIPTIONS

The information in this section is based on the following reports:

Economic and Design Factors for Flue Gas Desulfurization Technology, April 1980, EPRI CS-1428, prepared by Bechtel National, Inc. for Electric Power Research Institute;

Comparative Economics of Advanced Regenerable Flue Gas Desulfurization Processes, March 1980, EPRI CS-1381, prepared by M.R. Beychok for Electric Power Research Institute;

"Flue Gas Desulfurization and Fertilizer Manufacturing: Pircon-Peck Process" by R.B. Boyda of A.G. McKee & Company in *Proceedings: Symposium on Flue Gas Desulfurization, Las Vegas, Nevada, March 1979,* Vol II, EPA-60017-79-167b, compiled by F.A. Ayer of Research Triangle Institute for the U.S. Environmental Protection Agency (These proceedings will be hereinafter referred to as Symposium—Vol 2.);

"SO_2 and NO_x Removal Technology in Japan" by J. Ando of Chuo University in Symposium—Vol 2;

NATO-CCMS study, *Flue Gas Desulfurization Pilot Study—Phase I—Survey of Major Installations—Appendix 95-H, Flue Gas Desulfurization by Scrubbing with Dilute Sulfuric Acid,* January 1979, NTIS PB-295 009, prepared by F. Princiotta of U.S. Environmental Protection Agency and R.W. Gerstle and E. Schindler of PEDCo Environmental.

This section presents descriptions of advanced processes with only brief summaries included of those processes more fully described in subsequent chapters.

Process Summaries

Double Alkali Process: As applied to U.S. FGD applications, a double alkali process utilizes a sodium sulfite solution to absorb SO_2 from flue gas. The spent absorbent is reacted with lime to precipitate calcium sulfite and regenerate the active sodium absorbent. The precipitated calcium salts are separated and dewatered for disposal.

There are two versions of the process, a concentrated mode and a dilute mode. The selection is dependent on the level of absorbent oxidation (sulfite conversion to sulfate) experienced during absorption. Oxidation is favored by high oxygen-to-sulfur-dioxide ratios and by any catalytic activity contributed by the traces of fly ash captured by the absorber.

The "concentrated" mode process is favored for high-sulfur-coal uses (low oxidation) and is the simpler of the two process versions. The "dilute" mode process is applicable to low-sulfur-coal cases (high oxidation) and is the more complex and inherently more costly process for the same scope.

Wellman-Lord Process: The Wellman-Lord process employs wet absorption of SO_2 from flue gas by reaction with sodium sulfite to form sodium bisulfite and some sodium sulfate. Desulfurized flue gas is reheated and released to the atmosphere. The primary reactions occurring in the absorber are:

(1) $$Na_2SO_3 + H_2O + SO_2 \rightarrow 2NaHSO_3$$

(2) $$Na_2SO_3 + \frac{1}{2}O_2 \rightarrow Na_2SO_4$$

The absorber effluent liquor is filtered to remove solids and divided into two streams, a portion going to the regeneration area and the remainder going to purge treatment to reject the unreactive sodium sulfate. Double-effect evaporator/crystallizers are used to convert the dissolved $NaHSO_3$ to crystalline Na_2SO_3 and liberated SO_2 by the reverse of reaction (1). The regenerated Na_2SO_3 crystals are dissolved and returned to the absorbers. The regenerated SO_2 stream is converted to elemental sulfur by reduction with natural gas in an Allied Chemical SO_2 reduction plant.

Magnesia Slurry Process: The magnesia slurry process uses an aqueous slurry of magnesium oxide to absorb and remove SO_2 from flue gases. The resulting magnesium sulfite and magnesium sulfate slurry is dried and calcined to yield recycle magnesium oxide and an SO_2-rich gas suitable for conversion to sulfuric acid.

The magnesia slurry process has been demonstrated in operation at oil- and coal-burning power plants in the United States and in Japan. The demonstration plants have operated on flue gas flows equivalent to those from power plants with outputs of up to 150 MW. The process design, engineering and construction of proprietary versions of the magnesia slurry process can be obtained in the United States from the Chemico Air Pollution Control Company and from United Engineers and Constructors.

Citrate Process: The absorption/steam stripping process as marketed by Flakt, Inc., the United States subsidiary of AB Svenska Flaktfabriken (SD) of Sweden, utilizes a solution of sodium citrate to absorb SO_2 from flue gas. The scrubber effluent liquor is stripped with low-pressure steam to liberate SO_2. The dilute SO_2 stream is then reduced to elemental sulfur using anthracite coal in a Foster Wheeler RESOX plant. A portion of the stripped effluent liquor is treated to remove sulfates and impurities and then recombined with the balance of the liquor before being returned to the scrubber sprays. The overall chemical reactions occurring in this process may be written as follows:

Absorption
$$Na_3Ci + SO_2 + H_2O \rightarrow NaHSO_3 + Na_2HCi$$
$$Na_2HCi + SO_2 + H_2O \rightarrow NaHSO_3 + NaH_2Ci$$
$$NaH_2Ci + SO_2 + H_2O \rightarrow NaHSO_3 + H_3Ci$$

Oxidation
$$2NaHSO_3 + O_2 \rightarrow Na_2SO_4 + H_2SO_4$$
$$H_2SO_4 + Na_3Ci \rightarrow Na_2SO_4 + NaH_2Ci$$

Desorption
$$NaHSO_3 + Na_2HCi \xrightarrow{\text{Heat}} H_2O + SO_2 + Na_3Ci$$

Reduction
$$C + SO_2 \rightarrow CO_2 + S$$

Lime Slurry Spray Dryer/Fabric Filter Process: The lime slurry spray dryer/fabric filter process selected for evaluation in this report uses a slurry of hydrated lime $[Ca(OH)_2]$, calcium sulfite, and fly ash to sorb SO_2 from flue gas by contact with atomized droplets of the alkali slurry. These droplets are typically generated by high-speed centrifugal atomizers.

The following reactions occur in the spray dryer reaction chamber:

(1) $$Ca(OH)_2 + SO_2 \rightarrow CaSO_3 + H_2O$$

(2) $$CaSO_3 + \tfrac{1}{2}O_2 \rightarrow CaSO_4$$

Reaction (1) is the primary reaction for SO_2 removal. The dry salt mixture produced by these reactions is usually about 70% anhydrous sulfite and 30% anhydrous sulfate.

The atomized droplets rapidly evaporate in the spray dryer to produce a cooled and partially humidified particulate-laden gas from which most of the SO_2 has been removed. The mixture of fly ash, reaction products, and unreacted absorbent is removed from the gas stream in fabric filters. As the flue gas passes through the particulate-laden fabric, some additional SO_2 is absorbed by the remaining unreacted alkali, reducing the total amount of unreacted alkali in the filter cake.

Clean flue gas discharges from the fabric filters (baghouses) to the ID fans which discharge directly to the chimney. Individual groups of filter bags are periodically isolated from the flue gas stream and back-blown with clean flue gas to dislodge the filter cake, which is discharged from the baghouse hoppers and conveyed to storage. The total waste product stream is divided, the bulk of the material being trucked to a landfill site, and a varying amount being recycled to the absorbent preparation section to utilize the unreacted alkali.

Dry Bicarbonate Process: The dry bicarbonate process, proposed by Atomics International (since renamed the Environmental & Energy Systems Division of Rockwell International), injects dry sodium bicarbonate into flue gases to react with and remove SO_2 from the gases. The dry reaction products (a mixture of sodium sulfate and unreacted sodium bicarbonate) are removed from the flue gases by passage through a baghouse. Subsequent processing of the reaction products yields regenerated sodium bicarbonate and hydrogen sulfide gas. The hydrogen sulfide gas is converted to elemental sulfur as an end product. The dry bicarbonate process was offered by Atomics International with United Engineers and Constructors participating as an architect-engineering subcontractor.

Carbon Adsorption Process: The dry adsorption process, developed by Bergbau Forschung GmbH, removes SO_2 from flue gases by adsorption in a moving bed of activated char. Subsequent heating of the saturated char yields concentrated gaseous SO_2 and regenerates the char for reuse. The gaseous SO_2 is reduced and converted to elemental sulfur with the RESOX process, developed by the Foster Wheeler Energy Corporation, which uses anthracite coal to reduce the SO_2.

The overall combined system was jointly proposed by Bergbau Forschung GmbH, Foster Wheeler, STEAG AG and Deutsche Babcock AG with program support by the Federal Republic of Germany's environmental agency (Umweltbundesamt).

Chiyoda Thoroughbred 101 Process

General Process Description: The CT-101 process uses dilute sulfuric acid solution to absorb the SO_2 in the flue gas. The absorbed SO_2 gas is oxidized with air and forms additional sulfuric acid. The acid thus produced is reacted with limestone ($CaCO_3$) to produce gypsum ($CaSO_4 \cdot 2H_2O$) as a saleable by-product. Ferric ion was chosen as a catalyst for the oxidation step because it is harmless, available, and resistant to poisoning by contaminants. Figure 3.1 presents a simplified hypothetical flow diagram of a 500-MW, coal-fired boiler with a dilute sulfuric acid desulfurization system.

Figure 3.1: Flow Diagram for a Dilute Sulfuric Acid Desulfurization Process

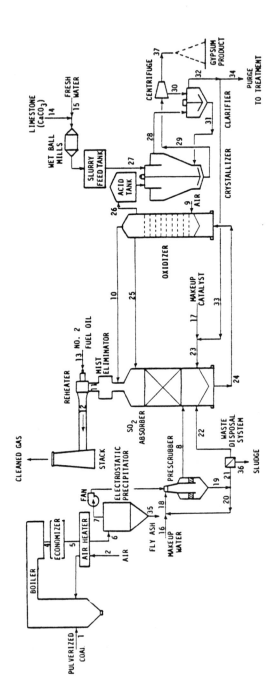

Process Chemistry: The CT-101 process includes three major process areas: absorption, oxidation and crystallization. The overall reactions that take place in the absorber are:

$$SO_2 \text{ (g)} + H_2O \rightarrow H_2SO_3$$

At the flue gas–scrubbing solution interface in the absorber, the SO_3 anion forms in the following manner:

$$H_2SO_3 \rightarrow H^+ + HSO_3^- \quad \text{and} \quad HSO_3^- \rightarrow H^+ + SO_3^=$$

The catalytic oxidation reaction proceeds as follows:

$$SO_2 \text{ (dis)} + 2H_2O + Fe_2(SO_4)_3 \rightarrow 2FeSO_4 + 2H_2SO_4$$

$$SO_2 \text{ (dis)} + O_2 \text{ (dis)} + 2FeSO_4 \rightarrow Fe_2(SO_4)_3$$

$$2H^+ + SO_3^= + H_2O + Fe_2(SO_4)_3 \rightarrow 2FeSO_4 + 2H_2SO_4$$

In this stage, oxidation is aided by dissolved catalyst of ferric sulfate, $Fe_2(SO_4)_3$.

The sulfuric acid is sent to the crystallizer and neutralized with limestone to a concentration of 1% H_2SO_4, which generates a gypsum by-product in the following manner:

$$H_2SO_4 \cdot H_2O + CaCO_3 \rightarrow CaSO_4 \cdot 2H_2O\downarrow + CO_2\uparrow$$

As a result of heat loss, the operating temperature is approximately 5°C lower than in the absorption and oxidation sections.

According to the stoichiometry, 1 mol of limestone is required to remove 1 mol of SO_2. Assuming 10% excess, actual limestone required is 1.72 kg/kg of SO_2 and the gypsum produced is 2.84 kg/kg of SO_2. The gypsum by-product contains about 10% free water.

Chiyoda Thoroughbred 121 Process

Process Description: The Chiyoda Thoroughbred 121 (CT-121) process is a second-generation limestone slurry process in which SO_2 absorption, sulfite/bisulfite oxidation, and precipitation of calcium sulfate as gypsum are accomplished in a single reaction vessel. The overall chemical reactions are the same as those occurring in the limestone slurry process, except that virtually all of the sulfur is removed as an easily dewatered gypsum slurry rather than as mixed sulfite/sulfate sludge which is difficult to dewater. Figure 3.2 is a simplified flow diagram developed by Bechtel on the basis of previously published descriptions of the process.

A summary of the basic process design parameters for the CT-121 FGD system is presented in Table 3.2. The corresponding raw material and utility requirements are presented in Table 3.3. These data were provided by Chiyoda International Corporation, and were reproduced in EPRI CS-1428 with their concurrence.

Description of Major Process Subsystems: *Slurry Preparation* — Limestone is received by rail in uncovered bottom-dump rail cars. On-site storage consists of an uncovered reserve storage pile and a ball mill feed bin. Limestone is ground in wet ball mills and diluted with recycled process wastewater to produce slurry for makeup to the absorption section.

Flue Gas Desulfurization Technology

Figure 3.2: Chiyoda Thoroughbred 121 Process—Flow Diagram

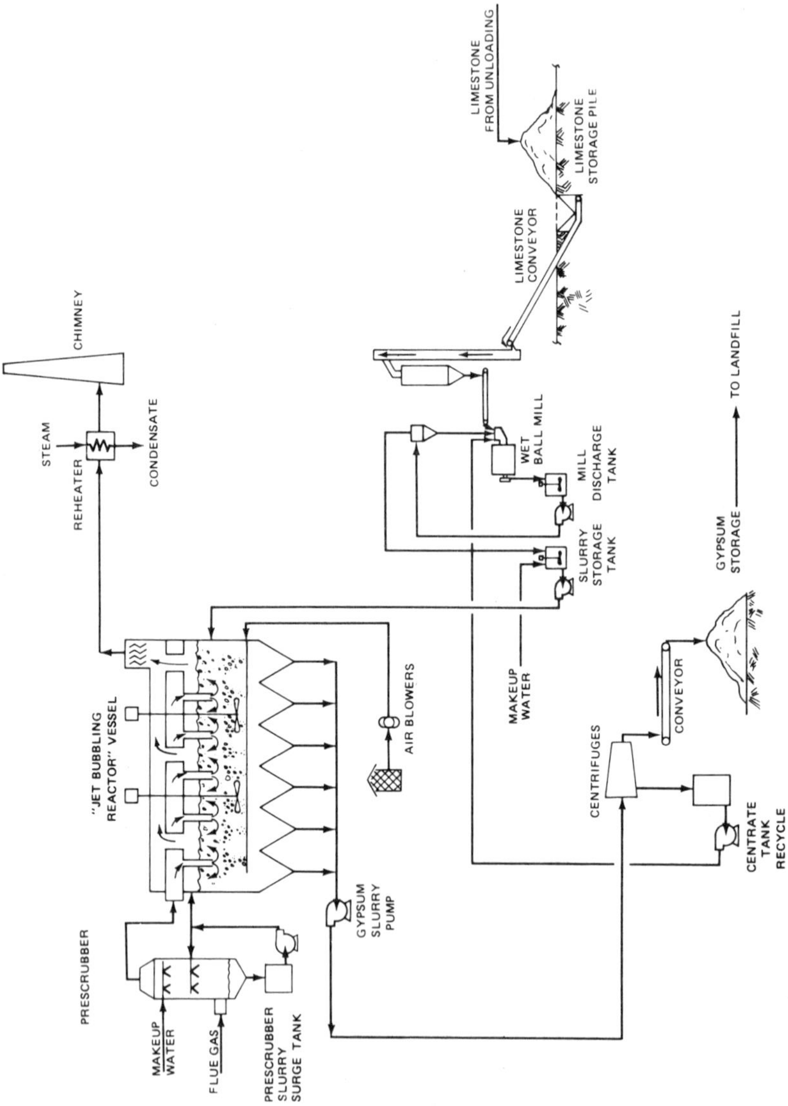

Source: EPRI CS-1428

Table 3.2: Chiyoda Thoroughbred 121 Process System Parameters

Data for 1 x 500 MW Unit
(Maximum Load Conditions)

	Eastern Coal	Western Coal
Flue gas flow rate, m^3/sec (10^3 acfm)	856 (1,814)	942 (1,995)
Absorption trains per unit	4	4
Capacity per train, %	33⅓	33⅓
Gas flow per absorber, saturated, m^3/sec (10^3 acfm)	239 (507)	261 (552)
Presaturator type	. . . . spray tower.	
Presaturator L/G, ℓ/m^3 (gal/10^3 acf)	0.936 (7)	0.936 (7)
Absorber type	. . . . . .jet bubbling reactor	
Mist elimination	. . . . .chevron	
System total pressure loss, pascals (inch wg)	4,982 (20)	2,740 (11)
Absorbent stoichiometric ratio*	1.01	1.01
Oxidation air stoichiometric ratio*, %	4	10
Flue gas reheat required, °C (°F)	31.1 (56)	26.7 (48)

*Mols alkali per mol SO_2 removed.

Source: EPRI CS-1428

Table 3.3: Chiyoda Thoroughbred 121 Process Raw Materials and Utility Requirements

Data for 1 x 500 MW Unit
(Maximum Load, Average Sulfur)

	Eastern Coal	Western Coal
Raw materials		
Limestone, 10^3 kg/hr (tons/hr)	25.6 (28.2)	3.9 (4.3)
Utilities		
Electricity		
Raw material handling and preparation, kW	730	110
Pretreatment and absorption, kW	2,510	510
Waste dewatering and storage, kW	300	55
Fans, kW	5,190	3,160
Total, kW	8,730	3,835
Steam, 10^6 Btu/hr	96.6	89.4
Process water, 10^3 ℓ/hr (10^3 gal/hr)	122.3 (32.3)	111.7 (29.5)
Products		
Gypsum (10 wt % moisture),		
10^3 kg/hr (tons/hr)	44.8 (49.4)	6.9 (7.6)
m^3/hr (yd^3/hr)	33.0 (43.1)	5.1 (6.6)
Incremental heat rate penalties		
Steam (basis gross heat rate), %	2.14	1.91
Electricity (basis unit net rating), %	1.75	0.77
Corrected net heat rate, Btu/kWh	9,931	10,168
Variation from base, %	-0.55	-1.21

Source: EPRI CS-1428

SO₂ Absorption — Hot flue gas discharges from the ID fans into a plenum from which it is distributed to the operating absorption trains. Before entering the absorbers the hot flue gas passes through low-pressure-drop spray tower pre-scrubbers where it is adiabatically cooled by sprays of recirculating fly ash slurry which also remove residual fly ash and some SO_2, SO_3 and chlorides. A slip stream of the recirculating slurry is sent to the absorber vessels where it is neutralized.

This bleed stream is controlled to maintain the suspended solids level in the recirculating slurry at about 5%. Make-up water is added to the prescrubber vessels to offset evaporation and bleed losses.

The cooled, saturated gas enters a distribution plenum within the absorber vessels, and is sparged into a tank containing a slurry of limestone and gypsum. The slurry is moderately stirred both by mechanical agitators and by oxidizing air which enters through distribution headers near the bottom of the vessel. The sparged gas creates a froth layer above the liquid in which SO_2 absorption and sulfite oxidation take place.

The sulfates formed in the froth layer precipitate as gypsum crystals which grow as they circulate in the slurry bath until they reach sufficient size to settle to the bottom of the reaction vessel. Gypsum slurry is continually withdrawn from the bottom of the reaction vessel. A portion of this slurry is recycled to the reaction tank to provide seed crystals, while the balance is sent to centrifuges for dewatering.

The treated flue gas passes through chevron mist eliminators and reheaters before being discharged through the chimney.

Waste Slurry Dewatering and Storage — Gypsum slurry from the absorption section is centrifuged to produce a wet gypsum cake, which is stockpiled and subsequently trucked off-site for landfill disposal. Centrate liquor is recycled to the slurry preparation area to minimize makeup water requirements.

Development Status: Development of the CT-121 process was initiated in Japan for oil firing in 1975 in an effort to reduce the cost and complexity of the commercial CT-101 process. Tests were initiated at the bench and laboratory scale, and finally at the 0.354 m³/sec (750 scfm) level. A conceptual design of a 59 m³/sec (125,000 scfm) plant has been completed, and a 23 MW pilot plant at Gulf Power Company's coal-fired Scholz Power Station started in mid-1978 and is currently undergoing testing.

Ammonia Scrubbing Process

The ammonia scrubbing process proposed jointly by Catalytic, Inc. (a subsidiary of Air Products and Chemicals, Inc.) and Institut Francais du Petrole (IFP) uses an aqueous ammonia solution to absorb and remove SO_2 from flue gases. The resulting ammoniacal salt brine is then regenerated to yield elemental sulfur as an end product.

The SO_2 absorption section of the overall process involves ammonia scrubbing technology developed by Catalytic. The brine regeneration is based on technology developed and licensed by IFP.

Process Description: A schematic flow diagram of the Catalytic/IFP process is presented in Figure 3.3. Flue gas from an electrostatic precipitator (ESP) in a coal-burning power plant is first washed with water in a venturi scrubber to condition the flue gas and to remove chlorides, some sulfur trioxide and residual fly ash. The flue gas then enters the absorber.

Figure 3.3: Catalytic/IFP Ammonia Scrubbing Process

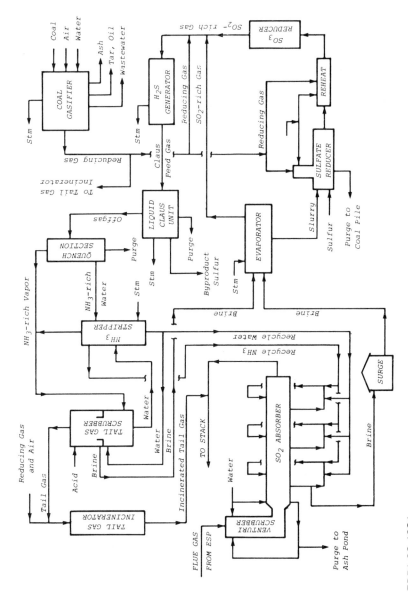

Absorption Section — The Catalytic/IFP process uses multiple scrubbing stages in a horizontal absorber. The SO_2 absorption stages contain Koch Flexipac packing and the final stage is a spray chamber in which the flue gas is scrubbed with water to remove and recover any residual ammonia. Mist eliminators are included after the venturi and after each stage. Each stage of the absorber includes a liquid storage sump and circulation pump, and the circulation of the scrubbing liquid within each stage is controlled so as to achieve the desired ratio of liquid flow to gas flow. The scrubbing liquid cascades stage-by-stage to the first SO_2 absorption stage. In effect, the flow of the scrubbing liquid through the absorption stages is countercurrent relative to the flue gas flow, except that ammonia solution is fed to the individual stages as required to control the scrubbing liquor concentration.

SO_2 is removed from the flue gas by absorption in the ammonia solution which the gas contacts as it flows through the absorption stages of the absorber. The primary reactions occurring in the SO_2 absorption stages are:

$$SO_2 + 2NH_4OH \rightleftharpoons (NH_4)_2SO_3 + H_2O$$
$$SO_2 + (NH_4)_2SO_3 + H_2O \rightleftharpoons 2NH_4HSO_3$$
$$SO_3 + (NH_4)_2SO_3 \rightleftharpoons (NH_4)_2SO_4 + SO_2$$
$$2(NH_4)_2SO_3 + O_2 \rightleftharpoons 2(NH_4)_2SO_4$$

Only about 10% of the sulfur values are oxidized to $(NH_4)_2SO_4$ in the scrubber.

The flue gas is then scrubbed with water to remove any residual ammonia and passes through the final mist eliminator before leaving the absorber.

The scrubbing liquid leaving the absorber is a concentrated brine of these ammoniacal salts: ammonium sulfite, ammonium bisulfite and ammonium sulfate.

Regeneration Section (Reduction of Ammoniacal Salts) — The concentrated brine from the absorption section is sent to an intermediate surge storage tank from which the brine is pumped into a steam-heated forced circulation evaporator. Approximately 60% of the brine is evaporated and thermally decomposed to yield an SO_2-rich gas which also contains ammonia and water vapor.

The reactions occurring in the evaporator are:

$$(NH_4)_2SO_3 \rightarrow 2NH_3 + H_2O + SO_2$$
$$NH_4HSO_3 \rightarrow NH_3 + H_2O + SO_2$$

Since the sulfates in the brine do not decompose in the evaporator, a concentrated sulfate slurry is withdrawn from the evaporator bottom and sent to a sulfate reducing reactor. In the presence of molten sulfur, a temperature of about 600° to 700°F and a reducing atmosphere, the sulfates are decomposed in the sulfate reducer to yield an SO_2-rich gas which also contains ammonia, water vapor and a small amount of SO_3. The reactions occurring are:

$$(NH_4)_2SO_4 \rightarrow NH_4HSO_4 + NH_3$$
$$2NH_4HSO_4 + S \rightarrow 3SO_2 + 2NH_3 + 2H_2O$$
$$(NH_4)_2SO_4 \rightarrow SO_3 + 2NH_3 + H_2O$$

Reaction heat and the reducing atmosphere for the sulfate reducer are provided by the submerged exhaust of gas combusted within the reactor.

The gas yielded from the sulfate reducer passes through an in-line fired reheater and into an SO_3 reactor where the reducing atmosphere catalytically converts the small amount of SO_3 in the gas to SO_2.

The SO_2-rich gas streams from the brine evaporator and from the SO_3 reducer are combined with a reducing gas containing hydrogen and carbon monoxide. The combined gases are then sent to a hydrogen sulfide (H_2S) generator.

Regeneration Section (H_2S Generator) — To utilize the Claus reaction for converting SO_2 into elemental sulfur, two-thirds of the SO_2 must first be converted to H_2S since the Claus reaction requires a feed gas with approximately a 2:1 molar ratio of H_2S to SO_2. To accomplish this, a portion of the SO_2 in the SO_2-rich gas is reacted with the carbon monoxide and/or hydrogen in any ratio in the reducing gas as follows:

$$SO_2 + 3CO + H_2O \rightarrow H_2S + 3CO_2$$
$$SO_2 + 3H_2 \rightarrow H_2S + 2H_2O$$

After conversion of the required portion of the SO_2 to H_2S, the gases are cooled in a waste heat boiler and sent to a liquid phase Claus reactor.

The reducing gas required for generating the H_2S is provided by a commercial air-blown coal gasifier. The gasifier also provides gas for: the submerged combuster supplying hot exhaust for use in the sulfate reducer, the in-line fired reheater ahead of the SO_3 reducer and the final tail gas incinerator. The gasifier includes facilities for separation and removal of gasifier bottom ash as well as for cleansing the reducing gas of fly ash, tar and oil, and wastewater.

Regeneration Section (Liquid Phase Claus Unit) — The liquid phase Claus unit consists of a vertical tower about 75% full of a polyethylene glycol (PEG) and sulfur solution. The conversion of H_2S and SO_2 to elemental sulfur occurs in the liquid phase via the following reaction:

$$2H_2S + SO_2 \rightarrow 3S + 2H_2O$$

Most of the molten sulfur end product from the Claus unit is sent to storage and a small percentage is recycled to the sulfate reducer.

Regeneration Section (Ammonia Recovery and Recycle) — The Claus unit off-gas is quenched with water to reduce its temperature and to remove sulfur vapor and PEG from the gas. The gas then goes through a water-cooled condenser and into a condenser flash drum. The uncondensed vapor from the flash drum (which contains ammonia) is sent to a tail gas scrubber. The NH_3-rich aqueous solution from the condenser flash drum is routed through a steam-reboiled NH_3 stripper from which the following streams are produced:

> An overhead, NH_3-rich vapor which is sent to the tail gas scrubber along with vapor from the quench condenser flash drum.

> An overhead solution of aqueous ammonia which is recycled for reuse in the flue gas SO_2 absorber.

> A bottoms stream of water, containing some ammonia, which is also recycled for reuse in the flue gas SO_2 absorber. A part of this water is used as scrubbing liquid in the tail gas scrubber.

The tail gas scrubber is a two-stage unit. In the lower section, water from the NH_3 stripper bottoms is used to absorb NH_3 from the NH_3-rich tail gases and the water recycles back to the NH_3 stripper. In the upper section, a sulfuric acid solution removes the last traces of NH_3 from the tail gases and the resulting ammonium sulfate brine is sent to the evaporator and the sulfate reducer.

The scrubbed tail gases are then incinerated and blended with the flue gas from the SO$_2$ absorber to provide about 20°F of flue gas reheat.

Purge Streams — A slip-stream of the wash water from the flue gas inlet venturi scrubber is neutralized with lime and then filtered. The filtrate is recycled for reuse in the venturi scrubber. The sludge from the filter is purged to the power plant's ash pond.

The sulfate reducer is purged on a periodic basis to discharge any buildup of metallic salts which may originate with fly ash carryover into the SO$_2$ absorber.

A "sulfate cream" is periodically skimmed from the PEG-sulfur interface in the Claus reaction tower. The skimmed cream is separated to recover PEG and sulfur for recycle to the Claus reactor. The sulfate solution which remains is then purged to the sulfate reducer.

A slip-stream from the Claus off-gas quenching water is similarly separated to recover PEG for recycle to the Claus reactor. The sulfur compounds which have accumulated and combined to form a thiosulfate solution are also purged to the sulfate reducer.

As discussed above, the coal gasifier which produces reducing gas also discharges ash, tar and oil, and wastewater which will require appropriate handling and disposal.

Conoco Process

The Consol Flue Gas Desulfurization Process, developed by Conoco Coal Development Company (a subsidiary of Continental Oil Company), uses an aqueous solution of potassium carbonate and other potassium salts to absorb and remove SO$_2$ from flue gases. The resulting potassium bisulfite solution is converted to a potassium thiosulfate solution which is then reduced with carbon monoxide to yield a regenerated solution of potassium carbonate and hydrogen sulfide gas. The hydrogen sulfide gas is concentrated and converted to elemental sulfur as an end product.

The Consol flue gas desulfurization process was offered by Conoco with The Rust Engineering Company participating as Conoco's engineering and construction subcontractor. The overall process includes two licensed subprocesses: the Selexol Process for concentrating hydrogen sulfide which is licensed by the Allied Chemical Company, and the Partial Oxidation Process for producing carbon monoxide (from heavy fuel oil) which is licensed by the Texaco Development Corporation.

Process Description: A schematic flow diagram of Conoco's process is presented in Figure 3.4. Flue gas from an electrostatic precipitator (ESP) in a coal-burning power plant is first washed with water in a venturi scrubber to remove chlorides, sulfur trioxide and residual fly ash. The flue gas then enters the SO$_2$ absorber.

Absorption Section — The flue gas flows upward through a packed SO$_2$ absorber where it is contacted by the downward flowing potassium carbonate solution which absorbs and removes SO$_2$ from the flue gas. The SO$_2$ reacts with the potassium carbonate (K$_2$CO$_3$) to form potassium bisulfite (KHSO$_3$) as shown below:

$$4SO_2 + 2K_2CO_3 + 2H_2O \rightarrow 4KHSO_3 + 2CO_2$$

Figure 3.4: Conoco Process

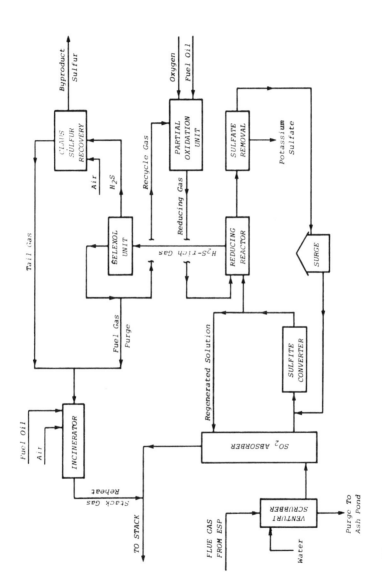

Source: EPRI CS-1381

The absorbent solution leaves the absorber, is joined by regenerated solution returning to the absorption section and the combined stream passes through the sulfite converter. The regenerated solution contains almost equimolar amounts of potassium carbonate and potassium hydrosulfide (KHS). The KHS in the regenerated solution reacts with the $KHSO_3$ in the absorbent solution to yield potassium thiosulfate $(K_2S_2O_3)$:

$$2KHS + 4KHSO_3 \rightarrow 3K_2S_2O_3 + 3H_2O$$

The potassium carbonate present in the sulfite converter does not enter into the reaction. The absorbent solution then recirculates from the sulfite converter back to top of the absorber for reuse. Thus, the absorbent solution entering the absorber is predominantly a potassium thiosulfate solution with a small amount of the active component (potassium carbonate) and a very small amount of potassium bisulfite.

Since potassium thiosulfate is a stable, nonoxidizable salt, the absorbent solution can be continuously recirculated through the SO_2 absorber. A small portion of the solution is withdrawn to be regenerated by reducing the potassium thiosulfate to potassium carbonate and potassium hydrosulfide so as to maintain the desired composition of the circulating absorbent solution.

Regeneration Section (Thiosulfate Reduction and Sulfate Removal) — The thiosulfate-rich solution from the absorption section is reacted with a carbon monoxide-rich reducing gas in a noncatalytic, thermal reductor. The reaction yields gaseous H_2S and regenerates the solution to potassium carbonate and potassium hydrosulfide:

$$3K_2S_2O_3 + 12CO + 5H_2O \rightarrow 2KHS + 2K_2CO_3 + 4H_2S + 10CO_2$$

The reaction occurs at about 450°F and 600 psig in an agitated reductor vessel.

After cooling and pressure letdown, the regenerated solution passes through a crystallizer and a filter. This removes any fly ash picked up in the absorber as well as a small amount of by-product potassium sulfate formed and accumulated in the absorbent solution. The regenerated solution then returns to the absorption section via an intermediate surge storage tank.

H_2S Concentration — The selexol process, licensed by the Allied Chemical Company, selectively absorbs H_2S from gases high in CO_2 content. The reductor off-gas entering the Selexol unit absorber contains about 11 and 48 volume percent of H_2S and CO_2, respectively. The H_2S is selectively absorbed in the lean Selexol absorbent (a solution of mixed polyethylene glycol dimethyl ethers) along with some CO_2. The remainder of the reductor off-gas passes through the absorber and exits the Selexol unit containing less than 125 ppm of H_2S.

The rich Selexol absorbent is then stripped to yield an off-gas which contains about 50 and 48 vol % of H_2S and CO_2, respectively. Thus, the H_2S has been separated from the reductor off-gas and has been concentrated by a factor of almost five.

The stripped Selexol absorbent is recycled to the Selexol absorber for reuse. The concentrated H_2S is sent to a conventional, vapor-phase Claus unit for conversion to elemental sulfur as an end product.

The sulfur-free reductor off-gas (less than 125 ppm of H_2S) from the Selexol absorber is divided into two streams. About 20% is burned as fuel gas for the incineration of the Claus unit tail gas. The remaining 80%, which is rich in CO_2, is recycled to the Partial Oxidation Unit for conversion of the CO_2 into CO-rich reducing gas.

Partial Oxidation Unit (Production of Reducing Gas) — The carbon mon-oxide-rich reducing gas, required for regenerating the potassium thiosulfate solu-tion, is produced by the partial oxidation of fuel oil using the proprietary Texaco Partial Oxidation Process.

In the partial oxidation unit, heavy fuel oil is reacted with oxygen of 95% purity to produce a reducing gas containing carbon monoxide, carbon dioxide and hydrogen. After cooling in a waste heat boiler and water scrubbing for removal of carbon fines, the product reducing gas is sent to the potassium thio-sulfate reductor.

Conditions in the partial oxidation unit are such that the reverse water gas reaction occurs:

$$CO_2 + H_2 \rightarrow CO + H_2O$$

Thus, the CO_2 recycled to the partial oxidation unit from the Selexol absorber off-gas acts to increase the yield of CO in the reducing gas from the partial oxida-tion unit. The flow of recycle CO_2 to the partial oxidation unit is controlled to maintain a 2:1 volume ratio of CO to CO_2 in the product reducing gas.

The carbon fines scrubbed from the product reducing gas are recycled to extinction within the partial oxidation unit.

The sulfur content of the fuel oil feedstock used in the partial oxidation unit is not a limiting constraint. Any H_2S produced in the reducing gas joins the H_2S produced in the potassium thiosulfate reductor and is subsequently con-verted to elemental sulfur in the Claus unit.

Claus Unit — The concentrated H_2S from the Selexol stripper is converted to elemental sulfur in a conventional, vapor-phase Claus unit. The overall reaction occurring in the Claus unit is:

$$2H_2S + O_2 \rightarrow 2S + 2H_2O$$

The Claus unit includes three converter stages to achieve a 95% conversion of H_2S to elemental sulfur.

The residual tail gas from the Claus unit is incinerated and blended with the flue gas from the SO_2 absorber to provide gas reheat.

Purge Streams — A slip-stream of the wash water from the flue gas inlet venturi scrubber is neutralized with lime and discharged to the power plant's ash pond.

The by-product potassium sulfate from the regeneration section will require appropriate handling and disposal.

Aqueous Carbonate Process

The Aqueous Carbonate Process, developed by Atomics International (since renamed the Environmental & Energy Systems Division of Rockwell International), contacts hot flue gases with an aqueous sodium carbonate solu-tion in a spray dryer. The sodium carbonate reacts with and removes SO_2 from the flue gases, and the solution is evaporated to dryness by the hot flue gases. The dry reaction products (a mixture of sodium sulfite, sodium sulfate and unreacted sodium carbonate) are removed from the flue gases by passage through cyclones and an electrostatic precipitator. Subsequent processing of the reaction products yields regenerated sodium carbonate and hydrogen sulfide gas. The hydrogen sulfide gas is converted to elemental sulfur as an end product.

Process Description: A schematic flow diagram of Atomics International's Aqueous Carbonate Process is presented in Figure 3.5. Flue gas from the combustion air preheaters of the boilers in a coal-burning power plant first passes through cyclones for removal of about 80% of the fly ash. The flue gas then enters the spray dryer at a temperature of 250° to 350°F.

Reaction Section — The flue gas flows through a spray dryer where it is contacted in cross-flow with atomized droplets of an aqueous sodium carbonate solution. The reaction between the SO_2 in the flue gas and the atomized sodium carbonate solution (Na_2CO_3) yields sodium sulfite (Na_2SO_3) and sodium sulfate (Na_2SO_4):

$$Na_2CO_3 + SO_2 \rightarrow Na_2SO_3 + CO_2$$
$$Na_2CO_3 + SO_2 + 0.5O_2 \rightarrow Na_2SO_4 + CO_2$$

Any SO_3 in the flue gas also reacts with Na_2CO_3 to yield Na_2SO_4:

$$Na_2CO_3 + SO_3 \rightarrow Na_2SO_4 + CO_2$$

The sensible heat of the hot flue gases evaporates the reaction product sodium salts and unreacted sodium carbonate to dryness. The flue gas leaves the spray dryer at a temperature of 150° to 250°F, passes through cyclones and an electrostatic precipitator, and the cleansed flue gas then exits to the plant stack. Passage of the flue gas through the cyclones and precipitator removes the dry reaction products along with unreacted sodium carbonate and the residual fly ash. The mixture of fly ash and spent carbonate (i.e., sodium sulfite, sodium sulfate and unreacted sodium carbonate) collected in the cyclone and precipitator hoppers is pneumatically conveyed to a storage hopper. From there, it is fed to the regeneration section along with crushed coal from a coal storage hopper.

Regeneration Section (Sulfate Reduction) — The mixture of fly ash, spent carbonate and crushed coal enters the sulfate reducer which contains a pool of molten sodium salts and which operates at about 1800°F. At that temperature, the reductant carbon in the coal reacts with the sodium sulfate and sodium sulfite in the spent carbonate and reduces them to molten sodium sulfide as shown below:

$$2C + Na_2SO_4 \rightarrow Na_2S + 2CO_2$$
$$1.5C + Na_2SO_3 \rightarrow Na_2S + 1.5CO_2$$

The unreacted sodium carbonate in the feed to the reducer does not enter into the reaction and goes through unchanged. The CO_2-rich gas yielded from the reducer is cooled by water-scrubbing and is sent to a carbonator in the carbonate recovery system.

The reducer melt, containing the molten Na_2S, is continuously discharged from the reducer. The melt is quenched, shattered and dissolved by recycle liquor. Off-gas vented from the quenching of the melt is cooled and sent to incineration. The Na_2S solution from the melt quench is filtered in a rotary vacuum filter which removes and discharges fly ash and any unreacted coal. The filtered solution is sent to a precarbonator tower where it is contacted with an H_2S and CO_2-rich gas derived from the carbonator in the carbonate recovery system. In the precarbonator, the Na_2S solution is converted to a solution of sodium hydrosulfide (NaHS) and sodium carbonate:

$$Na_2S + H_2S \rightarrow 2NaHS$$
$$2Na_2S + H_2O + CO_2 \rightarrow 2NaHS + Na_2CO_3$$

Figure 3.5: Atomics International Aqueous Carbonate Process

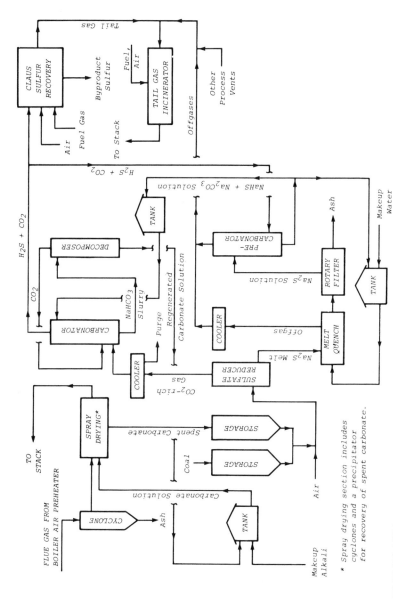

Source: EPRI CS-1381

A portion of the precarbonated solution is recycled, via an intermediate surge storage tank, for reuse in quenching the reducer melt. The remainder of the stream (a solution of NaHS and Na_2CO_3) goes to the carbonator in the carbonate recovery system.

Off-gas from the precarbonator is sent to incineration along with the off-gas from the quenching of the reducer melt.

Regeneration Section (Carbonate Recovery) — The CO_2-rich gas from the sulfate reducer and the solution of NaHS and Na_2CO_3 from the precarbonator are contacted in the carbonate tower to produce a slurry of sodium bicarbonate:

$$NaHS + CO_2 + H_2O \rightarrow NaHCO_3 + H_2S$$
$$Na_2CO_3 + CO_2 + H_2O \rightarrow 2NaHCO_3$$

The off-gas from the carbonator contains H_2S and CO_2. A portion of the gas is sent to the precarbonator. The remainder is sent to the Claus unit for conversion to elemental sulfur.

The sodium bicarbonate slurry from the bottom of the carbonator is sent to a decomposer which produces the regenerated sodium carbonate solution for reuse in the flue gas spray dryer. The off-gas from the decomposer is rich in CO_2 and is recycled to the carbonator.

Claus Unit — A conventional, vapor-phase Claus unit converts the H_2S in the carbonator off-gases into elemental sulfur as an end product. The overall reaction occurring in the Claus unit is:

$$2H_2S + O_2 \rightarrow 2S + 2H_2O$$

The residual tail gas from the Claus unit is incinerated along with the off-gases from the melt quench, the precarbonator and other process vents. If the hot flue gas from the electrostatic precipitator requires any reheat to compensate for the heat removed in the spray dryer, the incinerated tail gas can be blended with the exit flue gas.

Purge Streams — There are no liquid purge streams from the Aqueous Carbonate Process. The discharge of ash and unreacted coal, from the filtering of the precarbonator feed solution, will require appropriate handling and disposal.

Two Advanced Processes in Use in Japan

Nippon Kokan Ammonia Scrubbing Process: Nippon Kokan (NKK) recently completed two large FGD plants to treat flue gas from iron ore sintering machines to produce ammonium sulfate utilizing ammonia in coke oven gas (1). Flue gas (1,120,000 Nm^3/hr) is treated with ammonium sulfite liquor to remove over 95% of SO_2. The liquor discharged from the scrubbers (two in parallel) containing ammonium bisulfite is contacted with coke oven gas to absorb ammonia. A portion of the resulting ammonium sulfite liquor is oxidized to produce ammonium sulfate.

Hydrogen sulfide in the coke oven gas is removed prior to the ammonia absorption. By-products from the sulfide removal unit are also oxidized and converted to ammonium sulfate.

The flue gas leaving the scrubber is passed through wet electrostatic precipitators (eight in parallel) and heated to 110°C by two Ljungstrom-type heat exchangers installed in parallel. No plume at all is observed from the stack.

Nippon Kokan has a similar plant at Fukuyama, with a capacity of treating 760,000 Nm^3/hr of flue gas. At this plant, the treated flue gas is heated by conventional oil firing using neither wet electrostatic precipitator nor heat exchanger. An appreciable plume is observed at the stack.

Electron Beam Process: Flue gas at about 100°C is mixed with NH_3 and exposed to electron beam radiation (2). About 80% of both SO_x and NO_x are removed with 2 Mrad of the beam forming fine crystals of ammonium nitrate sulfate double salt which are caught by an electrostatic precipitator for fertilizer use. Tests are in progress at a pilot plant (1 MW equivalent) of Nippon Steel.

Pircon-Peck Process

The Pircon-Peck flue gas desulfurization process was developed by Mr. Ladd Pircon and Dr. Ralph Peck. The process utilizes patented heterogeneous reactor technology developed by Mr. Pircon as cooling and absorber towers in conjunction with chemistry developed and demonstrated by Mr. Pircon and Dr. Peck at the Illinois Institute of Technology. The process is unique in that in addition to providing a means of controlling SO_2 emissions, it provides the owner with an opportunity to earn a profit on the invested capital.

Process Description: The Pircon-Peck FGD process, as shown in Figure 3.6a, utilizes "activated" phosphate rock, ammonia, and flue gas as raw materials to produce ammoniated phosphate fertilizers. Laboratory and pilot plant testing indicate that in the process of producing fertilizer, sufficient SO_2 can be removed from the flue gas to provide compliance with all EPA regulations, both existing and proposed.

The chemistry utilized to produce the product material can be summarized by three chemical reactions. These are as follows:

$$SO_2 + (NH_4)_2HPO_4 \xrightarrow{+ H_2O} NH_4HSO_3 + NH_4H_2PO_4$$

$$Ca_3(PO_4)_2 + 3NH_4HSO_3 \rightarrow 3CaSO_3 + (NH_4)_2HPO_4 + NH_4H_2PO_4$$

$$NH_4H_2PO_4 + NH_3 \rightarrow (NH_4)_2HPO_4$$

An important feature of this chemistry is the pretreatment of the phosphate rock which facilitates the reaction of the rock with the weak sulfurous acid produced in the scrubber. Standard phosphate fertilizer chemistry requires concentrated sulfuric acid to acidulate the rock and liberate the desired phosphate molecule.

In many ways, the Pircon-Peck process is similar to standard double alkali technology. In each system, a soluble alkali is utilized in the scrubber portion of the process and in each system this alkali is regenerated by a calcium source. In the Pircon-Peck process, the soluble alkali is diammonium phosphate and the calcium source is phosphate rock. However, where alkali regeneration with lime produces water as a by-product, diammonium phosphate regeneration with phosphate rock produces additional phosphate. Neutralization of this by-product with ammonia produces the final diammonium phosphate product.

In addition to producing diammonium phosphate, the regeneration reaction also produces calcium sulfite and sulfate. This material can be included in the fertilizer product or separated for disposal. If included in the fertilizer product, the material would be similar to superphosphate fertilizers presently produced by the fertilizer industry. These fertilizers contain the calcium sulfate solids produced by the acidulation of phosphate rock with sulfuric acid. While the final product form will be dictated by the local market conditions, the advantage of including the calcium sulfite/sulfate solids in the product is obvious.

Figure 3.6: Pircon-Peck Process

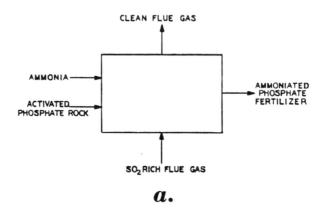

Source: Symposium—Vol 2

Figure 3.6b provides a view of the scrubber portion of the process. Flue gas enters the system and is first passed through a high efficiency particulate control device. The use of a particulate removal system minimizes the possibility of product contamination from fly ash. After fly ash removal, the gas is sent to a cooling tower where it is quenched and saturated to its adiabatic saturation temperature. The cooled and saturated flue gas is next processed through a heterogeneous reactor tower where the flue gas is contacted with a saturated solution of diammonium phosphate and the SO_2 is removed. The SO_2 free flue gas, after passing through a demister, is sent to the stack. If necessary, reheat can be added to aid in plume dispersion.

The SO_2-rich absorption liquor is taken from the heterogeneous reactor system and sent to an agitated attack tank where it is contacted with ammonia and phosphate rock. The slurry generated in the attack tank is next sent to a product separator (clarifier) where solid diammonium phosphate and calcium sulfite/sulfate crystals are withdrawn as a slurry. The product slurry is sent to a fertilizer plant where the product is converted to its final form. The overflow from the product separator is a saturated solution of diammonium phosphate and is sent back to the heterogeneous reactor tower for further reaction with SO_2.

The material from the product separator, on a dry basis, has an approximate analysis of 7-20-0 ($N-P_2O_5-K_2O$). This analysis can be adjusted as desired by the addition of ammonia, phosphoric acid, or other N-P-K materials.

As indicated, it is possible to produce product in various forms. Some alternates to granules with N and P_2O_5 are: pure diammonium phosphate, granules with various N, P and potassium values, suspensions, and flakes.

References

(1) Ando, J., SO_2 Abatement for Stationary Sources in Japan, EPA-600/7-78-210, November 1978. U.S. EPA.

(2) Japan Steel Federation, Status of Development of NO_x Removal Technology by Steel Industry (April 1978) (in Japanese).

COMPARISON OF ENERGY REQUIREMENTS OF SELECTED FGD SYSTEMS

Energy Consumption: Figure 3.7 presents a comparison of the energy consumptions of FGD systems in terms of millions of Btu per hour for the high- and low-sulfur cases evaluated. Electrical energy consumption is represented in this figure by the equivalent heat input to the boiler required to produce this amount of energy. For purposes of comparison, the heat input to the model magnesia slurry systems provided by fuel oil has beeen treated as though an equivalent heat input were provided by steam. While the two energy sources may not be directly comparable on a per-Btu basis because of the higher temperatures attainable with fuel oil, the total heat input approach provides an indication of the overall energy consumption of each process for each type of coal.

Analysis of the individual energy inputs for each wet throw-away process and coal case indicates that the bulk of the electrical energy for all cases goes for fan power and absorbent pumping, both of which are related to gas flow volume. The other energy input, steam, is used entirely for flue gas reheat in all wet throw-away cases. Because the flue gas densities and reheat temperature differentials for both the high- and low-sulfur cases are comparable, this energy input is also primarily a function of gas volume.

Thus, it may be concluded that, for wet throw-away FGD processes employing flue gas reheat, total energy consumption is determined primarily by gas volume, and is only weakly influenced by total SO_2 removal requirements. Low-sulfur applications are therefore relatively more energy-intensive than high-sulfur applications for a given generating unit size and heat rate.

The general characteristic of the wet throw-away processes examined indicates that the total excess combustion air and total air in-leakage in the boiler draft system can have a significant effect on actual total FGD energy consumption. It also appears that there may be a basic level of energy consumption associated with the creation of the turbulent environment required for the absorption

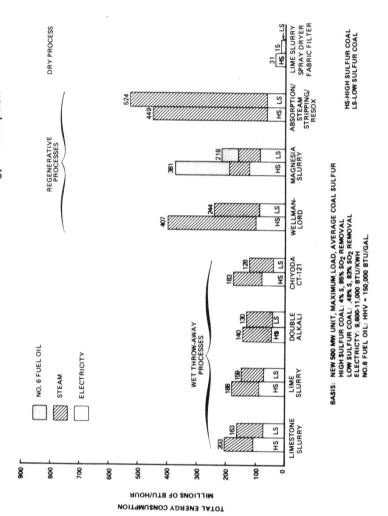

Figure 3.7: Flue Gas Desulfurization—Total Energy Consumption

Source: EPRI CS-1428

of SO_2 by slurries of calcium compounds. Comparison of the total energy consumptions of the Chiyoda 121 and limestone slurry processes suggests that the Chiyoda 121 process absorption vessel design and process chemistry reduce the energy input required for absorption when using limestone slurry as the absorbent. This observation must, however, be considered tentative due to the relatively immature state of development of the Chiyoda process.

The total energy consumptions for all regenerative process cases are substantially higher than for the throw-away process cases due to the energy required to regenerate and reduce the absorbed SO_2. However, the requirements of the high- and low-sulfur cases still differ by less than a factor of two with the heat demands of the regeneration and reduction sections accounting for the bulk of both the total consumptions and the difference for each process.

In contrast to all of the other processes examined, the total energy consumption of the low-sulfur absorption steam stripping case is higher than that of the high-sulfur case. This difference is due to the fact that the specific steam consumption (pounds of steam per pound of SO_2 regenerated) of the stripping process increases as the sulfur loading of the absorbent liquor decreases.

The lime slurry spray dryer/fabric filter process employs no flue gas reheat and requires the lowest total energy input of all of the FGD processes examined.

Energy Intensity: Figure 3.8 presents a comparison of the energy intensity of FGD systems in terms of thousands of Btus expended to remove one pound of SO_2 from the flue gas of the model generating units. Comparison of Figure 3.8 with Figure 3.7 indicates that for the four wet throw-away processes the absolute energy consumptions for the high- and low-sulfur cases differ by less than a factor of two, but that the energy intensities differ by approximately a factor of five with the low-sulfur case requiring the higher energy input per pound of SO_2 removed. Low-sulfur applications are therefore relatively more energy-intensive than high-sulfur applications for a given generating unit size and heat rate.

Comparison of the energy intensity characteristics of the three regenerative processes examined shows the same general trend found in the throw-away processes. The Wellman-Lord and magnesia slurry processes show somewhat reduced differences between the energy intensities for the high- and low-sulfur cases as compared with the wet throw-away processes. This is due largely to the fact that the energy demands of the regeneration and reduction sections are determined primarily by SO_2 removal rate rather than flue gas flow rate.

The absorption/steam stripping process shows a different trend from that of the other two regenerative processes because of the high steam consumption associated with stripping the relatively dilute absorber effluent liquor of the low-sulfur case. Further development of this process may reduce the specific steam consumption, and therefore the total energy intensity for all applications.

As would be expected, the spray dryer/fabric filter cases show the lowest total energy intensities of all of the FGD processes examined.

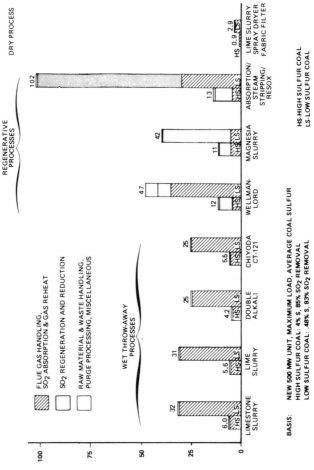

Figure 3.8: Flue Gas Desulfurization—Energy Intensity

Source: EPRI CS-1428

DOUBLE ALKALI PROCESS

The information in this chapter is based on *Definitive SO$_X$ Control Processes Evaluations: Limestone, Double Alkali and Citrate Processes,* August 19, 1979, EPA-600/7-79-177, by S.V. Tomlinson, F.M. Kennedy, F.A. Sudhoff and R.L. Torstrick of TVA, Office of Power Emission Control Development Projects, for the U.S. Environmental Protection Agency; *Economic and Design Factors for Flue Gas Desulfurization Technology,* April 1980, EPRI CS-1428, prepared by Bechtel National, Inc. for the Electric Power Research Institute; and the following papers from Symposium—Vol. 2: "Summary of Utility Dual Alkali Systems," by N. Kaplan of Industrial Environmental Research Laboratory, Office of Research and Development, EPA; "Operating History on Present Status of The General Motors Double Alkali SO$_2$ Control System," by T.O. Mason, E.R Bangel, E.J. Piasecki, and R.J. Phillips of GMC and "Environmental Assessment of the Dual Alkali: FGD System Applied to an Industrial Boiler Firing Coal and Oil," by W H. Fischer of Gilbert Associates, W.H. Ponder of U.S EPA, Industrial Environmental Research Laboratory and R. Zaharchuk of Firestone Tire and Rubber Co.

GENERIC DOUBLE ALKALI PROCESS

As in the limestone slurry system, double-alkali processes dispose of removed SO$_2$ as throwaway calcium sludge. Unlike limestone, however, absorption of SO$_2$ and production of disposable waste are separated—the addition of limestone or lime occurring outside the scrubber loop. The scrubbing step utilizes an aqueous solution of soluble alkali. The absorption reaction depends on gas/liquid chemical equilibrium and mass transfer rates of sulfur oxides (SO$_x$) from flue gas to scrubbing liquid instead of limestone dissolution, the limiting factor in limestone scrubbing. Therefore, SO$_x$ absorption efficiency in a double-alkali system is potentially higher than in a limestone system with the same physical dimensions and liquid-to-gas (L/G) flow rates (1). Scaling and plugging in the absorption area are reduced because calcium slurry is confined to the regeneration and disposal loop and soluble calcium is minimized in the scrubber liquor. The process has been described by Kaplan (2) and LaMantia, et al (3)(4).

Technically, the use of any combination of alkaline compounds, organic or inorganic, for SO_2 removal and disposal can be classified as a double-alkali process. The process described in this section is a sodium sulfite absorbent-lime reactant system.

Sodium sulfite in solution absorbs SO_2 in the scrubbing step represented by equation (1).

(1) $$SO_3^= + SO_2 + H_2O \rightarrow 2HSO_3^-$$

Sodium hydroxide formed in the regeneration step and sodium carbonate added as sodium makeup react with SO_2 as shown below. The absorption reactions actually involve reaction of SO_2 with an aqueous base such as sulfite, hydroxide, or carbonate rather than sodium ion which is present only to maintain electrical neutrality.

(2) $$2OH^- + SO_2 \rightarrow SO_3^= + H_2O$$

(3) $$CO_3^= + SO_2 \rightarrow SO_3^= + CO_2$$

The use of lime for regeneration allows the system to be operated over a wider pH range which in turn includes the complete range of active alkali hydroxide/sulfite/bisulfite. Limestone regeneration operates only in the sulfite/bisulfite range.

(4) $$Ca(OH)_2 + 2HSO_3^- \rightarrow SO_3^= + CaSO_3 \cdot \tfrac{1}{2}H_2O + \tfrac{3}{2}H_2O$$

(5) $$Ca(OH)_2 + SO_3^= + \tfrac{1}{2}H_2O \rightarrow 2OH^- + CaSO_3 \cdot \tfrac{1}{2}H_2O$$

(6) $$Ca(OH)_2 + SO_4^= \rightleftharpoons 2OH^- + CaSO_4$$

(7) $$Ca(OH)_2 + SO_4^= + 2H_2O \rightleftharpoons 2OH^- + CaSO_4 \cdot 2H_2O$$

Total oxidizable sulfur (TOS) is the total concentration of sulfite and bisulfite in solution. Oxidation of TOS to sulfate may occur in any part of the system and is affected by composition of the scrubbing liquor, oxygen content of the flue gas, impurities in the lime, and design of the equipment.

(8) $$SO_3^= + \tfrac{1}{2}O_2 \rightarrow SO_4^=$$

(9) $$HSO_3^- + \tfrac{1}{2}O_2 \rightarrow SO_4^= + H^+$$

The sum of the concentrations of $NaOH$, Na_2CO_3, $NaHCO_3$, Na_2SO_3, and $NaHSO_3$ in the scrubbing solution is termed active alkali. The active alkali concentration in a system can be dilute or concentrated; a concentrated mode (active sodium concentration greater than 0.15 M) was chosen for this discussion. In this mode high sulfite levels prevent the precipitation of calcium sulfate ($CaSO_4$) as gypsum ($CaSO_4 \cdot 2H_2O$), equation 7. However, $CaSO_4$ is precipitated along with calcium sulfite ($CaSO_3 \cdot \tfrac{1}{2}H_2O$) as shown in equations 4-6. In this way the system can keep up with sulfite oxidation at the rate of 25 to 30% of the SO_2 absorbed without becoming saturated with $CaSO_4$. Usually, soluble calcium levels are less than 100 ppm in the regenerated liquor of a concentrated mode double-alkali process.

Several U.S. and Japanese companies have developed double-alkali FGD processes. Unlike the Japanese processes which generally result in the production of gypsum, U.S. processes are of the waste-producing type, producing a calcium sludge that is primarily $CaSO_3 \cdot \tfrac{1}{2}H_2O$.

The first U.S. patent for a double-alkali system was awarded to FMC Corporation in October 1975. FGD investigation was started at FMC in 1956 with limestone and lime scrubbing (5)(6), but by the 1960s FMC was testing a sodium-based scrubbing process on a pilot-plant scale.

The process produced sodium sulfite (Na_2SO_3) and sodium sulfate (Na_2SO_4) and when efforts to sell these products failed, FMC began a search to find a method of recovering the sodium values from the system while producing an acceptable solid waste for disposal. The resulting concentrated double-alkali process was demonstrated as an equivalent 30-MW prototype installed on a reduction kiln at FMC's Modesto (California) Chemical Plant in 1971. Since 1971 several installations have been constructed or are planned for industrial boilers of up to 150 MW equivalent size. Removal of SO_2 and unit availability have been 90% or greater. An FMC system, scrubbing flue gas from coal of 3.75% sulfur, is planned for the 250-MW A.B. Brown Unit No. 1, Southern Indiana Gas and Electric Company.

Envirotech Corporation developed its Buell double-alkali SO_2 control process for both dilute and concentrated mode operation (7). A joint R&D effort with Utah Power and Light Company was undertaken at Gadsby station in Salt Lake City. The 1-MW pilot plant began testing in January 1972. Envirotech research has also focused on high-chloride coals and the acceptable disposal of chlorides in a throwaway system. Envirotech is constructing a 575-MW FGD system at Newton station unit 1, Central Illinois Public Service. Unit 1 will burn coal containing 4% sulfur and 0.2% chloride.

In 1972 Arthur D. Little, Inc., (ADL) was awarded a $1.1M EPA contract to develop double-alkali technology. In the ADL laboratory program, tests were conducted to develop process chemistry, to study regeneration of sodium scrubbing solutions, and to characterize double-alkali waste products (3)(4). ADL, in conjunction with Combustion Equipment Associates, Inc., (CEA), conducted pilot-plant work involving both concentrated and dilute modes of operation. A 20-MW prototype double-alkali system using lime in a concentrated mode was designed and developed by CEA and ADL for installation at Gulf Power Company's Scholz Steam Plant at Sneads, Florida. The test program, a part of the EPA contract, ran from May 1975 to July 1976. CEA and ADL have been awarded an EPA contract for a full-scale double-alkali demonstration unit under construction at Cane Run Unit 6, Louisville Gas and Electric Co. (8). Coal sulfur content at Unit 6 is 3.5 to 4%. Although the entire cost of the installation $16.3M, is being borne by Louisville Gas and Electric, EPA will provide additional funding of $4.5M to cover performance testing and a 1-year operational study. SO_2 removal efficiencies greater than 95% have been guaranteed.

Other U.S. companies have developed double-alkali processes or have conducted experimental programs to study process feasibility. A dilute mode, limestone regeneration double-alkali system, developed by General Motors Corporation and installed at its Parma, Ohio, steam plant, was put into operation in March 1974 (9). Under contract to EPA, ADL conducted a 2-year test program at the Parma site to study operating characteristics and waste by-product properties of the system. The Zurn double-alkali process (Zurn Industries, Inc.) is in use at the Caterpillar Tractor Company plant in Joliet, Illinois (10). The process is dilute mode, lime regeneration, and scrubs gas from two industrial boilers burning 4% sulfur coal. The CALSOX system, a Monsanto Enviro-Chem Systems, Inc., development (11), absorbs SO_2 in an aqueous solution of ethanolamine and regenerates scrubbing liquor with lime. Chemico and Bechtel have also conducted pilot-plant tests of double-alkali systems.

Double-alkali process chemistry was studied in the laboratories of EPA at Research Triangle Park, North Carolina (12), in the early 1970s as a part of an EPA program which included work contracted to Radian Corporation and ADL. Radian designed a mathematical model of the double-alkali system which included certain chemical species not already considered in the laboratory. A bench-scale study was undertaken by ADL to find optimum equipment arrangement, to develop mass transfer coefficients, and to use process knowledge to develop economic information about the system. In the early 1970s laboratory- and bench-scale studies of sodium and ammonia sorbents were conducted by TVA (13)(14), using limestone or lime as regenerant. A pilot-plant study at TVA's Colbert Steam Plant under EPA contract used an ammonia system (15).

The double-alkali process described in this section (Figure 4.1) has been generalized from the several processes available in the United States. In this design, an ESP is used for removal of fly ash and a common plenum and booster fans are included downstream from the ESP and the power plant ID fans for distribution of the gas. Flue gas is cooled and saturated in a presaturator with a recycle stream of scrubber effluent. In the absorber tower SO_2 is removed using a mixture of a regenerated sodium sulfite solution and recycle scrubber effluent (pH about 6.0). The outlet gas from the scrubber passes through a chevron-type entrainment separator with provisions for upstream wash with fresh makeup water. The cleaned flue gas is reheated to 175°F by indirect steam heat before entering the stack.

Incoming pebble lime, from an across-the-fence limestone calcination plant, is received in a silo with a 10-day capacity and conveyed to two 4-hour feed bins that supply the slakers. The lime is processed in two parallel slakers to a slurry concentration of 15% solids. A slurry feed tank with a residence time of 8 hours is provided for in-process storage.

Lime slurry is reacted with a bleedstream of absorber effluent in agitated tanks. The reaction product, predominately calcium sulfite, is pumped to a thickener where the slurry is concentrated to 25% solids. This stream is further dewatered using drum filters to produce a cake of about 55% solids. The filter is designed with two wash sections to minimize sodium loss. The filter cake is reslurried to 15% solids with supernate from the pond and fresh makeup water for pumping to disposal. The solids settle to a 40% concentration in the pond.

Makeup soda ash is pneumatically conveyed from a rail hopper car to the storage silo and fed to an agitated tank where it is slurried in fresh makeup water. The slurry is added to the regenerated scrubber liquor at the thickener overflow storage tank. The material balance for the system is shown in Table 4.1.

Major Process Areas: The generic double-alkali process has been divided into the following operating areas.

Materials Handling — This area includes facilities for receiving pebble lime from an across-the-fence limestone calcination plant, lime storage silo, and in-process storage for supply to the slakers. Soda ash storage is also provided.

Feed Preparation — Included in this area are two parallel slaking systems and the facilities for dissolving makeup soda ash in water before feeding to the absorption system.

Gas Handling — Fan location and duct configuration are the same as in the limestone slurry process.

SO₂ Absorption — Four tray tower absorbers with presaturators, recirculation tanks, and pumps are included.

Figure 4.1: Generic Double Alkali Process

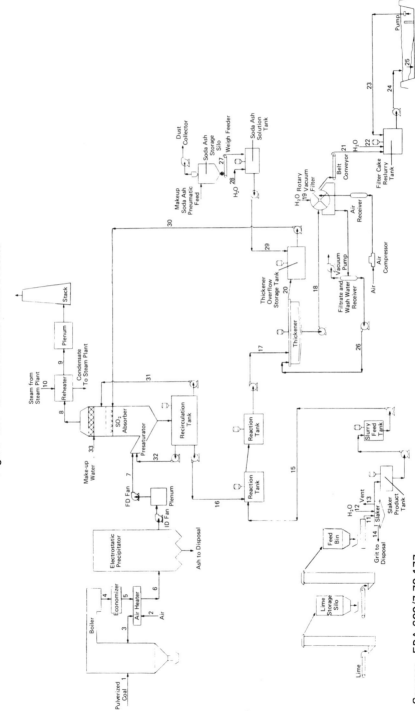

Source: EPA-600/7-79-177

Table 4.1: Generic Double-Alkali Process—Material Balance

Stream*	Total Stream lb/hr	Volume scfm, 60°F	Temp °F	Pressure psig	Gallons per Minute	Specific Gravity	pH	Undissolved Solids, %
1	428,600							
2	4,546,200	1,005,000	80					
3	4,101,800	906,700	535					
4	4,516,100	958,000	890					
5	4,516,100	958,000	705					
6	4,960,400	1,056,000	300					
7	4,905,800	1,056,000	300					
8	5,094,000	1,125,000	127					
9	5,094,000	1,126,700	175					
10	92,810		470	500				
11	18,170							
12	135,150				270			
13	1,056				2			
14	182							
15	152,080				279	1.09		15
16	1,898,900				3,547	1.07	6.0	
17	2,051,000				3,726	1.1	11.0	2
18	167,800				271	1.24		25
19	59,700				120			
20	2,034,400				3,800	1.07		
21	76,270							55
22	28,600				57			
23	174,780				349			
24	279,650				508	1.1		15
25	104,870				161	1.3		40
26	151,230				283	1.07		
27	1,730							
28	9,800				20			
29	11,530				20	1.14		
30	2,045,900				3,821	1.07		
31	2,727,500				5,094	1.07	5.8	
32	2,727,500				5,094	1.07		
33	41,100				82			

*Streams are identified as follows:

1 - Coal to boiler
2 - Combustion air to air heater
3 - Combustion air to boiler
4 - Gas to economizer
5 - Gas to air heater
6 - Gas to electrostatic precipitator
7 - Gas to presaturator
8 - Gas to reheater
9 - Gas to stack
10 - Steam to reheater
11 - Lime to slaker
12 - Water to slaker
13 - Vent from slaker
14 - Grit to disposal
15 - Lime slurry to reaction tank
16 - Bleedstream to reaction tank
17 - Slurry to thickener
18 - Thickener underflow to filter
19 - Filter wash water
20 - Thickener overflow to storage tank
21 - Filter cake to reslurry tank
22 - Water to reslurry tank
23 - Recycle pond water
24 - Slurry to pond
25 - Settled sludge
26 - Filtrate to thickener
27 - Soda ash to solution tank
28 - Water to solution tank
29 - Makeup soda ash stream
30 - Scrubbing liquor to absorber
31 - Recycle liquor to absorber
32 - Recycle liquor to presaturator
33 - Makeup water to absorber

Source: EPA-600/7-79-177

Stack Gas Reheat — Equipment in this area includes indirect steam reheaters and soot blowers for the coal variations. The oil-fired unit is designed with one direct oil-fired reheater per duct which discharges hot combustion gases directly into the duct.

Reaction — Reaction tanks with agitators and pumps are provided in this area.

Solids Separation — Separation of calcium salts is accomplished by thickener and filters

Solids Disposal — Filter cake is reslurried in this area and purged to the disposal pond. A pond return pump is included.

Storage Capacity: Storage requirements for raw materials and allowances for in-process streams are listed below.

Raw materials:
Lime storage silo — 10 days (from across-the-fence calcination plant)
Soda ash storage silo — 4 months (purchased in bulk quantity by rail)

In-process storage:
Lime feed bins — 4 hours each
Slaker product tank — 5 minutes
Slurry feed tank — 8 hours
Soda ash solution tank — 8 hours
Recirculation tanks — 10 minutes each (includes sufficient surge
  capacity for shutdown of scrubbers)
Thickener — 4 hours
Thickener overflow storage tank — 20 minutes
Filter cake reslurry tank — 5 minutes

Solids Disposal: Waste solids in the generic double-alkali process are handled by ponding or piling a dewatered filter cake. Pond areas for various cases are listed in Table 4.2.

Table 4.2: Acreage Required for Waste Solids Disposal

Case	Years Remaining Life	Acres
Coal-Fired Power Unit		
1.2 lb SO_2/MBtu heat input		
allowable emission; onsite		
solids disposal (ponding)		
200 MW E 3.5% sulfur	20	64
200 MW N 3.5% sulfur	30	116
500 MW E 3.5% sulfur	25	187
500 MW N 2.0% sulfur	30	127
500 MW N 3.5% sulfur	30	233
500 MW N 5.0% sulfur	30	329
1,000 MW E 3.5% sulfur	25	315
1,000 MW N 3.5% sulfur	30	393
Solids disposal by trucking		
500 MW N 3.5% sulfur	30	87
90% SO_2 removal; onsite		
solids disposal (ponding)		
500 MW N 3.5% sulfur	30	260

(continued)

Table 4 2: (continued)

Case	Years Remaining Life	Acres
Oil-Fired Power Unit 0.8 lb SO_2/MBtu heat input allowable emission; onsite solids disposal (ponding) 500 MW E 2.5% sulfur	25	97

Note: It is assumed that the coal has a heating value of 10,500 Btu/lb and contains 16% ash.

Source: EPA-600/7-79-177

References

(1) Kaplan, Norman, 1974. "An Overview of Double-Alkali Processes for Flue Gas Desulfurization." In: *Proceedings of Symposium on Flue Gas Desulfurization,* Vol. I, Atlanta, November 4-7, 1974. EPA-650/2-74-126a (NTIS PB 242 573), U.S. Environmental Protection Agency, Washington, D.C., pp 445-515.

(2) Kaplan, Norman, 1976. "Introduction to Double-Alkali Flue Gas Desulfurization Technology." In: *Proceedings of Symposium on Flue Gas Desulfurization,* Vol. I, New Orleans, March 8-11, 1976. EPA-600/2-76-136a (NTIS PB 252 317), U.S. Environmental Protection Agency, Washington, D.C., pp 387-421.

(3) LaMantia, C.R., et al., 1976. "Operating Experience—CEA/ADL Dual-Alkali Prototype System at Gulf Power/Southern Services, Inc." In: *Proceedings of Symposium on Flue Gas Desulfurization,* Vol. I, New Orleans, March 8-11, 1976. EPA-600/2-76-136a (NTIS PB 252 317), U.S. Environmental Protection Agency, Washington, D.C., pp 423-469.

(4) LaMantia, C.R., et al., 1977. *Final Report: Dual-Alkali Test and Evaluation Program, Vol. II, Laboratory and Pilot Plant Programs.* EPA-600/7-77-050b (NTIS PB 272 770), U.S. Environmental Protection Agency, Washington, D.C.

(5) FMC Corporation, 1976. *Capabilities Statement, Sulfur Dioxide Control Systems.* FMC Technical Report 100.

(6) Legatski, L.K., K.E. Johnson and L.Y. Lee, 1976. "The FMC Concentrated Double-Alkali Process." In: *Proceedings of Symposium on Flue Gas Desulfurization,* Vol. I, New Orleans, March 8-11, 1976. EPA-600/2-76-136a (NTIS PB 252 317), U.S. Environmental Protection Agency, Washington, D.C., pp 471-502.

(7) Bloss, H.E., J. Wilhelm, and W.J. Holhut, 1976. "The Buell Double-Alkali SO_2 Control Process." In: *Proceedings of Symposium on Flue Gas Desulfurization.* Vol. I, New Orleans, March 8-11, 1976. EPA-600/2-76-136a (NTIS PB 252 317), pp 545-562.

(8) Van Ness, R.P., 1978. "Louisville Gas and Electric Company, Scrubber Experiences and Plans." In: *Proceedings of Symposium on Flue Gas Desulfurization.* Vol. I, Hollywood, Florida, November 8-11, 1977. EPA-600/7-78-058a, U.S. Environmental Protection Agency, Washington, D.C., pp 235-245.

(9) Interess, Edward, 1977. *Evaluation of the General Motors' Double-Alkali SO_2 Control System.* EPA-600/7-77-005 (NTIS PB 263 469), U.S. Environmental Protection Agency, Washington, D.C.

(10) Lewis, P.M., 1976. "Operating Experience With the Zurn Double-Alkali Flue Gas Desulfurization Process." In: *Proceedings of Symposium on Flue Gas Desulfurization,* Vol. I, New Orleans, March 8-11, 1976. EPA-600/2-76-136a (NTIS PB 252 317), U.S. Environmental Protection Agency, Washington, D.C., pp 503-514.

(11) Barnard, R.E., R.K. Teague, and G.C. Vansickle, 1974. "CALSOX System Development Program." In: *Proceedings of Symposium on Flue Gas Desulfurization,* Vol. II, Atlanta, November 4-7, 1974. EPA-650/2-74-126b (NTIS PB 242 573), pp 1127-1149.

(12) Draemel, D C., 1972. *Regeneration Chemistry of Sodium-Based Double-Alkali Scrubbing Process.* EPA-R2-73-186 (NTIS PB 220 077), U.S. Environmental Protection Agency, Washington, D.C.

(13) TVA, 1973. *Applied Research Branch Annual Report.* U.S. Tennessee Valley Authority, Muscle Shoals, Alabama.

(14) TVA, 1974. *Applied Research Branch Annual Report.* U.S. Tennessee Valley Authority, Muscle Shoals, Alabama.

(15) Williamson, P.C., and E.J. Puschaver, 1977. *Ammonia Absorption/Ammonium Bisulfate Regeneration Pilot Plant for Flue Gas Desulfurization.* TVA Bulletin Y-116, U.S. Tennessee Valley Authority, Muscle Shoals, Alabama; EPA-600/2-77-149 (NTIS PB 272 304/AS) U.S. Environmental Protection Agency, Washington, D.C.

CONCENTRATED MODE-DILUTE MODE COMPARISON

A summary of the basic process design parameters for the double alkali process is presented in Table 4.3, the concentrated mode applies to Eastern Coals, and the dilute mode to Western Coals. The corresponding raw material and utility requirements are presented in Table 4.4.

Although similar to that given above, the description of the concentrated mode process that follows is included for comparison purposes.

Table 4.3: Double Alkali Process System Parameters

Data for 1 x 500 MW Unit
(Maximum Load Conditions)

	Units	Eastern Coal	Western Coal
Flue gas flow rate	m^3/sec	871	943
	(10^3 acfm)	(1846)	(1998)
Number of absorption trains per unit	each	4	4
Capacity per train	%	33-$\frac{1}{3}$	33-$\frac{1}{3}$
Gas flow per absorber, saturated	m^3/sec	239	261
	(10^3 acfm)	(507)	(552)
System operating mode	–	concentrated	dilute
Presaturator type	–	spray duct	spray duct
Presaturator L/G	$1/m^3$	0.267	0.267
	(gal/10^3 acf)	(2)	(2)
Absorber type	–	TCA	TCA
Number of contacting stages	each	2	2
Absorber L/G	$1/m^3$	1.07	1.34
	(gal/10^3 acf)	(8)	(10)
Mist elimination	–	chevron	chevron
Total system pressure drop	pascals	2491	2491
	(inches wg)	(10)	(10)
Absorbent oxidation ratio[*]	%	10	30
Number of regeneration trains per unit	each	1 at 120%[***]	2 at 60%[***]
Number of reaction stages	each	2	2
Stoichiometric ratio Ca in sludge[**]	–	1.00	1.00
Sodium loss rate[**]	%	2.2	2.2
Dewatered sludge solids content	wt %	55	50
Flue gas reheat required	°C	13.3	8.9
	(°F)	(56)	(48)

[*]Mols Na_2SO_4 formed/mol SO_2 removed.
[**]Mols/mol SO_2 removed.
[***]Capacity based on design requirements for maximum coal sulfur content.

Source: EPRI CS-1428

Table 4.4: Double Alkali Process Raw Materials and Utility Requirements

Data for 1 x 500 MW Unit
(Maximum Load, Average Sulfur)

	Units	Eastern Coal	Western Coal
Raw Materials			
Soda Ash	kg/hr	557	84.8
	(lb/hr)	(1,228)	(187)
Lime for regeneration	10^3 kg/hr	15.7	2.40
	(tons/hr)	(17.3)	(2.64)
Lime for sludge stabilization	10^3 kg/hr	1.8	0.51
	(tons/hr)	(2.0)	(0.57)
Total lime	10^3 kg/hr	17.5	2.91
	(tons/hr)	(19.3)	(3.21)
Utilities			
Electricity			
Raw material handling and preparation	kW	420	70
Absorption	kW	500	540
Absorbent regeneration	kW	130	380
Sludge treatment	kW	750	200
Fans	kW	2,610	2,820
Total	kW	4,410	4,010
Steam	10^6 Btu/hr	96.6	89.4
Process Water	10^3 ℓ/hr	138.9	116.2
	(10^3 gal/hr)	(36.7)	(30.7)
Waste Materials			
Sludge	10^3 kg/hr	62.5	11.2
	(tons/hr)	(68.8)	(12.3)
Fly Ash (Dry)	10^3 kg/hr	28.6	14.8
	(tons/hr)	(31.5)	(16.3)
Lime (Dry)	10^3 kg/hr	1.8	0.52
	(tons/hr)	(2.0)	(0.57)
Combined Wastes	10^3 kg/hr	92.9	26.5
	(tons/hr)	(102.3)	(29.2)
	m^3/hr	96.6	27.5
	(yd^3/hr)	(126.3)	(36.0)
Incremental Heat Rate Penalties			
Steam (basis gross heat rate)	%	2.14	1.91
Electricity (basis unit net rating)	%	0.88	0.80
Corrected Net Heat Rate	Btu/kWh	9,851	10,170
Variation from Base	%	-1.35	-1.19

Source: EPRI CS-1428

Concentrated Mode

During absorption, a portion of the reactive sodium sulfite is oxidized to unreactive sodium sulfate. Regeneration of this sulfate occurs by precipitation of a double salt of calcium sulfite/sulfate. Regeneration by this mechanism requires a high concentration of the sulfate in solution. A high dissolved salts concentration is also desirable to reduce oxygen solubility and minimize oxidation. An equilibrium salt concentration comfortably below the salt solubility limit (20 to 25%) can generally be maintained with 25% or less total oxidation. This is normally the case with coal sulfur of 2% or greater. If oxidation exceeds this level, a purge of absorbent is required to remove excess sodium sulfate. This results in a loss of sodium and a purge salt disposal requirement.

The chemical reactions associated with the various process steps include:

Absorption

$$Na_2SO_3 + SO_2 + H_2O \rightarrow 2NaHSO_3$$

$$Na_2SO_3 + \tfrac{1}{2}O_2 \rightarrow Na_2SO_4$$

Regeneration

$$Ca(OH)_2 + 2NaHSO_3 \rightarrow Na_2SO_3 + CaSO_3 \cdot \tfrac{1}{2}H_2O + \tfrac{3}{2}H_2O$$

$$2Ca(OH)_2 + 4NaHSO_3 + Na_2SO_4 \rightarrow CaSO_3 \cdot CaSO_4 \cdot 4H_2O + 3Na_2SO_3$$

(illustrative double salt)

Figure 4.2 is a simplified flow diagram for the concentrated mode double alkali process.

Description of Major Process Subsystems: The concentrated mode sodium/lime double alkali process includes the following major process subsystems.

Raw Material Receiving, Storage and Preparation — This area includes facilities for receiving and storing soda ash and dry lime and preparing makeup sodium carbonate solution and lime slurry.

Soda ash is conveyed pneumatically from self-unloading delivery trucks to storage silos. Soda ash and recycled process water are fed to a dissolving tank where a 20% sodium carbonate absorbent makeup solution is prepared. This solution is fed to the regenerated absorbent liquor storage tank to replace the sodium values lost in the waste sludge filter cake.

Dry pebble lime is delivered by rail car and conveyed mechanically to bulk storage bins, and then to smaller process surge bins. The lime is slaked with makeup water and the resulting slurry is transferred to a dilution tank. The raw lime slurry is diluted with recycled process wastewater and transferred to an agitated slurry storage tank from which it is fed to the first-stage absorbent regeneration reactor.

SO_2 Absorption — Hot flue gas discharges from the ID fans into a plenum from which it is distributed among the operating absorption trains. Before entering the absorber vessels, the hot gas is adiabatically cooled and saturated by absorbent liquor sprays located in specially designed duct sections (presaturators) just upstream of the absorber inlets.

The cooled, saturated flue gas is contacted with the sodium sulfite absorbent solution in a vertical, countercurrent absorber having two stages of mobile-ball packing. The cleaned flue gas passes through a chevron mist eliminator for removal of entrained liquor droplets and is then reheated before discharging to the chimney.

Absorbent Regeneration — A slip stream of the absorber effluent liquor is pumped to the first-stage reactor tank where it is mixed with makeup lime slurry. The first-stage reactor overflows to a second-stage reactor, where the regeneration reactions are completed. The resulting slurry of calcium salts in regenerated absorbent liquor is pumped to a thickener for separation of the waste solids. The thickener underflow sludge is pumped to a sludge storage tank, then to the sludge dewatering and stabilization section.

The clarified liquor overflows from the thickener into a regenerated absorbent storage tank. Makeup sodium carbonate solution is added to the regenerated absorbent to replace sodium lost in the waste sludge cake, and the resulting solution is injected into the absorbers' recycle liquor systems.

Figure 4.2: Sodium/Lime Double Alkali Process—Flow Diagram—Concentrated Mode

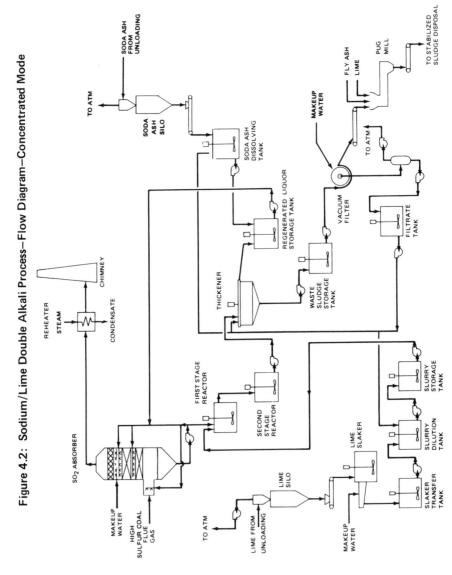

Source: EPRI CS-1428

Sludge Dewatering and Stabilization — Sludge from the waste sludge storage tank is fed to rotary drum vacuum filters where it is further dewatered. During filtration the sludge solids cake is washed with process makeup water to recover the sodium contained in the salt cake surface moisture. The filtrates liquor is collected and pumped back to the regeneration and raw material processing sections for reuse.

Development Status (Concentrated Mode): The concentrated mode system is offered by a number of vendors and is considered commercially developed. This particular process was extensively piloted by one vendor in coal-fired service at a scale of 25 MW. A major 227 MW utility application is nearing completion.

Dilute Mode

In order to regenerate the more fully oxidized absorbent from a low-sulfur coal application, it is necessary to dilute the absorbent to less than 0.1 molar (10 wt percent). The regenerated sodium is in the form of hydroxide rather than sodium sulfite.

Because of the large dilute volume of absorbent recycle liquor and low sulfite concentration, a large amount of dissolved calcium is contained in the regenerated absorbent. If recycled to the absorber, this calcium can produce absorber scaling. For this reason, the regenerated absorbent is normally softened by addition of sodium carbonate with clarification to remove the precipitated calcium carbonate.

The most significant chemical reactions occurring in each stage of the process are shown below.

Absorption

$$2NaOH + SO_2 \rightarrow Na_2SO_3 + H_2O$$
$$Na_2SO_3 + \tfrac{1}{2}O_2 \rightarrow Na_2SO_4$$

Regeneration

$$Na_2SO_4 + Ca(OH)_2 \rightarrow 2NaOH + CaSO_4$$

Absorbent Softening

$$CaSO_4 \text{ (dissolved)} + Na_2CO_3 \rightarrow CaCO_3 + Na_2SO_4$$
$$Ca(OH)_2 \text{ (dissolved)} + Na_2CO_3 \rightarrow CaCO_3 + 2NaOH$$

Figure 4.3 is a simplified flow diagram for the dilute mode double alkali system.

Description of Major Process Subsystems: The process systems for the dilute mode are similar to those described above for the concentrated mode. The major differences are the larger circulation rates required by the dilute absorbent and the addition of the regenerated absorbent softening equipment.

The calcium-rich regenerated absorbent liquor overflows from the thickener to the regenerated absorbent storage tank where it is mixed with makeup sodium carbonate solution and a small amount of recycled calcium carbonate sludge which provides seed crystals for the soda softening process. The resulting slurry of calcium carbonate crystals in regenerated absorbent liquor is pumped to a clarifier for separation of the solids. The bulk of the clarifier underflow sludge is pumped to the waste sludge storage tank, with a bleed stream going to the regenerated absorbent tank.

Development Status: The dilute mode system has been applied in the United States to an industrial boiler, and is considered commercially developed. Its com-

Figure 4.3: Sodium/Lime Double Alkali Process—Flow Diagram—Dilute Mode

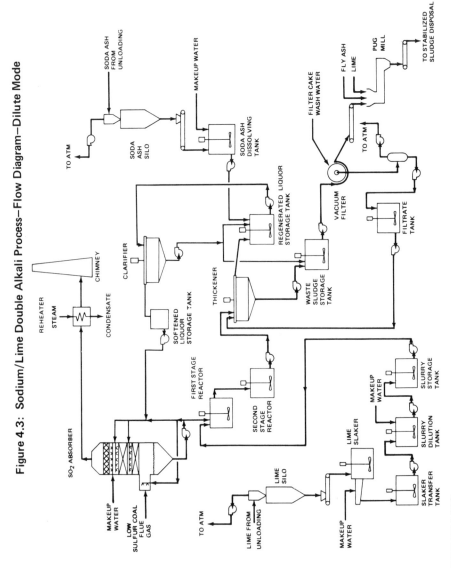

Source: EPRI CS-1428

plexity and the recent improvements made in the limestone slurry and lime slurry processes probably limit its future use to specialized small-scale applications.

UTILITY DUAL ALKALI SYSTEMS IN THE U.S.

Design Criteria

Based on the performance and reliability demonstrated in various dual alkali pilot and prototype plants in the U.S. and Japan, and in an increasing number of full-scale applications in Japan, several general design and performance criteria are of interest. Principal among them are: (1) SO_2 removal performance, (2) particulate matter removal performance, (3) sodium consumption, (4) calcium consumption, (5) energy consumption, (6) waste solids quality, and (7) system reliability.

SO_2 Removal: A commercial dual alkali system must remove the desired quantity of SO_2 to allow compliance with the applicable Federal and/or local standards. With the present state-of-the-art, it is reasonable to expect long-term average SO_2 removal capability on the order of 95% with moderate, 10-15 gal/10^3 acf, liquid-to-gas ratios, and high (up to 4,000 ppm) SO_2 inlet concentration.

Particulate Matter Removal: The scrubbed gas from the dual alkali flue gas desulfurization unit should not contain particulate matter in excess of applicable standards. In some cases, this could require particulate matter removal by the FGD unit; in others it could merely imply no net addition of particulate matter to the gas stream by the FGD unit.

Sodium Consumption: Sodium consumption is an important performance criterion for a dual alkali system not so much from the viewpoint of economics but rather from its potential for secondary pollution. Sodium consumption is only a minor factor in the operating cost of a system, representing about 2% of the annual operating cost. Thus, even if soda makeup to a system were to increase by 100% over its design value, operating cost will be increased by a factor of less than 1.02. On the other hand, the environmental consequences of higher sodium consumption may be significant if the sodium is leached from the waste product to the environment.

A logical way to measure sodium consumption is in mols of Na consumed per mol of sulfur removed by the system. A value of 0.05 mol of Na makeup per mol of sulfur removed (equivalent to 0.025 mol of Na_2CO_3 makeup per mol of sulfur removed) appears to be a reasonable design target based on present U.S. technology. This target is achievable in concentrated dual alkali systems burning relatively high sulfur coal (over 3% sulfur) and having a relatively low oxygen content in flue gas. In Japan, sodium makeup is reportedly as low as 0.02 mol Na/mol sulfur removed for some systems (1).

Calcium Consumption: A logical way to specify calcium consumption is as calcium stoichiometry; i.e., mols of calcium added per mol of sulfur removed (or collected). A calcium consumption of 0.98 to 1.0 appears to be a reasonable design target for concentrated dual alkali systems (values less than 1.0 are possible due to the alkali added with sodium makeup to the system).

Energy Consumption: Design targets for energy consumption can be in the range of 1-2% of power plant output without reheat, or under 4% even with 50°F of reheat. These figures are based on a system which has a scrubber pressure drop in the range of 6-8 inches H_2O and a scrubber system L/G ratio of about

10 gal/10^3 acf. This assumes that the power plant is equipped with some means of efficient particulate collection upstream of the FGD unit (e.g., an electrostatic precipitator).

Waste Solids Quality: The dual alkali process must produce easily handled, transportable, environmentally acceptable waste. Present state of technology indicates that a filter cake containing a minimum of 65 wt % solids and a low level of soluble salts, such as Na_2SO_3 and NaCl, is a reasonable expectation. Systems currently being built in the U.S. (concentrated mode) will produce a waste solid containing primarily calcium sulfite with some amount of sulfate in the crystal. The Japanese systems are primarily designed to produce saleable gypsum.

System Reliability: The FGD system should have a high availability. Based on existing vendor guarantees availability of 90% for 1 year, as defined by the Edison Electric Institute for power plant equipment, may be a reasonable target.

In 1979 three U.S. full-scale, utility boiler dual alkali systems, ranging in size from 250 to 575 MW, were scheduled for startup. The owners of these systems, their size, and the vendors providing them are: (1) Central Illinois Public Service Co., Newton, No. 1, 575 MW unit, Envirotech; (2) Louisville Gas & Electric Co., Cane Run No. 6, 277 MW unit, CEA/ADL; and (3) Southern Indiana Gas & Electric Co., A.B. Brown No. 1, 250 MW unit, FMC. All of these systems are of the concentrated sodium based dual alkali type and they service boilers firing high sulfur coal. Brief descriptions of these full-scale dual alkali facilities and their design bases, are reported below.

Central Illinois Public Service Co. (CIPSCO)(2)(3)

Boiler: Newton Station Unit No. 1, Newton, Illinois, utilizes a pulverized coal-fired boiler manufactured by Combustion Engineering Inc. It is designed for a maximum steam generating capacity of 4,158,619 lb/hr at 2620 psig and 1005°F at the steam outlet connection. Design excess air in the boiler is 22% with flue gas flow to the scrubbing system totalling 6,615,000 lb/hr. The boiler design includes two parallel induced draft fans with discharge into a common plenum which can feed flue gas to the FGD system or to the stack. The boiler is tangentially fired with coal from bowl-type pulverizing mills utilizing high-sulfur Illinois bituminous coals from several local sources.

FGD Facility: Process equipment of interest, in the "duplex-type" dual alkali FGD system, built to serve Newton No. 1, includes four booster fans; four precooler spray towers, including mist eliminators; four mobile-ball gas scrubbers, including mist eliminators; three spent absorbent causticizers (reactors); one precooler effluent neutralizer; one reactor/clarifier; two 100-ft diameter thickeners, for the dual alkali solids concentration system, with concrete bottoms and access to bottom discharge cone through a tunnel; one 50-ft diameter thickener, for the precooler loop, constructed of coated steel; and three horizontal Eimco-Extractor filters for dewatering and multistage washing of thickener underflow.

Figure 4.4 is the process flow diagram of the FGD system. An overview of the absorption, regeneration, and dewatering sections is given below.

Absorption Section: *Precooler/HCl Absorber* — Because of design criteria requiring the use of local Illinois coal, the system is designed to accommodate these high-chloride coals. The presence of chlorides in the coal may require a preconditioning step before the flue gas is treated for SO_2 removal in the main scrubbing loop. This preconditioning of the flue gas in the precooler spray towers also minimizes soda ash consumption.

Figure 4.4: Envirotech Flue Gas Desulfurization System on CIPSCO'S Newton No. 1 Unit

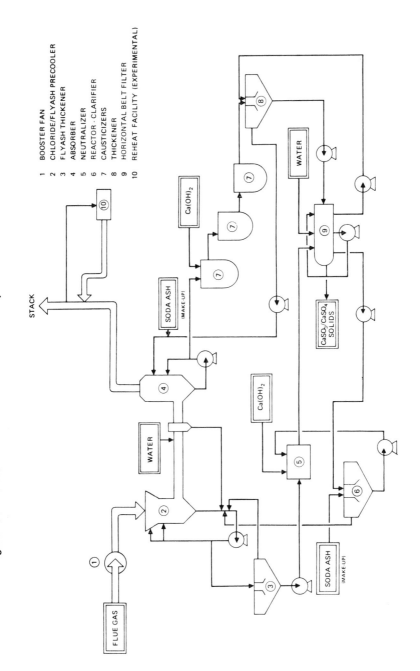

1 BOOSTER FAN
2 CHLORIDE/FLYASH PRECOOLER
3 FLYASH THICKENER
4 ABSORBER
5 NEUTRALIZER
6 REACTOR - CLARIFIER
7 CAUSTICIZERS
8 THICKENER
9 HORIZONTAL BELT FILTER
10 REHEAT FACILITY (EXPERIMENTAL)

The precooler spray towers treat and cool the flue gas. The towers are designed to remove 90% of the chloride in the form of HCl. To minimize any carryover of mist or particulate matter, vertical mist eliminators are installed in the horizontal precooler tower outlet ducts (liquid drains perpendicular to the gas flow). These mist eliminators control the acid mist within the precooler loop and the level of dissolved solids (chlorides) in the scrubbing liquor to reduce sodium consumption and minimize operational problems.

The spent liquor from the precooler is sent to the fly ash thickener for clarification and prevention of significant concentrations of abrasive slurry solids in the recirculating liquor. Also, to maintain steady sodium level, pH, and solids level in the precooler loop, the clarifier underflow is dewatered and treated with lime and sodium carbonate. The regenerated liquor is returned to the precooler loop.

SO_2 Scrubber — SO_2 emissions are reduced to less than 200 ppm utilizing a countercurrent, two-tray-stage mobile-ball bed scrubber operating at 8.3 ft/sec design gas velocity with a L/G of less than 10 gal/10^3 acf. A mist eliminator mounted vertically in the horizontal duct (liquid drains perpendicular to the gas flow) controls sodium liquor carryover. The eliminator vanes are intermittently washed with service water, which is collected and returned to the precooler loop to maintain a favorable water balance.

In the initial operation of the unit, two types of reheat systems are used (half the gas volume for each): (1) recycle reheat; and (2) bypass reheat.

With the recycle reheat mode of operation, the reheat is accomplished by blending a side stream of steam-heated flue gas with the saturated flue gas exiting the scrubber trains. The blending occurs across a high-alloy perforated-plate diffuser section in the main duct. The steam heated flue gas is maintained above the saturation point during normal operation by recycling and preheating gas on the downstream side of the diffuser. The gas entering the reheater will be above the saturation point to reduce corrosion potential. The gas in the recycle reheat section during normal operation will range between 156° and 300°F, as compared to the saturated gas temperature of 131°F. The system is designed to achieve an exit gas temperature of 156°F to the stack.

In the bypass reheat method, a portion of the unscrubbed flue gas is bypassed around the scrubber and mixed with the cleaned, saturated, scrubbed gases exiting the scrubber. A diffuser section, as described above, is used to disperse and mix the two gas streams. This system is designed to achieve a maximum of 43°F reheat because of the limitation on allowed SO_2 and particulate emission levels. The quantity of bypass gas is controlled by damper control to realize the desired degree of reheat. In addition, the gas is also monitored in regard to opacity and SO_2 level as an additional means of control.

The overall design philosophy in regard to the reheat systems was to provide sufficient flexibility to allow either type of reheat system to be installed later if one of the systems proved to be more effective. Space has been reserved to achieve up to 50°F of reheat, based on space requirements for the larger system.

Regeneration Section: A portion of the spent liquor from the absorber is pumped to the first stage of a three-stage causticizer (reactor) system, where slaked lime slurry is mixed with spent liquor.

The primary causticizer overflows by gravity into the secondary causticizer which in turn overflows into the tertiary causticizer. The slurry from the tertiary causticizer is pumped to the dewatering section. It is not certain whether one, two, or three reactor stages will be operated. If one is needed, there will be two spares. Likewise, if two are needed one will act as a spare.

Dewatering Section: After thickening of the sulfur oxides solids collected in the mobile-ball scrubber, the solids are dewatered and fresh-water washed in a top-loading horizontal-belt-type extractor filter. The unique geometry of this equipment permits countercurrent multistage washing of the raw cake with limited quantities of service water. Over 80% of the entrained sodium in the vacuum filter cake is recovered. This sodium is recycled to the absorption loop for reuse. In addition, the liquid purge stream from the precooler loop, containing collected chlorides, residual fly ash, and trace elements, flows to a lime-neutralization tank and then is utilized as the wash medium for a final "cake-impregnation wash." Thus high-chloride, low-pH liquor bypasses the SO_2 absorber and is discarded as surface moisture in the final waste cake.

Louisville Gas & Electric Co (LG&E)(4)

Boiler: Cane Run Station Unit No. 6, Louisville, Kentucky, utilizes a pulverized coal-fired boiler manufactured by Combustion Engineering Inc. It is designed for a maximum steam generating capacity of 1,854,217 lb/hr at 2600 psig and 1005°F at the steam outlet connection. Design flue gas flow to the scrubbing system totals 3,372,000 lb/hr. The boiler is tangentially fired with coal from bowl-type pulverizing mills utilizing high-sulfur, midwest bituminous coals.

FGD Facility: Process equipment of interest, in the dual alkali FGD system for Cane Run No. 6, includes two booster fans; two dual-tray absorbers; two pairs of spent absorbent regeneration reactors; one 125-ft diameter thickener with concrete bottom and access to bottom discharge cone through a tunnel; three rotary-drum vacuum filters with water wash; and two external combustion, scrubbed gas reheaters.

Figure 4.5 is a process flow diagram of the FGD System. An overview of the absorption, regeneration, and dewatering sections is given below.

Absorption Section: The flue gas from the existing electrostatic precipitator (ESP) induced draft fan is forced by a booster fan into an absorber. There are two absorber modules, each equipped with a booster fan. A common duct connects the two inlet ducts to the booster fans.

The hot flue gas is adiabatically cooled and saturated by sprays of absorber solution directed at the underside of the bottom tray. These sprays keep the underside of the tray and the bottom of the absorber free of buildup of fly ash solids. The cooled gas then passes through two sieve trays, where SO_2 is removed, and leaves the absorber through a chevron type mist eliminator. Prior to entering the stack, the saturated gas from the mist eliminator is heated 50° to 175°F, by hot combustion gas from a grade-mounted reheater fired with No. 2 oil.

Each absorber is designed for 9.0 ft/sec gas velocity, a rate consistent with good mass transfer, low pressure drop, and minimal entrainment. Each absorber is sized to handle 60% of the design gas flow rate and the overall system can be turned down to 20% of the design flow rate by shutting down one absorber. At levels less than 50% of the design capacity, the system can be operated with one absorber module by use of the common inlet duct.

For control of tray feed liquor pH, regenerated scrubbing liquor from the thickener hold tank is mixed in-line with absorber recycle liquor for each unit; the mixture is then fed to the top tray in each absorber. The absorber recycle liquor is used in the spray section below the trays. A bleedstream of the absorber

Figure 4.5: CEA/ADL Flue Gas Desulfurization System on LG & E'S Cane Run No. 6 Unit

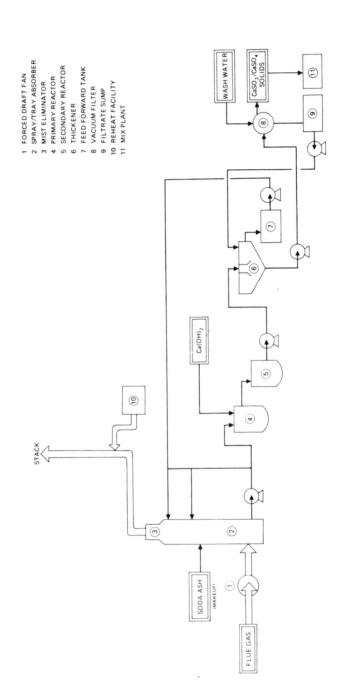

1   FORCED DRAFT FAN
2   SPRAY/TRAY ABSORBER
3   MIST ELIMINATOR
4   PRIMARY REACTOR
5   SECONDARY REACTOR
6   THICKENER
7   FEED FORWARD TANK
8   VACUUM FILTER
9   FILTRATE SUMP
10 REHEAT FACILITY
11 MIX PLANT

recycle liquor is withdrawn and sent to the reactor system for regeneration. The bleed rate is controlled by the liquid level in the absorber. The feed forward rate of regenerated liquor from the thickener hold tank to the scrubber trays corresponds to an L/G of 4.0 gal/10^3 acf (saturated) at design conditions. The total recirculation rate for each absorber (sprays plus trays but excluding the feed forward regenerated liquor) corresponds to an L/G of 5.7 gal/10^3 acf. Thus, the total L/G is about 10 gal/10^3 acf.

Regeneration Section: The spent liquor (absorber bleed) is fed to the primary reactor of the two-stage reactor system along with slurried carbide lime. The primary reactor has a nominal liquor holdup time of 4.5 minutes at design flow. The primary reactor overflows by gravity into the secondary reactor, which has a nominal holdup of 40 minutes at design flow, where the reaction between lime and sodium salts is completed. The reaction product is a slurry containing 2-4% insoluble calcium salts and the regenerated sodium salt solution. The reactor slurry is pumped to the solid/liquid separation section. The pumping rate is controlled by the liquid level in the reactor.

Two reactor trains are provided, each train consisting of a primary reactor, a secondary reactor, and a reactor pump. At design conditions, both of the reactor trains would normally be in service, with each reactor train regenerating the spent liquor from the corresponding absorber. The reactor trains are identical, and each can be operated on liquor from either absorber or combined liquor from both the absorbers. For short-term, a reactor train may handle the total liquor from both the absorbers operating at design conditions. Thus, maintenance can be performed on one reactor train while the other is operating.

Dewatering Section: The reactor effluent streams are fed to the thickener. Clarified liquor overflows to the thickener hold tank from which the regenerated solution is pumped automatically to the absorbers to maintain the pH of the absorber liquor. The total volume in the system is maintained by controlling the liquid level in the thickener hold tank using process makeup water.

The thickener underflow slurry, controlled at about 25 wt % solids, is pumped to the filter system where solids separation is completed. The filter cake is washed with fresh water to recover the sodium salts in the liquor. Combined filtrate and wash water are returned to the thickener.

There are three rotary drum vacuum filters, each rated to handle 50% of the total solids produced at the design conditions. Each filter can be operated independently. For optimum performance (to obtain cake containing high solids content and low soluble salts content) it is desirable to operate the filters at fixed conditions (constant drum speed, submergence, wash ratio, etc.). Therefore, the cake rate is controlled by changing the number of filters in operation. The number of filters in operation is determined by the amount of solids accumulated in the thickener, which is reflected in the solids concentration in the underflow slurry. The density of the underflow slurry is measured and thickener hold tank liquor is added as required to maintain the percent solids in the underflow slurry at about 25%. The number of filters in operation is changed if the concentration of solids in the underflow slurry cannot be controlled using the dilution liquor.

Southern Indiana Gas & Electric Co. (SIGECO)(5)

FGD Facility: Process equipment items of interest, in dual alkali system for Unit 1, include two booster fans; two, three-stage disc contactors; one lime reactor; one, 100-ft diameter, thickener for dual alkali solids concentration; and three, rotary-drum vacuum filters with water wash.

Figure 4.6 is the process flow diagram. A description of the absorption, regeneration, and dewatering sections is given below.

Absorption Section: The FGD system receives flue gas from two points on the discharge side of the electrostatic precipitator. The flue gas is drawn into two fans to increase the static pressure to the level required to force the gas through the absorbers and to the stack.

The SO_2 absorbers are low pressure drop, three-stage, countercurrent disc contactors. Liquid/gas contacting is accomplished with fixed position flow diverters. The units are 30 ft-5 inches in diameter and have an overall height to the discharge of 70 ft. The top of the scrubber contains a chevron type mist eliminator to prevent liquid entrainment. This mist eliminator is washed approximately once per shift to remove any accumulation of salts and/or particulate matter.

Four manways are provided on each scrubber to give access to the mist eliminator, scrubber internals, and liquid sump. The bottom of the vessel is used as a reservoir for the scrubbing solution. During normal operation, the flue gas provides adequate heat to prevent freezing of the scrubbing solution. During system shutdown and subsequent startup, a steam sparger is provided to heat the solution. The bottom 8 ft of the contactor is insulated with fiberglass and covered with an aluminum jacket.

Two rubber-lined recirculation pumps circulate scrubbing liquor to the top stage of each scrubber. Each pump is capable of supplying 5200 gal/min of solution. One pump is the primary operating pump; the second pump provides 100% spare capacity. The pumps are piped into the system with check valves to permit remote selection and startup without manual valve changes. A crossover starting system automatically starts the spare pump when flow fails. The flow of recirculated solution, not modulated for scrubbing load, is set for maximum gas flow.

Regeneration Section: Scrubbing solution to be regenerated is pumped to the lime reactor where it is neutralized with the lime slurry. The pH of the lime reactor is controlled by two separate pH control loops, one in operation and one spare.

The system has high and low pH alarms. A selector switch allows the operator to convert from one pH analyzer element to the other. The pH analyzer probes in this service are cleaned periodically. The regenerated slurry overflows from the lime reactor to the thickener where the calcium sulfite solids settle out of the regenerated solution. The rate of soda ash solution makeup to the scrubber modules controls the regenerated solution density. Density is an indication of the total solution concentration which must be maintained for optimum operation of the dual alkali process. The thickener underflow is pumped by diaphragm pumps to three rotary drum vacuum filters. The slurry flow rate to each filter is controlled by a low level switch. The regenerated scrubbing solution is collected in the regeneration surge tank and pumped back to the scrubbers based on pH demand.

Liquid level in the regeneration surge tank is controlled by addition of makeup water. The temperature of the surge tank is monitored and, in the event of low temperature, the tank is heated with steam supplied through a sparger.

Dewatering System: The underflow from the thickener is pumped to three rotary drum vacuum filters where the calcium sulfite precipitate is separated from the sodium sulfite solution. Cake removal is facilitated by low pressure air blown through the filter cloth just before the scrapers at the discharge. The filtrate is

Figure 4.6: FMC Flue Gas Desulfurization System on SIGECO'S A.B. Brown No. 1 Unit

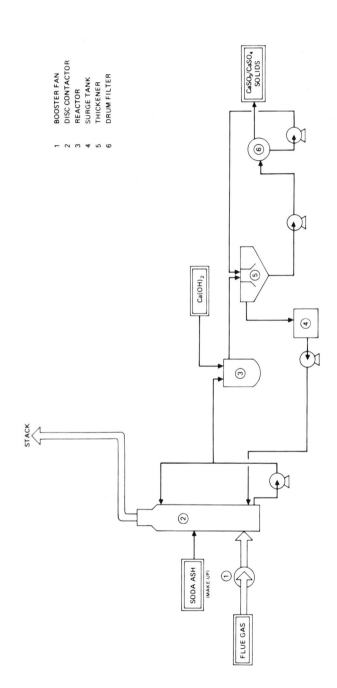

1 BOOSTER FAN
2 DISC CONTACTOR
3 REACTOR
4 SURGE TANK
5 THICKENER
6 DRUM FILTER

Source: Symposium – Vol. 2

drawn into a filtrate receiver and is pumped back to the thickener. The vacuum is produced by a mechanical vacuum pump using water flushing. The water for this service is used only once.

Design Bases

The design bases and design features for the three, utility dual alkali FGD systems are presented in Tables 4.5 and 4.6, respectively. Typical analyses of the coals fired are summarized in Table 4.7.

Information in these tables leads to several general observations:

> All installations service boilers firing high-sulfur coal. Design coal sulfur range is 4.0 to 5.0 wt %. Inlet SO_2 entering the absorbers ranges between 2600 and 3500 ppm.

> Design coal chloride content varies widely from 0.04 to 0.20 wt %.

> Excess air including preheater leakage varies considerably as indicated by the 3.2 to 7.6% flue gas O_2 content range.

> Flue gas flow rate per scrubber module ranges between 395,000 and 522,000 acfm at 300°F.

> The number of regeneration reactors varies between designs, indicating differing design philosophies.

> All facilities use primary (thickener) and secondary (vacuum filters) dewatering devices. Sodium is recovered by wash schemes ranging from single- to multistage.

> Planned ultimate disposal of filter cake is to be accomplished, in all cases, by landfill. At the Newton No. 1 and Cane Run No. 6 plants, stabilization efforts may range from simple mixing with fly ash to mixing with fly ash and lime.

Table 4.5: FGD System Design Bases (2)(4)(5)

	CIPSCO Newton No. 1	LG&E Cane Run No. 6	SIEGCO A.B. Brown No. 1
Coal (Dry Basis):			
Sulfur, wt %	4.0	5.0	4.5
Chloride, wt %	0.20	0.04	0.05
Heat content, Btu/lb	10,900	11,000	13,010
Inlet Gas:			
Flow rate (volumetric), acfm	2,163,480*	1,065,000	790,036
(weight), lb/hr	6,615,000	3,372,000	2,415,764
Temperature, °F	327	300	290
SO_2, ppm	2590*	3471	2810
O_2, %	7.7	5.7	3.2*
Particulate, lb/10^6 Btu	not available	0.10	0.10
Outlet Gas:			
SO_2, ppm	200	200	520
Particulate, lb/10^6 Btu	0.10**	0.10	0.10

*Calculated from other data
**System designed to accommodate ESP upsets.

Table 4.6: FGD System Design Features (2)(4)(5)

	CIPSCO Vendor: Envirotech	LG&E Vendor: CEA/ADL	SIGECO Vendor: FMC
FGD unit rating, MW	575	277	250
Process	Sodium/calcium hydroxide dual alkali/concentrated	Sodium/carbide lime dual alkali/concentrated	Sodium/calcium hydroxide dual alkali/concentrated
Fly ash collection	ESP	ESP	ESP
No. of modules	Four	Two	Two
Module design	High velocity cocurrent spray tower, countercurrent two-tray-stage polysphere absorber	Two-stage tray tower absorber	Three-stage disc contactor
Regeneration reactor	Three reactors in series for four modules*	Two reactors in series per module	One reactor per two modules
Reactor residence time	No information	Primary – 4.5 minutes Secondary – 40 minutes	No information
Dewatering	Three horizontal belt filters	Three rotary drum vacuum filters	Three rotary drum vacuum filters
Filter cake disposal	Fly ash stabilization/onsite landfill	Fly ash stabilization/onsite landfill	Transported to landfill without stabilization
Additional equipment	Chloride/fly ash precooler; fly ash thickener; neutralizer; reactor clarifier; and two versions of experimental reheat facilities	Scrubbed gas reheat facility; mix plant for sludge stabilization	None

*In actual operation, one, two or three reactors may be used.

Source: Symposium – Vol. 2

Table 4.7: Typical Coal Analyses (2)(4)(5)
(Dry Basis)

Species	CIPSCO Newton No. 1 Value, wt %	LG&E Cane Run No. 6 Value, wt %	SIGECO A.B. Brown No. 1 Value, wt %
Ultimate			
Carbon	66.70	67.15	72.19
Hydrogen	4.92	4.72	5.01
Nitrogen	1.68	1.28	1.56
Chlorine	0.22	0.04	0.05
Sulfur	4.47	4.81	3.97
Ash	14.19	17.06	9.88
Oxygen	7.82	4.94	7.34
Proximate			
Moisture	10.50	8.95	11.34
Heat content, Btu/lb	10,900	11,000	13,010

Source: Symposium — Vol. 2

References

(1)　Kaplan, N., "Introduction to Double Alkali Flue Gas Desulfurization Technology," presented at the EPA Flue Gas Desulfurization Symposium, New Orleans, LA, March 8-11, 1976.

(2)　Bloss, E.H., Private Communication, Envirotech.

(3)　Holhut W.J., et al., "Zero-Effluent Throwaway SO_2 System Design for High-Chloride, High-Sulfur Coal," *Proceedings of the American Power Conference,* Vol. 38, 1976.

(4)　Van Ness, R.P., et al., *Project Manual for Full-Scale Dual Alkali Demonstration at Louisville Gas and Electric Co. — Preliminary Design and Cost Estimate,* EPA Report No. 600/7-78-010, (NTIS PB 278 722), January 1978.

(5)　Ramirez, A.A., Private Communication, FMC.

SUMMARY OF STATUS OF DUAL ALKALI TECHNOLOGY IN JAPAN (1)

About 42 major dual alkali FGD plants, having a combined capacity of about 4,000 MW equivalent, were operational in Japan at the beginning of 1978. Table 4.8 summarizes these systems by process developer/constructor and capacity. Approximately 45% of the total capacity represents utility boiler application (primarily oil-fired) while the remainder includes industrial boilers, sintering plants, smelters, and sulfuric acid plants.

An interesting facet of emerging dual alkali technology in Japan is processes having unlimited oxidation tolerance and utilizing limestone as a regenerant. This is generally accomplished by circulating absorbents other than sodium sulfite. Examples include Dowa's basic aluminum sulfate process, Kawasaki's magnesium-gypsum process, and Kureha's sodium acetate-gypsum process. Some of these processes do not use a clear solution as the absorbent. The current EPA funded program at Gulf Power's Scholz Station is also investigating the feasibility of limestone regeneration with sodium sulfite systems. A brief discussion of the Dowa, Kawasaki, and Kureha processes follows.

Table 4.8: Numbers and Capacities of Dual Alkali FGD Plants by Major Process Developers/Constructors in Japan (1)

Process Developer/Constructor	Number	Capacity MW
Dowa Engineering	8	200
Kawasaki Heavy Industries	2	180
Kobe Steel	6	745
Kurabo Engineering	5	215
Kureha Kawasaki	5	1685
Kureha Chemical	1	2
Nippon Kokan (NKK)	1	45
Tsukishima Kikai (TSK)	2	160
Showa Denko	3	360
Showa Denko-Ebara	9	370
Total	42	3,962

Source: Symposium—Volume 2

Dowa Basic Aluminum Sulfate Process (2)

This process has been developed by Dowa Mining Co. of Japan. In the U.S. this process is licensed by Universal Oil Products (UOP) and is being tested at the TVA Shawnee Test Facility to evaluate its applicability to coal-fired boilers.

Process Description: As shown in Figure 4.7 this process consists of three operations: absorption, oxidation, and neutralization.

Absorption — SO_2 is absorbed in a clear solution of basic aluminum sulfate of pH 3 to 4 to form $Al_2(SO_4)_3 \cdot Al_2(SO_3)_3$;

(1) $Al_2(SO_4)_3 \cdot Al_2O_3 + 3SO_2 \rightarrow Al_2(SO_4)_3 \cdot Al_2(SO_3)_3$

Oxidation — The aluminum sulfite in the spent liquor is oxidized by air to aluminum sulfate:

(2) $Al_2(SO_4)_3 \cdot Al_2(SO_3)_3 + \frac{3}{2} O_2 \rightarrow Al_2(SO_4)_3 \cdot Al_2(SO_4)_3$

Neutralization — The oxidized liquor is treated with powdered limestone to precipitate gypsum and to regenerate the basic aluminum sulfate solution:

(3) $Al_2(SO_4)_3 \cdot Al_2(SO_4)_3 + 3CaCO_3 + 6H_2O \rightarrow Al_2(SO_4)_3 \cdot Al_2O_3$
$$+ 3(CaSO_4 \cdot 2H_2O) + 3CO_2$$

Commercial Applications: One of the earlier applications (~1 MW) was at the Mobara Works of Taenake Mining. This plant started operation in October 1972 to treat waste gas from a molybdenum sulfide roaster containing 7500 ppm SO_2 at 100°C. No problems have been reported to date. Two commercial units, each with a capacity of treating 150,000 Nm^3/hr (50 MW) of tail gas from a sulfuric acid plant were built at Okayama Works of Dowa. For these systems, inlet SO_2 averages 600 ppm. They are designed to remove at least 95% of SO_2 at a liquid-to-gas ratio of 15-20 gal/10^3 acf. These plants have operated successfully since startup in July 1974, with an average SO_2 removal efficiency of 99% during 20 months of testing. The only installation on a boiler, a small one at Naikai, also continues to perform well.

Other recent applications include two on iron ore sintering plants, one on a sulfuric acid plant, and one on a smelter.

Figure 4.7: Dowa Basic Aluminum Sulfate-Gypsum Flue Gas Desulfurization Process

1   BOOSTER FAN
2   ABSORBER
3   MIST ELIMINATOR
4   OXIDIZER
5   THICKENER
6   CENTRIFUGE
7   BASIC ALUMINUM SULFATE
    REGENERATION TANKS
8   REHEAT FACILITY

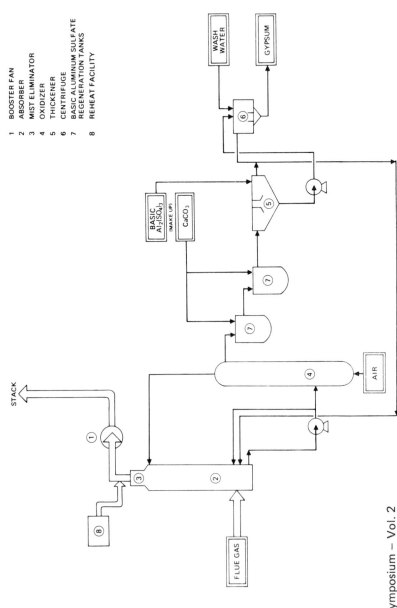

Source: Symposium – Vol. 2

Evaluation: The attractive features of this process, relative to the concentrated sodium-based processes, are unlimited oxidation tolerance and the use of limestone as a regenerant. However, this process operates at a higher liquid-to-gas ratio than the sodium-based processes.

Kawasaki Magnesium-Gypsum FGD Process (3)

Kawasaki Heavy Industries markets two versions of the magnesium-gypsum flue gas desulfurization process. In their "standard" process only a portion of the absorber bleed stream is oxidized and the mol ratio of calcium to magnesium in the absorbent liquor is maintained between 3 and 4. In the "new" process all the bleed stream is oxidized and the Ca/Mg mol ratio is held below 1.0.

Even though both the known commercial applications use the "standard" process the description presented herein relates to the new process because of its greater potential. Figure 4.8 is a simplified process flow diagram of this "new" process.

Process Description: Unlike the Dowa process, this process does not employ clear liquor scrubbing.

Its three unit operations (absorption, oxidation and gypsum recovery, and magnesium hydroxide regeneration) are described below.

Absorption — Flue gas containing SO_2 is contacted with a mixed slurry of calcium (as $CaSO_4$) and magnesium compounds. SO_2 is absorbed and removed according to reactions (4) and (5).

(4) $Mg(OH)_2 + SO_2 \rightarrow MgSO_3 + H_2O$

(5) $MgSO_3 + SO_2 + H_2O \rightarrow Mg(HSO_3)_2$

Oxidation and Gypsum Recovery — The spent liquor from the absorption section is oxidized by air to convert the magnesium sulfite to sulfate by the following reactions:

(6) $MgSO_3 + \frac{1}{2} O_2 \rightarrow MgSO_4$

(7) $Mg(HSO_3)_2 + \frac{1}{2} O_2 \rightarrow MgSO_4 + SO_2 + H_2O$

Since magnesium sulfate has a high solubility in water relative to calcium sulfate (calcium sulfate is recycled for seeding), separation of the calcium and magnesium sulfates is easy. Gypsum slurry from the oxidizer is dewatered to less than 10 wt % moisture with a centrifuge. Magnesium sulfate solution is forwarded to the magnesium hydroxide regeneration section.

Magnesium Hydroxide Regeneration — Since magnesium sulfate has no capacity to absorb SO_2, it is converted back to magnesium hydroxide. The regeneration is accomplished by adding lime or limestone. The following reaction occurs with lime.

(8) $MgSO_4 + Ca(OH)_2 + 2H_2O \rightarrow Mg(OH)_2 + CaSO_4 \cdot 2H_2O$

Magnesium hydroxide thus regenerated is returned to the absorber along with the precipitated gypsum.

Commercial Applications: Kawasaki Heavy Industries has built two commercial plants which have been operational since the beginning of 1976. Both plants use the partial oxidation mode (standard process). The first commercial plant utilizing this process is the Okazaki Works of the Unitika Co.; the second is the Saidaji Works of Japan Exlan Co. For the latter plant, inlet SO_2 averages 1400 ppm. For 95% SO_2 removal, a liquid-to-gas ratio of 35-45 gal/10^3 acf is required. The mol ratio of calcium to magnesium is maintained between 3.5 and 4.

Figure 4.8: Kawasaki Magnesium-Gypsum Flue Gas Desulfurization Process (New)

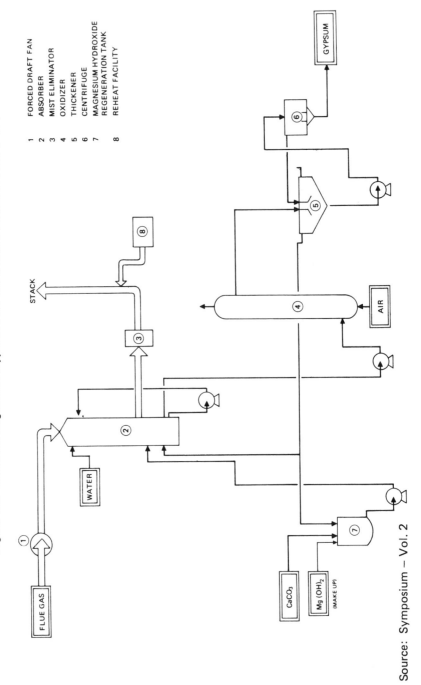

1   FORCED DRAFT FAN
2   ABSORBER
3   MIST ELIMINATOR
4   OXIDIZER
5   THICKENER
6   CENTRIFUGE
7   MAGNESIUM HYDROXIDE
    REGENERATION TANK
8   REHEAT FACILITY

Source: Symposium – Vol. 2

Lime is used in the plant for Japan Exlan Co., whereas limestone is used as the main absorbent (with a little lime) for the Unitika Co. For the Unitika application, lime is added to the reactor, and limestone is added to the absorber; the ratio of limestone to lime is about 5 to 1 (1). The use of limestone is a Unitika Co. refinement. Gypsum is recovered as a by-product in both plants.

Evaluation: Like the Dowa process, the major attractions of this process are unlimited oxidation tolerance and the option of limestone as a regenerant. However, the reported liquid-to-gas ratio for the "standard" version is significantly higher than that of the sodium-based processes. However, for the "new" version the liquid-to-gas ratio is comparable to that of the sodium-based process because most of the scrubbing is done by magnesium.

Kureha Sodium Acetate-Gypsum FGD Process (4)

Two limitations of the sodium sulfite dual alkali process marketed by Kureha were its inability to accommodate high oxidation levels and its difficulty in using limestone as a regenerant. In response to these shortcomings, Kureha developed the sodium acetate process described below.

Process Description: The process is composed of three unit operations: absorption, oxidation, and gypsum recovery.

Absorption — As illustrated in Figure 4.9, the absorption tower consists of two sections: SO_2 absorption and an acetic acid recovery section linked in series. In the SO_2 absorption section, SO_2 is removed by circulating sodium acetate solution. The following reaction takes place:

$$(9) \qquad 2CH_3COONa + SO_2 + H_2O \rightarrow Na_2SO_3 + 2CH_3COOH$$

Part of the acetic acid formed volatilizes in the scrubbed flue gas, and is recovered. Fresh limestone slurry is added to the top chamber and flows down countercurrently from chamber to chamber to the bottom to remove the acetic acid vapor as well as any remaining SO_2.

Oxidation — Sodium sulfite formed in the absorber is oxidized to sodium sulfate in the oxidation tower in which perforated plates facilitate fine dispersion of air bubbles and promote oxidation. Sulfite oxidation to sulfate takes place as follows:

$$(10) \qquad Na_2SO_3 + \tfrac{1}{2}O_2 \rightarrow Na_2SO_4$$

After oxidation, the liquor is sent to the gypsum recovery section, where gypsum is produced by the addition of limestone slurry.

Gypsum Recovery — In this operation, calcium carbonate reacts very rapidly with sodium sulfate in the presence of acetic acid; the reaction of calcium carbonate with sodium sulfate is very slow in the absence of acetic acid. It is believed that calcium carbonate reacts first with acetic acid present in the liquor to form calcium acetate, which reacts further with sodium sulfate to form calcium sulfate and sodium acetate by a double-decomposition reaction. The reaction mechanism can be expressed as in (11) and (12):

$$(11) \qquad CaCO_3 + 2CH_3COOH \rightarrow (CH_3COO)_2Ca + H_2O + CO_2$$

$$(12) \qquad (CH_3COO)_2Ca + Na_2SO_4 \rightarrow CaSO_4 + 2CH_3COONa$$

Evaluation: Like the Dowa and Kawasaki processes, unlimited oxidation tolerance and the use of limestone as a regenerant are attractive. The claimed liquid-to-gas ratio is on the order of 10 gal/10^3 acf. Kureha also offers the "lime-regenerant version" which is claimed to be simple and cheaper due to elimination

Flue Gas Desulfurization Technology

Figure 4.9: Kureha Sodium Acetate-Gypsum Flue Gas Desulfurization Process

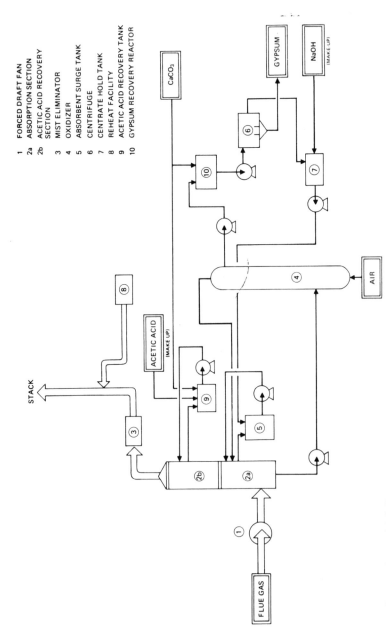

1   FORCED DRAFT FAN
2a ABSORPTION SECTION
2b ACETIC ACID RECOVERY
    SECTION
3   MIST ELIMINATOR
4   OXIDIZER
5   ABSORBENT SURGE TANK
6   CENTRIFUGE
7   CENTRATE HOLD TANK
8   REHEAT FACILITY
9   ACETIC ACID RECOVERY TANK
10 GYPSUM RECOVERY REACTOR

of the acetic acid recovery section, a smaller effluent hold tank, and the consolidation of the oxidation and the regeneration operations in one vessel. Reportedly, the process can also be modified by adding catalysts to the scrubbing liquor for simultaneous removal of SO_2 and NO_x.

Commercial Applications: Unlike the Dowa and Kawasaki processes, there are no commercial applications of this process. However, this process has been tested in a bench-scale plant (100 Nm^3/hr) with flue gas from an oil-fired boiler for 2 years since 1973. Furthermore, on the basis of the results in the bench-scale plant a 5,000 Nm^3/hr pilot plant was constructed in 1975. This pilot plant has operated smoothly and data necessary for scale-up have been obtained.

References

(1)   Ando, J , et al, *SO₂ Abatement for Stationary Sources in Japan,* EPA Report No. 600/7-78-210, (NTIS PB 290 198), November 1978.

(2)   Yamamichi, Y. and Nagao, J , "The Dowa's Basic Aluminum Sulfate-Gypsum Flue Gas Desulfurization Process," The Dowa Mining Co., Ltd., Okayama Research Laboratory, Okayama, Japan.

(3)   Tsugeno, H., et al., "Operating Experiences with Kawasaki Magnesium-Gypsum Flue Gas Desulfurization Process," *Proceedings: Symposium on Flue Gas Desulfurization,* Hollywood, Florida, November 1977, EPA Report No. 600/7-78-058b, (NTIS PB 282 091), March 1978.

(4)   Saito, S., et al., "Kureha Flue Gas Desulfurization-Sodium Acetate-Gypsum Process," Kureha Chemistry Industry Co., Ltd., Tokyo, Japan.

DESIGN AND PROCESS CONSIDERATIONS (1)

A commercial dual alkali system must be designed to remove the desired quantity of sulfur oxides from a given flue gas stream, while operating in a reliable manner and discharging environmentally acceptable solid waste product.

SO₂ Removal

For some time it has been known that small quantities of sulfur dioxide can be removed from large amounts of relatively inert gas by cyclic processes involving absorption into aqueous solutions of sodium sulfite/bisulfite. Johnstone et al (2) published a paper in 1938 giving data on the vapor pressure of SO_2 over solutions of sulfite/bisulfite and methods of calculating these equilibrium values under various conditions. The equilibrium partial pressure of SO_2 over sulfite/bisulfite solutions, the theoretical limit which a practical design can approach, is generally a function of solution temperature, pH, concentration of sulfite/bisulfite, and total ionic strength. Since Johnstone's work, a number of organizations have pursued this technology with laboratory, pilot plant, and full-scale applications for flue gas desulfurization, and many have demonstrated its ability for high removal efficiencies. (Note that, although Johnstone's work was aimed at cyclic processes with thermal regeneration, such as the Wellman-Lord system, the vapor pressure data are also applicable to dual alkali systems which use chemical regeneration.)

Once methods have been established to determine equilibrium SO_2 vapor pressure over scrubbing solutions of the various concentrations to be encountered in an operating system, it becomes a matter of standard chemical engineering practice to design adequate gas absorption equipment to accomplish the desired SO_2 removal in a system. For comparison, note that the design of lime/limestone slurry absorption equipment is further complicated by the kinetics of dissolution of the lime or limestone, the particle size of the suspended material, and the crystal morphology of the lime or limestone.

Reliable Operation

System reliability can be adversely affected by two classes of problems: mechanical and chemical.

Mechanical problems include malfunction of instrumentation and mechanical and electrical equipment such as pumps, filters, centrifuges, and valves. These problems in a commercial FGD system can be minimized by careful selection of materials of construction and equipment and by providing spares for equipment items such as pumps and motors which are expected to be in continuous operation and are prone to failure after a relatively short period of operation. Another important consideration in minimizing mechanical problems is the institution of a good preventive maintenance program.

Chemical (or physical/chemical) problems which may be associated with a dual alkali system include scaling, production of poor-settling solid waste product, excessive sulfate buildup, water balance, and buildup of nonsulfur solubles which enter the system as impurities in the coal or lime. Each factor is associated with reliable system operation, or production of an environmentally acceptable solid waste.

Scaling: One of the primary reasons, and probably the most important, for development of dual alkali processes was to circumvent the scaling problems associated with lime/limestone wet scrubbing systems. Therefore, a dual alkali system should be designed to operate in a nonscaling manner.

Scaling is caused by precipitation of calcium compounds (from process liquors) on the surfaces of various system components. In the scrubber, this is particularly troublesome since the flue gas path through the scrubber, if affected, could shut down the boiler/scrubber system and lower reliability.

Since scrubbing in dual alkali systems employs a clear solution rather than a slurry, there is a tendency to ignore potential scaling problems. Testing experience with dual alkali systems has indicated, however, that scaling can occur and indeed the problem should be a legitimate concern in the design of any system. Both gypsum and carbonate scale buildup has been recognized in these systems. Gypsum scaling is caused by the reaction of soluble calcium ion with sulfate ion formed in the system through oxidation of the absorbed SO_3 or from absorbed SO_3 according to the reaction:

$$(1) \qquad Ca^{2+} + SO_4^{2-} + 2H_2O \rightarrow CaSO_4 \cdot 2H_2O$$

In dilute systems, gypsum scaling is controlled by softening the regenerated liquor prior to recycling to the scrubber. In concentrated systems, gypsum is not a problem since the high sulfite concentration keeps the Ca^{2+} ion concentrations low. Softening ensures that the liquor recycled to the scrubber system is unsaturated with respect to gypsum; therefore, with proper softening even if some sulfate is formed in the scrubber, the liquor will not be saturated with gypsum and cause scaling on the inside surfaces of the scrubber. In concentrated active alkali systems, a special softening step is not necessary since high sulfite concentration is maintained throughout the system. This sulfite maintains a low Ca^{2+} ion concentration (sulfite softening), and thus maintains the scrubbing solutions unsaturated with respect to gypsum.

Based on experience gained in lime/limestone scrubbing testing, a certain factor of safety in the prevention of gypsum scaling probably exists in dual alkali systems. Gypsum has been found to supersaturate easily to about 130% saturation. Thus, scaling will occur only if supersaturation is excessive.

Carbonate scaling usually occurs as a result of localized high pH scrubbing liquor in the scrubber where CO_2 can be absorbed from the flue gas to produce CO_3 ions. These ions subsequently react with dissolved calcium to precipitate calcium carbonate scale according to the following series of reactions:

Carbon dioxide absorption by high pH liquor:

(2) $$CO_2 + 2OH^- \rightarrow CO_3^{2-} + H_2O$$

Calcium carbonate scaling:

(3) $$CO_3^{2-} + Ca^{2+} \rightarrow CaCO_3$$

Based on experience with the General Motors, full-scale, dual alkali system, carbonate scaling could occur with scrubber liquor pH above 9. At lower pH, the carbonate/bicarbonate equilibrium system tends to limit the free carbonate ion and thus prevent precipitation of calcium carbonate:

(4) $$H^+ + CO_3^{2-} \rightarrow HCO_3^-$$

Thus, carbonate scaling can be eliminated by control of pH in the scrubber.

Solids Quality: Under certain conditions, the waste solids produced in the regeneration sections of various dual alkali systems have a tendency not to settle from the scrubber liquors. This creates problems in the operation of settlers, clarifiers, reactor clarifiers, filters, and centrifuges. Although observed in the laboratory testing conducted by the EPA on dilute systems and in the laboratory and pilot plant work conducted by ADL on dilute and concentrated systems, this phenomenon is not completely understood, but is thought to be a function of reactor kinetics (3).

Some of the factors thus far identified which appear to affect the solids settling properties are reactor configuration, concentration of soluble sulfate, concentration of soluble magnesium and iron in the liquor, concentration of suspended solids in the reaction zones, and use of lime vs limestone for reaction. Based on laboratory work in dilute systems (about 0.1 M active sodium) using limestone, it appears that solids settling characteristics degraded significantly at soluble sulfate levels above 0.5 M. Based on laboratory work with concentrated systems (about 0.45 M TOS, 5.4 pH, 0.6 M sulfate) using limestone, marked degradation of solids settling properties occurred at a magnesium level of 120 ppm and virtually no settling of solids occurred at the 2000 ppm magnesium level. Equal degradation of solids settling properties also occurred in concentrated systems when the sulfate level was raised to 1.0 M while maintaining low magnesium level (about 20 ppm) and keeping other variables constant (3)

Envirotech (4) advocates the recycle of precipitated solids from the thickener underflow to the reaction zones in an effort to grow crystals which settle faster and are more easily filtered.

ADL cites reactor configuration as being important in the production of solids with good settling and filtration characteristics (3). Their basis for this is comparative tests of a simple continuous flow stirred tank reactor (CFSTR) with the ADL/Combustion Equipment Associates (ADL/CEA) designed reactor system under similar conditions. The ADL/CEA reactor system appeared to give better settling solids over a greater range of conditions than a simple CFSTR. The ADL/CEA reactor system consists of a low residence unit followed by a high residence unit.

Sulfate Removal: In dual alkali systems, some of the sulfur removed from the flue gas takes the form of soluble sodium sulfate due to oxidation in the system, thus changing some of the active sodium to the inactive variety. When sodium in the system is converted to the inactive form (Na_2SO_4), it is relatively difficult to convert back to active sodium. To convert inactive sodium to active sodium, sulfate ion must be removed from the system in some manner, while leaving the sodium in solution. The alternative to this is to remove the sodium sulfate from the system at the rate it is being formed in the system. This alternative is not desirable since it wastes sodium and generally is carried out by allowing the sodium sulfate to be purged from the system in the liquor which is occluded in the wet solid waste product (3). The solid waste product can then potentially contribute to water pollution by leaching. Water runoff can contaminate surface water, while leaching and percolation of the leachate into the soil can contaminate ground water near the disposal site. Failure to allow for sulfate removal from dual alkali systems will ultimately result in a) precipitation of sodium sulfate somewhere in the system if active sodium is made up to the system, or b) in the absence of makeup, eventual deterioration of the SO_2 removal capability due to the loss of active sodium from the system.

Equations (1), (2), (3), and (4) describe several sulfate removal techniques which have been used in FGD system pilot tests.

The first equation depicts the sulfate removal technique used in dilute active alkali systems:

(1) $$Na_2SO_4 + Ca(OH)_2 + 2H_2O \rightarrow 2NaOH + CaSO_4 \cdot 2H_2O$$
                                                                                    (gypsum)

Concerning the full-scale dilute alkali system installed and operating at the Parma, Ohio, transmission plant of General Motors, and dilute systems in general, Phillips (5) stated:

> The presence of Na_2SO_4 in the scrubber effluent is the prime factor influencing the design of the regeneration system. Na_2SO_4 is not easily regenerable into NaOH using lime, the reason being that the product, gypsum, is relatively soluble . . . Na_2SO_4 cannot be causticized in the presence of appreciable amounts of SO_3^{2-} or OH^- because Ca^{2+} levels are held below the $CaSO_4$ solubility product. To provide for sulfate causticization, the system must be operated at dilute OH^- concentrations below 0.14 molar. At the same time, SO_4^{2-} levels must be maintained in the system at sufficient levels to effect gypsum precipitation . . . We selected 0.1 molar OH^- and 0.5 molar SO_4^{2-} as design criteria.

In a previous paper, Phillips (6) showed a plot of equilibrium caustic formation in $Ca(OH)_2$-Na_2SO_4 solutions at 120°F which is the basis for selection of the design criteria. The essence of this discussion is that, if the active sodium concentration is sufficiently dilute, sulfate can be removed from the system by simple precipitation as gypsum by reaction of lime with sodium sulfate.

Since, as explained above, this reaction will not proceed to a great extent in concentrated active alkali systems, other techniques must be employed to effect sulfate removal in these systems.

The second equation depicts a technique which is used in the full-scale dual alkali systems in Japan, and which has been pilot-tested by ADL under contract with EPA:

(2) $Na_2SO_4 + 2CaSO_3 \cdot \frac{1}{2}H_2O + H_2SO_4 + \frac{3}{2}H_2O \rightarrow 2NaHSO_3 + 2CaSO_4 \cdot 2H_2O$
(gypsum)

This technique is used to precipitate gypsum by dissolving calcium sulfite in acidic solution, thus increasing the Ca^{2+} in solution enough to exceed the solubility product of gypsum. Ideally according to equation (2) 2 mols of gypsum should be precipitated for each mol of sulfuric acid added. In practice, however, this is not the case since any material which functions as a base can consume sulfuric acid and reduce the efficiency of this reaction for its intended purpose (3). Unreacted lime or limestone, sulfite ion, and even sulfate ion can consume sulfuric acid, thus lowering sulfate removal from the system.

Conceivably, this method of sulfate removal may be economically unattractive in applications with very high oxidation rates, and where the gypsum produced must be discarded. The economic picture is considerably changed where this system is used merely as a slipstream treatment to supplement other sulfate removal methods and/or where the solid product gypsum is saleable as is the case in Japan.

The third equation describes a phenomenon which has been referred to as mixed crystal or solid solution formation:

(3) $xNaHSO_3 + yNa_2SO_4 + (x+y)Ca(OH)_2 + (z-x)H_2O \rightarrow$
$(x+2y)NaOH + xCaSO_3 \cdot yCaSO_4 \cdot zH_2O$
(mixed crystal or solid solution)

This phenomenon is described by EPA/IERL-RTP's R.H. Borgwardt (7) as it applies to lime/limestone wet scrubbing based on pilot plant investigations. A similar phenomenon has been observed by ADL in some of their early pilot testing of dual alkali systems in conjunction with CEA, and later in the EPA/ADL dual alkali test program.

Under certain conditions the solids precipitated in lime/limestone and dual alkali systems contain sulfate, sulfite, and calcium; however, the liquor from which these solids precipitate appears to be subsaturated with respect to gypsum. This is based on the fact that pure gypsum crystal could be dissolved in the mother liquor from which the mixed crystal/solid solution was precipitated. In addition, the solid material, examined by x-ray diffraction contained no gypsum; infrared analysis confirmed the presence of sulfate.

Borgwardt found that the molar ratio of sulfate to sulfite in these solids was primarily a direct function of sulfate ion activity in the mother liquor. In pilot test work with lime/limestone scrubbing, with little or no chlorides present and normal magnesium level (below 1000 ppm) in solution, the sulfate to sulfite molar ratio in the mixed crystal solids reached a maximum level of 0.23. This is equivalent to a (SO_4^{2-})/total (SO_x) ratio in the solids of 0.19.

In pilot test work with concentrated dual alkali systems, ADL observed the simultaneous precipitation of sulfate and sulfite with calcium in lime and limestone treatment of concentrated dual alkali scrubbing liquors. This phenomenon was surprising at first, in light of the reasoning which led to the development of dilute dual alkali systems; i.e., gypsum cannot be precipitated from solutions containing high active alkali concentrations. It was a simple technique for sulfate removal in concentrated systems. The (SO_4^{2-})/total (SO_x) ratio observed in pilot dual alkali work was as high as 0.20 (3). Coincidentally, this was the same value observed by Borgwardt in lime/limestone testing. This led to the belief that the same phenomenon was occurring in both processes. The mother liquor from which these solids were precipitated was also found to be subsaturated in gypsum, and when the solids were examined, pure gypsum was not found.

Based on the observed data, it appears reasonable to design a concentrated active alkali system for a particular situation in which the system oxidation rate is below about 20% (3). In this case, sulfate can be removed at the desired rate, without purging Na_2SO_4 or supplementing the system with other complex method of sulfate removal.

The fourth equation shows sulfate removal as sulfuric acid in an electrolytic cell:

$$(4) \quad Na_2SO_4 + 3H_2O \xrightarrow{\text{Electrolytic cell}} 2NaOH + H_2SO_4 + H_2 + \tfrac{1}{2}O_2$$

This method is the basis for operation of the Stone and Webster/Ionics process sulfate removal technique. In Japan, Kureha/Kawasaki has pilot-tested the Yuasa/Ionics electrolytic process for sulfate removal in conjunction with their dual alkali process. They claim that this process is less expensive overall than the presently used sulfuric acid addition method. In addition, they claim that sodium losses from the system can be cut in half through the use of this method: from 0.018 mol Na loss/mol SO_2 absorbed to 0.009 mol Na loss/mol SO_2 absorbed (8).

Another approach to sulfate control is to limit oxidation. With sufficient limitation of oxidation, by process and equipment design, it may be possible to control sulfate by a small unavoidable purge of Na_2SO_4 with the solid waste product. To design for minimum oxidation, there should be minimum residence times in equipment where the scrubber liquor is in contact with oxygen-containing flue gas; all reactors, mixers, and solids separation equipment should be designed to minimize absorption of oxygen from air. In addition, it has been reported (13) that oxidation of scrubber liquors can be minimized by maintaining very high ionic strength. One possible explanation for this is that high ionic strength liquors are poor oxygen absorbers and that oxidation in these systems is oxygen absorption rate limited.

Water Balance and Waste Product (Cake) Washing: To operate a closed system to avoid potential water pollution problems, system water balance is a primary concern. Water cannot be added to the system at a rate greater than normal water losses from the system.

Generally fresh water is added to a D/A system to serve many purposes, including: saturation of flue gas, pump seal, mist eliminator washing, slurry makeup, waste product washing and tank evaporation.

On the other hand, water should only leave the system through: evaporation by the hot flue gas, water occluded with solid waste product and water of crystallization of solid waste product.

Careful water management, part of which is the use of recycled rather than fresh water wherever possible, is necessary in order to operate a closed system.

Disposal of wet solid waste containing soluble salts is ecologically undesirable. In addition, allowing active alkali or sodium salts to escape from the system is an operating cost factor. Sodium is usually made up to dual alkali systems by adding soda ash at some point in the system. Thus, both ecological and economic considerations dictate that waste product washing is desirable.

A rotary drum filter, belt filter, or centrifuge is usually where the final solids separation is made. This equipment can be designed for solids washing with fresh water.

One concern in waste product washing is the extent to which the cake should be washed, because of the effect it has on the concentration of solubles in the waste. Solubles in the waste of any FGD system (i.e., lime/limestone and dual alkali) have the potential to contaminate groundwater.

One of the major goals of the Resource Conservation and Recovery Act (RCRA), which requires Federal regulations for disposal of hazardous waste and guidelines (for use by the state) for disposal of nonhazardous waste, is to prevent groundwater contamination.

Although it will be some time before these standards are developed, an FGD system operator must take this potential for groundwater contamination into account in designing his disposal site.

One obvious consideration in waste product washing is system water balance. Unlimited waste product washing is not possible if a closed system operation with no liquid stream discharge is a goal. Another more subtle reason for limiting waste product washing is the potential problem of nonsulfur/calcium solubles buildup in the system. These nonsulfur/calcium solubles enter the system with the fly ash, flue gas, and lime and/or limestone and the makeup water. Of these, probably the soluble material in the highest concentration would be sodium chloride which results from the absorption of HCl from the flue gas by the scrubber solution. A material balance around the system at steady state necessitates that solubles leave the system at the rate they enter. Thus, depending upon how well the waste product is washed, a certain level of nonsulfur solubles will be established in the system.

Since the only mechanism for these solids to leave the system is as part of the wet solid waste, a certain purge is necessary. This purge also necessitates the loss of some sodium from the system. Practical limitations in filter design and water balance probably would limit a system to two or three "displacement washes" of the waste product (one displacement wash means washing with an amount of fresh water equivalent to the amount of water contained in the final wet waste product per unit of waste). Depending on the characteristics of the waste product and the design of the washing system, one displacement wash can reduce the solubles content of the waste product by as much as 80%.

Environmentally Acceptable Solid Waste

Dual alkali systems should be designed to produce an environmentally acceptable end product. Desirable solid waste product properties include: nontoxic, low soluble solids content, low moisture content, nonthixotropic and high compressive or bearing strength.

Gypsum vs Calcium Sulfite: One of the options available to some dual alkali processes is whether or not to oxidize the solid waste. Advantages accruing from oxidation include: gypsum has better handling properties than calcium sulfite because sludges containing a high ratio of gypsum to calcium sulfite are less thixotropic, faster settling, more easily filtered, and can be more completely dewatered than sludges containing a high proportion of calcium sulfite. Another important characteristic which has been attributed to high gypsum (as opposed to calcium sulfite) sludges is their higher compressive or bearing strength.

An explanation for the behavior of high sulfite sludges is given by Selmeczi and Knight (9). Although filter cakes appear dry, they still contain a considerable amount of water and thus, upon vibration or application of stress, have a tendency to again become fluid. This thixotropic property and high moisture content are both explained by the morphology of calcium sulfite clusters. Because of the highly open, porous, or spongelike nature of these clusters, a considerable amount of water is retained in the clusters. The calcium sulfite crystals are rather fragile and break under pressure, releasing some of the water, which results in fluid sludges.

In Japan, where by-product gypsum is saleable, the calcium sulfite solids produced are oxidized completely to gypsum in a separate oxidation process tacked onto the tail end of the system. In applications where high excess combustion air is present, where low-sulfur coal is burned, or a combination of these conditions, the oxidation rate in the system tends to be high (possibly about 90%) and the proportion of gypsum in the sludge tends to be high. In some dilute systems, the proportion of gypsum in the sludge can be increased by augmental aeration of the scrubbing liquor (10). Crystal seeding techniques used in conjunction with augmental aeration can produce relatively coarse grained gypsum crystals with good dewatering and structural properties in the final waste product.

Sludge Fixation Technology: Chemical or physical fixation of the sludge produced in a dual alkali system is another potentially important means of producing an environmentally acceptable solid waste product. This technology is commercially offered by I.U. Conversion Systems, Inc., Dravo Corporation, and Chemfix Corporation. Most of their efforts are concentrated on sludge produced from the more prevalent lime/limestone systems; however, there has been some evaluation of dual alkali sludges. The objective of sludge fixation technology is the production of a nontoxic, unleachable solid waste product which has reasonably high load bearing strength. If dual alkali sludges are amenable to this type of treatment, the need to reduce soluble sulfates in the solid waste product is mitigated. Some sodium sulfate has been found to be physically or chemically tied up in the solid calcium sulfate/sulfite crystal lattice (3), however the extent of this phenomenon is not generally considered to be adequate to remove all of the sodium sulfate produced by oxidation.

Sludge fixation technology may be a mechanism by which additional sodium sulfate can be removed from the system without adverse environmental effects. There is some concern that this is not viable, however, since sludge fixation chemistry involves pozzolanic reactions between calcium compounds and fly ash components in the sludge which may only involve multivalent ions rather than monovalent sodium. In other words, monovalent ions such as sodium may either (a) not take part in the pozzolanic reactions, or (b) inhibit or limit such reactions. Further investigation is needed in this area.

References

(1) Kaplan, N., "Introduction to Double Alkali Flue Gas Desulfurization Technology," presented at the EPA Flue Gas Desulfurization Symposium, New Orleans, Louisiana, March 8-11, 1976.

(2) Johnstone, H.F., et al., "Recovery of Sulfur Dioxide from Waste Gases," Ind. & Eng. Chem., Vol 30, No. 1, January 1938, pp 101-109.

(3) LaMantia, C.R., et al., "EPA/ADL Dual Alkali Program Interim Results," presented at EPA Symposium on Flue Gas Desulfurization, Atlanta, Georgia, November 4-7, 1974.

(4) Cornell, C.F. and Dahlstrom, D.A., "Performance Results on a 2500 ACFM Double Alkali Plant for SO_2 Removal," presented at the 66th Annual Meeting of AIChE, Philadelphia, Pennsylvania, November 11-15, 1973, condensed version of the paper appeared in December 1973 CEP.

(5) Phillips, R.J., "Operating Experiences with a Commercial Dual-Alkali SO_2 Removal System," presented at the 67th Annual Meeting of the Air Pollution Control Association, Denver, Colorado, June 9-13, 1974.

(6) Phillips, R.J., "Sulfur Dioxide Emission Control for Industrial Power Plants," presented at the Second International Lime/Limestone Wet-Scrubbing Symposium, New Orleans, Louisiana, November 8-12, 1971.

(7) Epstein, M., Borgwardt, R., et al., "Preliminary Report of Test Results from the EPA Alkali Scrubbing Test Facilities at the TVA Shawnee Power Plant and at Research Triangle Park," presented at Public Briefing, Research Triangle Park, North Carolina, December 19, 1973.

(8) Seamans, T., Private Communication, Ionics Incorporated, Watertown, Massachusetts.

(9) Selmeczi, J.G. and Knight, R.G., "Properties of Power Plant Waste Sludges," presented at the Third International Ash Utilization Symposium, Pittsburgh, Pennsylvania, March 13-14, 1973.

(10) Ellison, W., et al., "System Reliability and Environmental Impact of SO_2 Scrubbing Processes," presented at Coal and the Environment, Technical Conference, Louisville, Kentucky, October 22-24, 1974.

GENERAL MOTORS' DOUBLE ALKALI SYSTEM

The General Motors version of the double alkali set scrubbing process went on line in March of 1974 at the Chevrolet Motor Division facility in Parma, Ohio. This installation, was at that time, the latest development in a program started by General Motors engineers in 1968 to find an environmentally acceptable means of burning coal in an industrial size power plant. Most of the boilers in the Corporation range from steaming capacities of 50,000 to 150,000 lb/hr, with an average plant steaming capacity of 250,000 lb/hr. Generally, the pressure ranges below 200 psi, and the steam is used in process operations and building heat.

There are a number of operating characteristics associated with industrial boilers which have to be considered in the scrubber system design. These include boiler type, load fluctuation, high excess air, and dust load. In addition to these basic design characteristics, the system had to be simple to operate, reliable and economically competitive with other control alternatives.

With this in mind, the Chevrolet Parma, Ohio, plant was selected in 1972 for a prototype wet scrubbing system. The history and status of the installation follow.

Descriptive Layout of the Facility and Equipment: The Chevrolet Parma facility includes a sheet metal stamping plant, an automatic transmission manufacturing plant, and a propeller drive shaft manufacturing plant. Steam is supplied for process parts washers and building heat from four boilers, centrally located in a separated facility. Two 60,000 lb/hr boilers were installed in 1948 and two 100,000 lb/hr boilers were installed in 1966 and 1967. The boilers are all traveling grate, continuous-ash-dump type with stoker feeders. They are also equipped with mechanical dust collectors. The larger boilers have economizers, and stack discharge temperatures are approximately 350°F. The stack discharge temperature of the smaller boilers is approximately 600°F.

The double alkali type system was chosen because of its potential reliability over lime/limestone systems. The system is closed-loop with scrubbing being accomplished by use of a dilute caustic mixture and regeneration with lime. Each boiler has its own scrubber and caustic control. The lime regeneration system is common to all units.

Additional draft fans were necessary to overcome the 7- to 8-inch pressure drop of the scrubbers. These fans are installed upstream of the scrubbers to avoid corrosion problems.

The scrubbers have three bubble cap absorption trays and a mesh type mist eliminator. This type of scrubber was chosen to obtain the highest sulfur dioxide removal along with maximum turndown potential. All wetted surfaces are constructed of 316L stainless steel.

Caustic makeup to the scrubbers is determined by the pH of the liquor as it is recirculated. Liquid blowdown is diverted to two in-line reactor tanks where slaked lime slurry is added. These tanks overflow to a reactor clarifier where solids settling occurs and the lime reaction continues. The overflow from this clarifier goes to a second clarifier, where soda ash is added for softening and sodium makeup, and additional solids settling occurs. Overflow from the second clarifier is the makeup liquor used to replace blowdown at the scrubber.

Early Developments: The scrubbers were started up for the first time on February 28, 1974. A number of initial problems were encountered, many of these were due to lack of experience with the system. For example, just prior to an open house, an upset was experienced when a maintenance employee started cleaning the building walls with a soap solution. Since all spills or leaks drain back into the system via sump pumps, the soap entered the process and caused all the solids in the clarifiers to go into suspension. During the open house, scrubbing was done with slurry. This eventually shut down the system.

A number of functional problems with the system were also encountered. First, there was a problem with a reliable pH control. This was due to initial location of the pH sensor. Since the probe was in a pressure line, it would frequently break. In addition, because the response time was so short, pH readings were unreliable. An attempt was made to solve the problem by placing the probe in a gravity sample line off the top tray. This stopped the breakage, but soon the entire probe was coated with solids.

These experiences led to the present arrangement of sampling pH in a gravity sample line off the bottom tray. This lower pH location has eliminated the scaling problem and allowed maintenance of reliable pH control, which is maintained at a level of pH 6.

Eventually there was a problem due to the lack of intermixing of recycle liquor and scrubber feed which caused the top tray to plug. This was the result of a high pH condition on the top tray, with calcium carbonate scaling shutting down operations within 8 hours. An early fix was to have both lines feed into a mixing box above the top tray of the scrubber. This proved ineffective because sufficient time was not allowed for mixing. The next step was to put caustic feed into the recycle tank. This effectively reduced top tray pH and eliminated the carbonate scaling problem. Unfortunately, solids build-up of calcium sulfite occurred at a fantastic rate in the recycle tanks, causing them to fill with solids. This was primarily due to the long holdup time in the tank at a lower pH. Finally, it was determined that in order to minimize scaling, the proper place to introduce caustic makeup was in the recycle line just prior to entering into the top tray of the scrubber. This has greatly reduced the scrubber scaling problem.

An additional problem was encountered with settling calcium salts in the first clarifier. The problem was caused by hindered settling due to excess sludge recycling and trouble maintaining a constant sludge blanket in the bottom of the clarifier. This was resolved by eliminating sludge recycle and by not operating with a sludge blanket.

There has been a very long learning curve on the operation of vacuum filters. The first cloth was a polypropylene material that had good wear characteristics but poor cake release. This was replaced with nylon and improvements have resulted.

Also contributing to the problem of cake release was the high percentage of calcium sulfite in the cake. Cake characteristics were greatly improved by adding an oxidation step prior to lime addition. This produces a higher percentage of calcium sulfate. There is now approximately 60% solids in the cake.

Overdesign of the system has caused problems with control. The system was originally designed to handle 400,000 lb/hr of steaming capacity burning 3% sulfur coal. However, the maximum actual operations have been only 50% of design capacity. Since the controls were designed for maximum levels, there have been problems measuring flows and controlling the operation at these reduced conditions. This made it necessary to reduce line sizes in specific areas of the process in order to more accurately measure and control liquid flows.

There was also a serious problem with closed gravity lines carrying solids to the clarifiers. This caused numerous outages of the system due to plugging of these lines. The problem was finally resolved by installing removable covered troughs of a rectangular cross section which increased the flow rate in these lines and made them accessible for maintenance.

Current Operating Problems: The lime feed systems are being modified. The original system has never been able to accurately control the amount of lime added to the system. This was due to the number of variables in the control loop which made this lime addition so complex it could not be maintained. The system is being simplified by installing linear feed pumps and proportioning that feed to caustic flow to the scrubbers.

Mist carryover continues to be one of the most serious problems. Initial scrubber design did not allow for utilization of the entire mesh pad. This was corrected by modifying the transition from the scrubber to the exhaust stack and by relocating the mesh mist eliminator. However, this has not totally resolved the problem.

The vacuum filter agitators, slurry inlet, and cake support roll are also being modified to give better solids distribution over cloth and improve cake filtering characteristics.

System Performance Versus Time and Operability: The scrubbers have held up very well with little or no corrosion discernible. The rubber-lined recycle pumps have proved to be very maintenance-free. The recycle tanks are showing corrosion and will require relining. They are mild steel painted with a Bitumastic coating. They are presently being patched and relined with plastic. Corrosion of the stacks occurred almost immediately after startup. A stainless liner was added within the first year, and that has also failed. They have since been replaced with FRP stacks.

The sludge handling operation receives the most abuse and, therefore, requires the most service. The present sludge pumps, however, have provided good service. The original sump pumps failed early and were replaced with air operated diaphragm pumps.

Piping in the system has been mostly trouble-free. All piping in areas of the system where corrosion could occur is hand-laid, fiber-cast plastic.

The ratio of scrubber hours versus boiler hours has been increasing to a level now near 70%. One of the factors that must not be overlooked in this evaluation is that these hours include many scheduled shut-down periods for planned maintenance, evaluation programs, or system modifications. Typically, the system operates for extended periods of time without interruption.

ENVIRONMENTAL ASSESSMENT OF A DUAL ALKALI SYSTEM ON BOILER FIRING COAL AND OIL

In response to the need for a comprehensive environmental assessment of conventional combustion systems, EPA's Industrial Environmental Research Laboratory at Research Triangle Park (EPA/IERL-RTP), North Carolina, has established a unified Conventional Combustion Environmental Assessment (CCEA) program (1).

A major goal of the CCEA program is to evaluate the effects of implementation of the National Energy Plan, which calls for the increased use of coal to meet the Nation's energy requirements. Since fuel switching from oil to coal is an important facet of the NEP, the CCEA program initiated a study (2) to evaluate the environmental effects of oil and coal combustion in a controlled industrial boiler in order to compare environmental, energy, and societal impacts of firing coal vs firing oil. In order to conduct the comparative assessment, it was necessary to fully characterize feed streams, emissions, and effluents from the industrial boiler selected for study and all associated pollution control equipment.

Plant Description

The site chosen was the Pottstown, Pennsylvania, plant of the Firestone Tire and Rubber Company, with Firestone's agreement and cooperation. Boiler No. 4, one of four boilers which supply process and heating steam to the plant, was used in the assessment. The boiler burns either coal or oil and has a pilot FMC dual alkali flue gas desulfurization system designed to treat approximately one-third of the boiler flue gas.

Boiler No. 4 is a dry, bottom once-through integral furnace, Babcock and Wilcox (Type FH-18) unit. (See Figure 4.10 for a schematic of the boiler and associated equipment.) When it was installed in 1958, the boiler was designed as a coal-fired unit but was converted to fire either coal or oil in 1967. The changeover from one fuel to the other can be accomplished in less than 30 minutes.

Figure 4.10: Boiler System Schematic

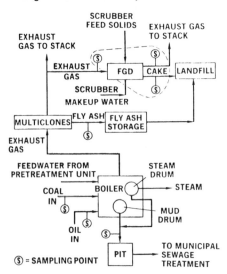

For test purposes, Firestone agreed to fire one fuel and then the other as long as required to conduct the appropriate sampling.

The two fuels are usually not burned simultaneously except when converting from oil to coal firing. The coal is ignited by continuing oil firing until a stable coal flame is obtained. Oil is fired simultaneously with coal to maintain acceptable steam generation rates when coal with a low heat content is burned.

Control Devices

The flue gases are treated by an air pollution system which consists of multiclone units and a pilot FGD unit. The multiclones are the primary particle control device. All of the flue gas passes through the multiclones after which the stream is split: two-thirds of the flue gas is ducted to the stack; and the other one-third is ducted to the pilot FGD system which removes SO_2 and additional particles. There are no NO_x controls on the system.

The collection efficiency of the multiclone varies as a function of the particle size distribution and grain loading. Typically, multiclones remove 90% of those particles with diameters of 10 μm and greater, and 50 to 80% of those particles with diameters of 3 μm and greater. The collection efficiency of multiclones drops off rapidly for particles less than 3 μm diameter.

The flue gas desulfurization system was designed and manufactured by FMC Corporation. The FGD system is a pilot unit designed to handle 280 acm/min (10,000 acfm) of flue gas, which is approximately one-third of the volume of the flue gas from the boiler. The pilot plant was placed on-line in January 1975. Figure 4.11 is the basic flow diagram of the FMC FGD system, as applied at this site.

The flue gas (stream 1) is withdrawn downstream of the boiler on the exit side of the multiclone dust collectors. Fly ash loading at the scrubber inlet is substantially higher during coal firing than during oil firing. To accommodate the wide variation in fly ash loading, the FGD system was designed to operate with or without fly ash, and can be operated on either fuel.

Upon entering the FGD unit, the flue gases are contacted with a slightly acidic scrubbing solution (stream 4) which removes SO_2 and particles. The SO_2 and particles are removed at the scrubber throat and carried away in the scrubbing solution. The process utilizes a sodium sulfite/sodium bisulfite solution as the absorbent. The basic reaction for SO_2 removal is:

(1) $\qquad Na_2SO_3 + SO_2 + H_2O \rightarrow 2NaHSO_3$.

A bleed stream (stream 5) of the scrubbing solution is removed from the system at a rate which keeps the pH of the solution in an acceptable range. The bleed stream is reacted with calcium hydroxide in a short retention time, agitated vessel to regenerate the sodium sulfite. The basic chemistry of sodium sulfite regeneration is:

(2) $\qquad 2NaHSO_3 + Ca(OH)_2 \rightarrow CaSO_3 \cdot \frac{1}{2}H_2O + \frac{3}{2}H_2O + Na_2SO_3$.

The slurry of precipitated sulfur compounds (stream 8) is concentrated and pumped to a rotary drum filter where the essentially clear liquid is separated from the solid waste products. The clear liquid (stream 10) is returned to the system for further utilization. The solid wastes, in the form of filter cake containing 40 wt % water (stream 3), are removed from the rotary drum filter and conveyed to a storage bin to await transportation to the dump site. Mainly due to the heavier particle loading, more filter cake is produced during coal firing than during oil firing.

Figure 4.11: FMC Unit at the Industrial Facility

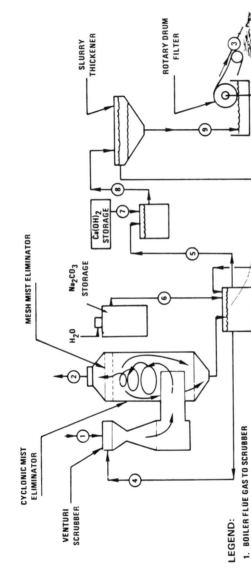

LEGEND:

1. BOILER FLUE GAS TO SCRUBBER
2. SCRUBBER OUTLET TO ATMOSPHERE
3. SOLID WASTE TO LANDFILL
4. ABSORBENT SOLUTION TO SCRUBBER
5. ABSORBENT SOLUTION TO REGENERATION
6. SODIUM CARBONATE MAKEUP
7. REGENERATION SOLUTION
8. REGENERATED SCRUBBER SOLUTION
9. CONCENTRATED SLURRY
10. RETURNED SCRUBBER SOLUTION

Source: Symposium—Vol 2

The on-site landfill, which is the final disposal facility for fly ash and scrubber cake generated at the facility, has several test wells from which samples are collected every 3 months and sent to an independent laboratory for analysis. Monthly tests are conducted by plant personnel to monitor Na^+ and specific conductivity. With permission from the Pennsylvania Department of Environmental Resources, this site is being used as an experimental disposal area for the filter cake from the FMC unit.

Test Description and Conditions

Multimedia emission tests were conducted on Boiler No. 4 of the Firestone Plant from September 27 through October 8, 1977. Gaseous and solid emissions were sampled during coal and oil firing to obtain data for the assessment. Flue gas was sampled before and after the scrubber to determine which pollutants are removed or modified by the control device. Sampling points used are indicated on the process diagram, Figure 4.10.

Emissions were characterized using EPA's phased approach. This approach utilizes two levels of sampling and analysis (Level 1 and Level 2). Level 1 screening procedures are accurate within a factor of 2 to 3. They provide preliminary assessment data and identify problem areas and information gaps. Based on these data, a site specific Level 2 sampling and analysis plan is developed. Level 2 provides more accurate and detailed information to confirm and expand on the information gathered in Level 1.

Normally all Level 1 samples are analyzed and evaluated before moving to Level 2. Because of the program time constraints, the Level 1 and Level 2 samples were obtained during the same test period; however, analysis of the samples did proceed in a phased manner except where sample degradation was of concern. In that case, Level 2 analysis was performed on the sample prior to Level 1 completion.

Gaseous Effluents: The boiler flue gas was sampled at the inlet and the outlet of the pilot flue gas desulfurization unit. Integrated bag samples were taken at both points during each test. On-site analyses of CO_2, O_2, N_2 and C_{1-6} organics were conducted on the bag samples. Continuous monitors were used to analyze CO, NO, NO_2, NO_x, SO_2 and total hydrocarbons (as CH_4).

Solid Effluents: Composite samples of the fly ash and scrubber filter cake were collected according to Level 1 procedures and returned to the laboratory for analysis. Grab samples of the scrubber feed solids were also obtained for laboratory analyses.

Laboratory Analyses: The samples from the various sampling trains were returned to the laboratory for analysis.

Test Conditions: Ten tests were performed with the industrial boiler, five each with coal and oil. Unit loadings ranged from 31,800 to 45,400 kg steam per hour (70,000 to 100,000 lb/hr), which corresponds to between 70 and 100% of full load operation.

The scrubber is a pilot unit which does not process the entire flue gas output of the furnace. From 11 to 32% of the total flue gas was processed through the scrubber during the tests. Typical inlet and outlet gas temperatures for the scrubber unit were 149°C and 52°C (300°F and 125°F). Only 51 to 57% of design loading, rather than full loading, was maintained during coal-fired testing because failure of the multiclone particle removal system upstream of the scrubber resulted in high solids loading at the scrubber and unacceptably high scrubber filter cake production rates. During oil-fired testing, 88 to 100% of full design flows were maintained. Analytical results were used to estimate total boiler emission on the basis of treatment of 100% of the flue gas from the boiler. That is, it was assumed that additional scrubber modules could be added in parallel to the system such that the total flue gas output would be processed with a mean scrubbing efficiency equal to that of the pilot scrubber. All stack emissions data are based on this assumption.

Emission of Sulfur Compounds

The average SO_2 emission rate from coal firing ahead of the scrubber was 1,112 ng/J and after scrubbing was 36.3 ng/J, for a mean scrubber efficiency of 96.7%. During oil firing, the SO_2 emission rates were 993 ng/J ahead of the scrubber and 26.8 ng/J after the scrubber, for a mean scrubber efficiency of 97%. In both cases, the SO_2 emissions after the FGD system were substantially below existing and proposed NSPS emission limit, see Table 4.9.

Table 4.9: Sulfur Dioxide Emissions ng/J (lb/10^6 Btu)

	Coal	Oil
Scrubber inlet	1,112 (2.59)	993 (2.31)
Scrubber outlet	36.3 (0.08)	26.8 (0.06)
NSPS	520 (1.2)	344 (0.80)
Efficiency, %	96.7	97

Source: Symposium—Vol 2

The scrubber removed 95 to 97% of the SO_2 during coal firing and 97 to 98% during oil firing. Only 32 to 33% of the SO_3 was removed during coal firing and 28 to 29% during oil firing. This relatively poor removal efficiency for SO_3 is an indication that SO_3 is either present as very fine aerosols in the scrubber inlet gas or is converted to very fine aerosols as the flue gas is rapidly cooled in the scrubber.

The removal rate for $SO_4^=$ was 88% during coal firing and 60% during oil firing, indicating that most of the $SO_4^=$ in the scrubber inlet is associated with larger particulates. However, combustion generated sulfates may not be simply passing through the scrubber. Because of the possibility that the $SO_4^=$ species from coal combustion may be changed by the scrubbing process, an analysis effort to determine the $SO_4^=$ species was initiated. Both the Fourier Transform IR (FTIR) analysis and the X-Ray Diffraction (XRD) analysis have confirmed the presence of sodium bisulfate ($NaHSO_4$) in the scrubber outlet, but not in the scrubber inlet. This is positive proof that sulfates are generated within the scrubber as the result of oxidation of sodium bisulfite ($NaHSO_3$) and sodium sulfite (Na_2SO_3) and emitted in the scrubber effluent gas. Also, tests on boilers with flue gas concentrations of 400 to 8,000 ppm SO_2 have shown that there is no correlation between initial SO_2 concentration and the net sulfate formation rate (3). This implies that the scrubber has a minimum sulfate emission rate that is virtually unaffected by inlet SO_2 concentration. These data must, however, be evaluated in the context of the potential for significantly increased sulfate loadings to the environment which would result from SO_2 emissions if the boiler flue gases were not controlled.

Based on the analysis of SO_3 and $SO_4^=$ emission data, it has been estimated that up to 40% of the fine particle emissions at the scrubber outlet could be attributed to scrubber generated $NaHSO_4$. The remaining portion of the net increase in fine particles across the scrubber may be attributable to the uncertainties associated with the assumptions used in converting the polarizing light microscope number size distribution data to weight size distribution, and to calcium sulfite hemihydrate $(CaSO_3 \cdot \frac{1}{2}H_2O)$ particles generated by the scrubber. This should be confirmed by further study since the unknown or unexplained portion amounts to more than half of the net increase.

Solid Emissions

Three solid waste streams are produced by the system: bottom ash, fly ash and scrubber cake. Table 4.10 shows the approximate quantities of bottom ash and scrubber cake that were produced. Only small quantities of fly ash were collected during the test because the multiclone malfunctioned.

The scrubber cake produced after filtration has the characteristics of silty soils, but its behavior closely resembles a clay in many respects. As obtained from the vacuum filter, the scrubber cake consists of small lumps and appears to be relatively dry; however, the water content generally ranges from about 30 to 50%.

Table 4.10: Generation Rate of Solid Waste from 10-MW Controlled Industrial Boiler

Waste	. . . Rate of Production, kg/hr. . . .	
	Coal Firing	Oil Firing
Bottom ash	80	1
Fly ash	240*	5*
Scrubber cake	700	400

*This is the amount of fly ash recovered by the cyclone collector. Approximately 25% of the fly ash is recovered in the scrubber and removed with the scrubber cake.

Source: Symposium—Vol 2

If it is assumed that calcium sulfite hemihydrate $(CaSO_3 \cdot \frac{1}{2}H_2O)$ is formed as a result of the SO_2 scrubbing and Na_2SO_3 regeneration processes, then the mass balance of coal firing in Table 4.11 shows that the scrubber cake is composed of 28.5% coal fly ash and 23.8% $CaSO_3 \cdot \frac{1}{2}H_2O$. However, if the multiclone had not malfunctioned during the test, more fly ash would have been removed upstream of the scrubber and the fly ash content of the scrubber cake would have been lowered proportionately. The amount of scrubber cake produced could be reduced to 600 to 750 kg/hr on wet basis, assuming approximately 60 to 80% multiclone efficiency.

Table 4.11 also shows the estimated composition of scrubber cake produced during oil firing. The cake is composed of 44 to 50% unbound water and at least 47% calcium sulfite hemihydrate. These data reflect the low particle emissions which are characteristic of oil firing. Only 1% of the scrubber cake during oil firing is estimated to be primary particle emissions, due to their small size and low removal efficiency.

Although the scrubber cake material is composed predominantly of relatively insoluble solids (calcium sulfite, calcium sulfate, and some calcium carbonate), the interstitial water does contain soluble residues of lime, sulfate, sulfite, and chloride salts. Trace elements in the fly ash may also contribute to the leachate from the disposed scrubber cake and are of special concern. Except for boron, the trace element concentrations in the scrubber cake from coal firing far exceed the MATE values for solids. Except for antimony, boron, molybdenum, and zinc, all trace elements in oil fired scrubber cake were found to exceed human health based MATE values for solids. Similarly, except for boron, all trace elements were found to exceed ecology based MATE values for solids. These results are consequence of reducing a high volume of low concentration wastes to a low volume of concentrated wastes. The high potential degree of hazard for most elements appears to warrant disposal of these solid wastes in specially designed disposal areas.

Table 4.11: Estimated Scrubber Cake Mass Balance

Component	. . . .Contribution to Scrubber Cake . . .			
	Coal	Oil	Coal	Oil
	. . . . . (kg/hr)		. . . (wt %) . . .	
Fly ash removed by scrubber	324	5	29	1
$CaSO_3 \cdot \frac{1}{2}H_2O$ formed from SO_2 scrubbing and Na_2SO_3 regeneration	262	210	24	47
$CaSO_4$, $CaCO_3$, Na_2SO_3, $Ca(OH)_2$, $NaHSO_4$, and Na_2SO_4 losses (estimated)	10-85	6-35	1-8	1-8
Water	429-504	193-222	39-46	44-50
Average	1,100	443	100	100

Source: Symposium—Vol 2

Scrubber Efficiency

Flue gas analyses indicate that the scrubber removes a significant percentage of input sulfur oxides (SO_2, SO_3 and particulate SO_4^{2-}), total particulates, and organics of the C_7 class and higher. The significance of data indicating NO_x and CO removal appears questionable. Therefore, these components are not included in this discussion. There is no NO_x control equipment.

Average removal efficiencies have been discussed; however, the C_7 and higher hydrocarbons are removed with 77% efficiency for coal firing and 85% efficiency for oil firing. These fractions comprise 32 to 96% of the total generated organics. Hence, based on the total generated organics, a removal efficiency of 25 to 53% was obtained for coal firing and 32 to 84% for oil firing.

Air Quality

Simplified air quality models were used to estimate the relative ground level air quality resulting from uncontrolled and controlled emissions. Worst case and typical (not average) weather conditions were considered. The worst case was assumed to be plume trapping, which was assumed to persist for as long as 3 hours. Typical conditions can reasonably be expected to occur almost anywhere in the country. It was further assumed that all species were inert and that no photochemical reactions occurred. Models for particles, SO_2, NO_x and CO were made. Keeping in mind the assumptions mentioned above, several observations can be made:

- Controlled emissions of particles for all cases are less than all particle emission standards.

- For controlled emissions of SO_2 during both coal and oil firing, no standards are exceeded.

- The NO_X standard is exceeded under both weather conditions during coal firing. During oil firing, the NO_X standard was exceeded under worst case weather conditions but not under typical weather conditions. Since the scrubber does not remove significant amounts of NO_X, there is no substantial difference between the air quality resulting from inlet and outlet emissions. (The boiler has no NO_X controls.)

- The CO standards are not exceeded under any conditions. As with NO_X there is no substantial difference between the inlet and outlet concentrations.

Comparative Assessment

Scrubbing removed 99% of the particles from coal firing and 75% of the particles from oil firing. The lower removal efficiency obtained during oil firing is attributed to the increased fraction of particles smaller than 3 μm; at least 21% of the uncontrolled particles from oil firing are less than 3 μm in diameter while substantially less than 1% of the uncontrolled particles from coal firing are under 3 μm.

There appears to be a net increase across the scrubber for particles less than 3 μm in diameter from coal firing. This net increase can be attributed to the poor removal efficiency of the scrubber for fine particles, and to the sodium bisulfate ($NaHSO_4$) and calcium sulfite hemihydrate ($CaSO_3 \cdot \frac{1}{2}H_2O$) particles generated by the scrubber. Both $NaHSO_4$ and $CaSO_3 \cdot \frac{1}{2}H_2O$ have been identified at the scrubber outlet but not at the inlet. Although a very slight increase in particles from oil firing in the 1 to 3 μm range was observed, a net decrease in particles less than 3 μm was observed during oil firing. Based on the results of coal firing tests, it appears reasonable that scrubber generated particles were present in the scrubber outlet stream during oil firing but that the high fine particle loading associated with oil firing masked detection of these particles. Particle emissions after scrubbing are below either existing or proposed NSPS limits.

Controlled SO_2 emissions for coal and oil firing are lower than either existing or proposed NSPS limitations. The overall uncontrolled sulfur balance indicates that over 92% of the fuel sulfur is emitted as SO_2, less than 1% as SO_3, and approximately 3% as SO_4^{2-} from coal burning and 1.5% from oil firing. The remaining input is in the bottom ash or is unaccounted for. Sulfates are more efficiently removed than SO_3 (60% removal for oil firing and 88% for coal firing). This indicates that SO_4^{2-} is probably associated with the larger particles, which are more efficiently removed than smaller particles. The higher sulfate removal from the coal flue gases is explained by the higher particulate loading during coal firing.

NO_X emissions increased with increasing load for both coal and oil firing, as expected. Available data indicate that for boiler loadings between 90 and 100%, NO_X emissions from coal firing are approximately 3 times greater than from oil firing. Observed reductions of NO_X emissions for coal firing and early oil firing tests appear to be due, at least in part, to air leakage into the scrubber outlet sampling line. Data from later oil firing tests, not known to be subject to leakage problems, indicate that maximum NO_x removal across the scrubber is on the order of 2%. Without controls, the emissions of NO_x from clustered coal-fired furnaces will cause the applicable NAAQS to be exceeded.

Uncontrolled CO emissions from coal firing were 3 times those from oil firing, although both emissions are very small. Apparent reductions in CO emissions across the scrubber are not considered significant due to air leakage in the sampling train and the low sensitivity of analysis at the measured CO concentrations.

Organic emissions for coal and oil firing were very similar, and appear to be primarily C_{1-6} hydrocarbons and organics heavier than C_{16}. While uncontrolled emission rates for both coal and oil firing are low, emissions of these organics were further reduced by about 85% in the scrubber unit. The organic compounds identified in the gas sample from both coal and oil firing were generally not representative of combustion generated organic materials, but were compounds associated with materials used in the sampling equipment and in various analytical procedures. This again confirms the low level of organic emissions. Polycyclic organic material (POM) was not found in the scrubber inlet or outlet at detection limits of 0.3 $\mu g/m^3$ for either coal or oil firing. Minimum Acute Toxicity Effluent (MATE) values for most POM are greater than this detection limit. However, since the MATE values for at least two POM compounds, benzo[a]pyrene and dibenz[a,h]anthracene, are less than 0.3 $\mu g/m^3$, additional GC/MS analyses at higher sensitivity would be required to conclusively determine the presence of all POM at MATE levels. Also a more accurate determination of oxygen in the flue gas at the furnace outlet could be important since POM levels decrease as excess air increases at constant temperature.

The air concentration of trace elements from plant clusters is expected to be approximately 4 orders of magnitude below the "allowed exposure levels" proposed for hazardous waste management facilities (4). They are also below typical urban ambient background, except for cobalt and selenium, which approach or slightly exceed endogenous levels. The concentrations are similar from coal and oil firing, except for cadmium, which is 40 times larger from oil firing than from coal firing.

Trace element concentrations in run-off water, which arise from deposition of emissions on soil and foliage, may be about 10^5 times the standards for livestock drinking and potable water. Concentrations due to oil firing are slightly lower than those due to coal firing; however, selenium and molybdenum concentrations in water are predicted to exceed their background levels. Mass closure for most trace elements from coal firing has been found to be in the 75 to 107% range. Mass closure for half of the trace elements from oil firing is in the 50 to 136% range; closure for the remainder of oil firing trace elements is poorer due to the extremely low elemental concentrations measured and/or contamination of the recycle scrubber solution during coal firing tests. These good closures instill confidence in the validity of the sampling and analysis data.

Beryllium emissions after scrubbing were less than or equal to the beryllium MATE value during coal and oil firing. At the measured emission concentrations, the National Standard for Hazardous Air Pollutants limitation of 10 g beryllium per day would only be exceeded by boilers of 50-MW capacity for coal firing and 100-MW capacity for oil firing.

Chlorides were removed with greater than 99% efficiency from coal flue gases and with about 51% efficiency from oil flue gases. This difference was attributed to the higher removal efficiency for the larger coal particles. Fluorides were removed with greater than 86% and about 87% efficiency for coal and oil firing, respectively. Nitrate emissions were removed from coal flue gases with at least 52% efficiency and from oil flue gases with 57% efficiency.

Scrubber cake production during coal firing was 3.3 times greater than during oil firing. If the multiclone had not malfunctioned, this ratio would have been reduced to 2.7, assuming 60% multiclone efficiency. Available data indicate that the principal difference between scrubber cake production rates from coal and oil firing is the particle loading and associated unbound moisture.

The scrubber cake produced from coal firing contains a significant amount of fly ash. Except for boron, trace element concentrations in the scrubber cake have exceeded their MATE values. This is the result of transferring an air pollution problem by scrubbing to an easier solid waste disposal problem. Because the trace elements may contribute to the leachate from the disposed scrubber cake, these solid wastes must be disposed of in specially designed landfills. In such landfills, leachate impact on groundwater is expected to be insignificant.

Conclusions

General: Several major conclusions have evolved from the environmental analysis. The difference in environmental insult expected to result between coal and oil combustion emissions from a single controlled 10-MW industrial boiler is insignificant. This is because: (1) there are only slight differences in the emissions levels of the pollutants, (2) the absolute impact of either fuel use is insignificant, and (3) the effectiveness of the control equipment makes environmental impacts small.

The environmental impacts of emissions from a cluster of five 10-MW industrial boilers at 200-m intervals aligned with the prevailing wind are potentially significant. The impacts include health effects, material damage, and ecological effects from high ambient levels of SO_2, NO_x, and suspended particulate matter; health effects and ecological damage due to trace metal accumulation in soils and plants; and degradation due to visibility reduction and the presence of waste disposal sites.

The environmental acceptability of a cluster of controlled industrial boiler emissions depends more on site specific factors (e.g., background pollution levels, location and number of other sources) than on the type of fuel used. Careful control of the site-specific factors can avert potential environmental damage and generally compensate for any differential effects arising between the use of coal or oil.

With the possible exception of ambient levels of NO_x, the risk of violating the NAAQS due to operation of clusters of controlled industrial boilers is essentially the same whether the fuel is coal or oil. Based on tests of the reference 10-MW boiler (which was not controlled for NO_x emissions), localized NO_x concentrations produced by coal firing are estimated to be twice those resulting from oil firing, and greater than those permitted by the NAAQS for 24-hr and 1-yr averaging periods.

Coal firing appears to produce a greater enrichment of trace elements in the flue gas desulfurization filter cake than does oil firing; however, the scrubber cake resulting from either coal or oil firing contains sufficient amounts of heavy metals and toxic substances to require specially designed disposal areas.

Health Effects: Regional emission levels of suspended sulfates from controlled oil or coal-fired industrial boilers would not be expected to cause a significant impact on regional health.

Sulfate emissions from clusters of controlled industrial boilers might be expected to cause significant adverse health effects in a localized area near the plant cluster. Oil firing would be expected to result in localized health effects about one-third less severe than those resulting from coal firing since oil firing produces only one-third the particle emission of coal firing.

The impact of solid waste generation on health is essentially the same for controlled coal firing and oil firing, if suitable land disposal techniques are employed to ensure sufficiently low leaching rates and migration of trace elements to groundwater and the terrestrial environment .

The concentration of metals in run-off waters due to controlled oil firing is predicted to be slightly less than that occurring from controlled coal firing; in either case, hazard to human health by drinking water is remote.

Trace element emissions from clusters of controlled industrial boilers may significantly increase local background levels in drinking water, plant tissue, soil, and the atmosphere; however, the expected increases in the levels of such elements are generally several orders of magnitude less than allowable exposure levels. Oil firing is estimated to cause cadmium burdens in plants approaching levels injurious to man. Because cigarettes contain significant cadmium levels, smokers are more apt to achieve thresholds of observable symptoms for cadmium exposure if they also consume additional cadmium via the food chain. Coal firing may produce plant concentrations of molybdenum which are injurious to cattle.

Ecological Effects: The potential for crop damage from either controlled coal firing or oil firing depends greatly on ambient levels of NO_x, SO_2, or trace element soil concentrations. If such levels are currently high, localized plant damage would be expected to occur within 1 to 2 km of a controlled boiler cluster. Leaf destruction from SO_2 exposure would be expected to be slightly more severe in the vicinity of a cluster of controlled boilers which are coal-fired as opposed to oil-fired. Plant damage may possibly occur even at levels below ambient air standards. (See pp. 5–29 to 5–32 of Volume 2 of Reference 2.)

For uncontrolled NO_x emissions, plant damage would be expected to be significantly greater in the vicinity of the coal-fired cluster, because of the higher levels of ambient NO_x produced. Emissions of CO and hydrocarbons will have negligible impacts on plants. The likelihood of damage occurring to plants due to emissions of trace elements from either controlled oil or coal firing is remote, with the possible exception of injury due to elevated levels of molybdenum and cadmium in plant tissue resulting from coal firing and oil firing, respectively.

The impact of fossil fuel combustion in controlled oil- or coal-fired boilers on plant damage via acid precipitation would be relatively insignificant. The levels of suspended sulfate (the precursor of acid rain) would be essentially the same whether the controlled boilers are coal- or oil-fired.

Measurements and analyses of leaching rates at experimental solid waste disposal sites indicate that landfills of untreated flue gas desulfurization system scrubber cake can be constructed without significant adverse impacts.

Societal Effects: The impact of boiler emissions on corrosion in the local area near a cluster of controlled industrial boilers will be significant (5). The corrosion rate will be slightly greater when the boilers are coal-fired; however, the extent of this overall impact (oil or coal) is minor compared to that which occurs when industrial boilers are uncontrolled.

The increase in annual total suspended particulate matter and the resulting soiling damage in the vicinity of a cluster of controlled industrial boilers results in additional cleaning and maintenance costs about 10 to 15% greater than that already experienced in a typical urban area (6). The cleaning costs may be slightly greater when the boilers are coal-fired.

Emissions of particulate matter from controlled industrial boilers will result in visibility reduction. This form of environmental degradation will occur in a localized area near the boiler cluster, and occurs to essentially the same extent whether the controlled boilers are oil- or coal-fired.

Total land disposal requirements for scrubber waste generated by controlled coal firing are 3 times greater than those for controlled oil firing. Disposal of the scrubber wastes may result in significant depreciation of property value and environmental degradation in the area of the disposal site. These impacts would be more severe if boilers use coal rather than oil.

Energy Effects: Obviously, the abundance of coal and the uncertainty of oil supplies are the driving forces for studies such as the coal versus oil comparative assessment study. Whether coal assumes a more significant role as an energy source by national choice or because there is no longer a choice, it is essential to be aware of and to be prepared to deal effectively with any environmental problems which result. Thus, the comparative impact of coal versus oil firing is indeed complex and involves consideration of all aspects of energy supply and use, including emissions characterization, multimedia environmental impacts identification, comparison of projected impacts with accepted levels of impacts, and evaluation of techniques for mitigating unacceptable levels of impacts.

While this study identified some potentially significant difference between coal and oil firing in clusters of boilers, the fuel choice of oil or coal may be a relatively minor factor in determining the environmental acceptability of controlled industrial boilers. Other site-specific and plant-design factors may exert greater environmental effects than fuel choice. As concern for environmental protection increases, the issue may become whether the increasing use of fossil fuels can be continued at the present levels of control technology without potential long-term dangers. If it is found that long-term effects of pollution (e.g., trace metals accumulation, lake acidity from acid rains) from fossil fuels combustion and other sources are environmentally unacceptable, it is clear that energy use may be affected. Increasing control requirements could result in energy cost increases to the level where the combustion of fossil fuels loses its economic advantage over other, cleaner sources of energy production.

References

(1) Ponder, W.H. and Kenkeremath, D.C., *Conventional Combustion Environmental Assessment Program,* MS78-44 Rev. 1, Mitre Corp., McLean, Virginia, September 1978.

(2) Leavitt, C., et al., *Environmental Assessment of Coal- and Oil-Firing in a Controlled Industrial Boiler,* 3 vol, EPA-600/7-77-164 a/b/c (NTIS PB289942/289941/291236), August 1978.

(3) *Capabilities Statement, Sulfur Dioxide Control Systems,* Tech. Report 100, FMC Corp., Environmental Equipment Div., Itasca, Illinois, November 1977.

(4) "Standards Applicable to Owners and Operators of Hazardous Waste Treatment, Storage and Disposal Facilities," draft of proposed rules obtained from EPA, Office of Solid Waste, March 1978.

(5) Upham, J., "Atmospheric Corrosion Studies in Two Metropolitan Areas," *J. Air Poll. Control Assn.,* June 1967.

(6) Michelson, I., "The Household Cost of Living in Polluted Air in the Washington, DC Metropolitan Area," a report to the U.S. Public Health Service.

MAGNESIUM OXIDE PROCESS

The information in this chapter is based on *Economic and Design Factors for Flue Gas Desulfurization Technology,* April 1980, EPRI CS-1428, prepared by Bechtel National, Inc. for Electric Power Research Institute and *Flue Gas Desulfurization Pilot Study–Phase I–Survey of Major Installations–Appendix 95-I–Magnesium Oxide Flue Gas Desulfurization Process,* January 1979, (NTIS PB-295010), a NATO-CCMS study, prepared by F. Princiotta of U.S. Environmental Protection Agency, R.W. Gerstle and E. Schindler of PEDCo Environmental.

BACKGROUND

Magnesium oxide flue gas desulfurization is a regenerable process. The sulfur dioxide (SO_2) in the flue gas is absorbed by a magnesium-based slurry, and the resulting magnesium sulfite ($MgSO_3$) is calcined to regenerate magnesium oxide (MgO), which is recycled back to the power plant for reuse in the scrubbing section. The sulfur value from the SO_2-rich by-product gas is recovered.

The chemistry and process design of this system are well developed. Similar processes have been used in the paper pulp industry for many years. One application of this FGD system was first developed in Germany in 1964 and tested in 1968. Full-scale demonstration programs in the United States were recently terminated, but one U.S. unit is still in operation. Two units also are operating in Japan. Initially, many mechanical problems were encountered, as well as design/sizing problems with the solids-handling equipment. Most have been identified and solved.

The magnesium oxide system can best be applied to boilers located in industrial areas where sulfuric acid can be marketed. This can become an additional advantage if other boilers in the plant or area choose this system because the regeneration facility can serve a number of scrubber systems. The system requires a minimal amount of fresh magnesium oxide and water; thus raw materials are not a limiting factor. Land requirements are relatively small because no sludge is generated. Unless the boiler is very large, the construction of an individual MgO regeneration facility and sulfuric acid plant is generally not feasible.

The advantages and disadvantages of the MgO system are summarized as follows.

Advantages: The regeneration of the absorbent (MgO) reduces the cost of raw materials. Disposal problems are relatively minor because the MgO is regenerated and the sulfur is recoverable in a usable form. The sulfur can be recovered as a high-grade acid or possibly elemental sulfur, both of which are marketable. Regeneration can be carried out at a site distant from the power plant, permitting the use of a central regeneration facility to service several FGD units. By maintaining adequate inventories of MgO, the boiler can tolerate extended outages of the regeneration facility without interrupting operation. Modifications and investigations of the first large-scale installations have resulted in improved process reliability. High solubility of $MgSO_4$, high circulation rates, control of slurry composition, high concentrations of crystallization nuclei, and a short residence time reduce scrubber plugging and scaling problems.

Disadvantages/Problems: Energy needs are relatively high because of the high temperatures (800° to 1000°C) required in the MgO regeneration step. Reliability at the demonstration plants in the United States has been relatively low since frequent equipment and processing problems have been encountered. Commercial plants in Japan, on the other hand, demonstrated high performance reliability. Experience with the application of the magnesium oxide process to coal-fired boilers is very limited. The centrifuge cake containing $MgSO_3$ can present handling difficulties. Very little data are available on the equilibrium of aqueous magnesium salts. Corrosion/erosion problems in the slurry-handling process of this system are common. Initial and makeup MgO is required in particle sizes of less than 150 μm (100 mesh) to increase reactivity. The scrubber requires liquid-to-gas (L/G) ratios of 5.3 ℓ/m^3 for absorption and 2.7 ℓ/m^3 for quenching. Because fly ash reduces the efficiency of the regeneration system, extensive particulate removal is required from a coal-fired boiler.

As mentioned previously, one of the early MgO processes was developed in 1964 by Grillo-Werke AG in Duisburg-Hamborn, Germany. In 1968 this process was tested in a 7.08 m^3/sec, oil-fired boiler at Union Kraft in Wesseling, Germany. The system was also developed and installed by the Mitsui Engineering and Shipbuilding Company in Japan. Although the results were encouraging, for economic reasons the system was later replaced with a lime scrubbing system.

The principal activity in connection with MgO scrubbing in the United States was initiated by a full-scale prototype test facility on oil-fired boiler Unit 6 at the Boston Edison Mystic Station in Everett, Massachusetts. The Chemical Construction Corporation (Chemico) built the scrubbing facility at Everett and the regeneration plant at Rumford, Rhode Island, in April 1972. During the 2-year test program which ended in 1975, operating problems (mostly mechanical) were identified and solved except for corrosion/erosion problems. Both plants indicated the potential for achieving 90% availability with over 90% SO_2 removal efficiency.

The MgO process is quite similar to the magnesium-based pulping system, generally called the "Magnefite" process. The main differences between the Magnefite and the MgO FGD system are that: the pulp mill gas contains more SO_2 (1% or more vs 0.1 to 0.4%); the object in pulping is to produce soluble $Mg(HSO_3)_2$, whereas in stack gas scrubbing $MgSO_3$ is desired; and the calcining step in the pulping operation involves decomposition of complex metal organic compounds rather than simple $MgSO_3$.

A 2-year test program at Dickerson Generating Station of the Potomac Electric and Power Company operated from September 1973 to August 1975 on a coal-fired utility boiler. This FGD system treated half of the flue gas from the 190-MW boiler while burning 2%-sulfur-content coal. The magnesium sulfate crystal from the coal-fired boiler was different from that obtained at the oil-fired Mystic Station and required a change in calciner operation.

GENERAL PROCESS DESCRIPTION

Process Chemistry

The principles of all MgO scrubbing systems are essentially the same. A magnesium-based slurry is contacted with flue gas containing SO_2. The SO_2 is absorbed and forms an insoluble product. After this product is removed from the scrubbing solution by a centrifuge, it is dried and shipped to a separate regeneration facility for the recovery of its sulfur value and the MgO. The regenerated MgO is recycled back to the power plant for reuse in the scrubbing system.

The overall reactions that take place in the absorber are:

(1) $MgO + SO_2 + 6H_2O \rightarrow MgSO_3 \cdot 6H_2O$

(2) $MgO + SO_2 + 3H_2O \rightarrow MgSO_3 \cdot 3H_2O$

Some $MgSO_4$ is also formed in the absorber as a result of partial oxidation of the $MgSO_3$ by the reaction:

(3) $MgSO_3 + 0.5O_2 + 7H_2O \rightarrow MgSO_4 \cdot 7H_2O$

The SO_2 is also absorbed by $MgSO_3$ according to the reactions:

(4) $MgSO_3 + H_2O + SO_2 \rightarrow Mg(HSO_3)_2$

and

(5) $Mg(HSO_3)_2 + Mg(OH)_2 + 10H_2O \rightarrow 2MgSO_3 \cdot 6H_2O$

Reactions (4) and (5) are favored in an acidic slurry, with a pH of about 6.

After the crystals are dried, they are calcined to generate SO_2 gas and regenerate MgO. Coke is added to reduce $MgSO_4$ to MgO and SO_2. The overall calcining reactions are:

(6) $MgSO_3 \rightarrow MgO + SO_2$

(7) $MgSO_4 + 0.5C \rightarrow MgO + SO_2 + 0.5CO_2$

The calcining temperature is generally between 800° and 1000°C.

According to the molecular weights of MgO and SO_2, the theoretical requirement is 1 mol of MgO per mol of SO_2 removed. Assuming a 1.1 excess stoichiometric requirement and 97% purity of MgO, the actual MgO required is:

$$\frac{40.3}{64} \times \frac{1.1}{0.97} = 0.71 \text{ kg MgO/kg of } SO_2.$$

Regenerated MgO is recycled back to the scrubbing system.

Flowsheet and Subsystem Descriptions

Figure 5.1 is a flow diagram of a model magnesia slurry process, developed by Bechtel on the basis of previously published data. While the process depicted in Figure 5.1 resembles the Philadelphia Electric Company's process in that it employs venturi-type scrubbers and a fluid bed calciner, the design parameters and the resulting cost data were developed by Bechtel on the basis of previously published data, and are not intended to represent any particular vendor's offering. A summary of the basic process design parameters for the magnesia slurry process model FGD system is presented in Table 5.1. The corresponding raw material and utility requirements are presented in Table 5.2.

Table 5.1: System Parameters

Data for 1 x 500 MW Unit
(Maximum Load Conditions)

	Units	Eastern Coal	Western Coal
Flue gas flow rate	m^3/sec	852	923
	(10^3 acfm)	(1,806)	(1,955)
Number of absorption trains per unit	each	5	5
Capacity per train	%	25	25
Gas flow per absorber, saturated	m^3/sec	179	195
	(10^3 acfm)	(380)	(414)
Prescrubber type	–	venturi	venturi
Prescrubber L/G	ℓ/m^3	2.00	2.00
	(gal/10^3 acfm)	(15)	(15)
Absorber type	–	venturi	venturi
Absorber L/G	ℓ/m^3	6.68	3.34
	(gal/10^3 acf)	(50)	(25)
System pressure drop	pascals	4,982	4,982
	(inches wg)	(20)	(20)
Mist elimination	–	chevron	chevron
Absorbent oxidation ratio*	%	10	30
Total system MgO loss**	%	3.5	3.5
Flue gas reheat required	°C	28.3	26.1
	(°F)	(51)	(47)
Number of solids separation trains per unit	each	4 at 30%	1 at 120%
Number of solids dryers per unit	each	1 at 120%	1 at 120%
Number of calcination trains per unit	each	1 at 100%	1 at 100%
Number of acid production trains per unit	each	1 at 100%	1 at 100%

*Mols MgSO$_4$/mols SO$_2$ removed.
**Mols MgO lost/mols SO$_2$ removed.

Source: EPRI CS-1428

Figure 5.1: Magnesia Slurry Process Flow Diagram

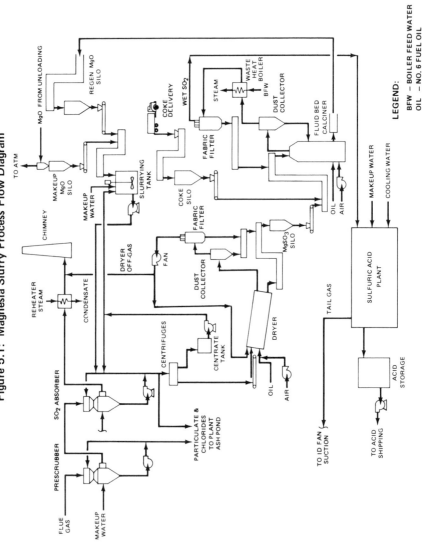

LEGEND:

BFW – BOILER FEED WATER
OIL – NO. 6 FUEL OIL

Source: EPRI CS-1428

Table 5.2: Raw Materials and Utility Requirements

Data for 1 x 500 MW Unit
(Maximum Load, Average Sulfur)

	Units	Eastern Coal	Western Coal
Raw materials			
Magnesium Oxide	kg/hr	343	52.2
	(lb/hr)	(756)	(115)
Coke	kg/hr	165	75.3
	(lb/hr)	(364)	(166)
Utilities			
Electricity			
Raw material handling and preparation	kW	520	80
Pretreatment and absorption	kW	2,690	3,270
Solids separation and drying	kW	1,280	195
SO_2 regeneration	kW	615	95
SO_2 reduction	kW	2,055	315
Fans	kW	5,800	6,260
Total	kW	12,960	10,215
Steam			
Calcination	10^6 Btu/hr	23.5	3.6
		credit	credit
Reheat	10^6 Btu/hr	88.0	87.5
Total	10^6 Btu/hr	64.5	83.9
Water			
Process makeup	10^3 ℓ/hr	111.3	110.1
	(10^3 gal/hr)	(29.4)	(29.1)
Cooling water	10^3 ℓ/hr	1,764.8	269.1
	(10^3 gal/hr)	(466.0)	(71.1)
Fuel Oil (No. 6)			
Solids drying	bbl/hr	14.3	2.2
	(gal/hr)	(600)	(91)
Calcination	bbl/hr	15.5	2.4
	(gal/hr)	(650)	(100)
Total	bbl/hr	29.8	4.6
	(gal/hr)	(1,250)	(191)
Products			
Sulfuric acid (98%)	10^3 kg/hr	23.9	3.64
	(tons/hr)	(26.3)	(4.01)
Incremental heat rate penalties			
Steam (basis gross heat rate)	%	1.43	1.80
Electricity (basis unit net rating)	%	2.59	2.04
Corrected net heat rate	Btu/kWh	9,939	10,277
Variation from base	%	−0.47	−0.16

Source: EPRI CS-1428

Description of Major Process Subsystems: The magnesia slurry FGD process includes the following major process subsystems.

Raw Material Receiving, Storage, and Preparation — This area includes facilities for receiving and storing makeup magnesium oxide (MgO) and coke.

Makeup magnesium oxide is conveyed pneumatically from self-unloading trucks to storage silos. Fresh MgO is added to the regenerated MgO to feed to the

MgO slurrying tank at a rate sufficient to offset system handling losses. MgO slurry is added to the SO_2 absorber recycle slurry loops as required to maintain the absorbent pH at the desired level.

Coke is delivered by truck and conveyed mechanically to a storage silo. Coke is fed from the storage silo to the regeneration area as required to reduce the magnesium sulfate ($MgSO_4$) formed by oxidation of magnesium sulfite ($MgSO_3$) in the absorbers.

Prescrubbers — Hot flue gas from the ID fans passes through venturi scrubbers where it is cooled and chlorides and residual fly ash are removed. Makeup water is added to the prescrubbers through wash sprays on chevron mist eliminators. A slip stream of the prescrubber slurry is pumped to the plant ash pond to purge fly ash and chlorides from the loop.

SO_2 Absorption — Saturated gas from the prescrubbers is contacted with a recirculating slurry of magnesium sulfite ($MgSO_3$) in a second venturi scrubber. SO_2 is absorbed by reaction with the soluble magnesium sulfite producing a slurry of insoluble magnesium sulfite and sulfate in the form of hydrate crystals. A slip stream of the absorber slurry is sent to the solids separation and drying area to remove sulfur compounds from the absorbent loop. Makeup MgO slurry from the MgO slurrying tank and absorbent liquor from the solids separation area are added to the absorber recycle loop to maintain the recycle slurry pH and specific gravity at the required levels.

The cleaned flue gas passes through chevron mist eliminators for removal of entrained liquor droplets, and is then reheated before being mixed with the dryer offgas and discharged to the chimney.

Solids Separation and Drying — Absorber effluent slurry is centrifuged to separate the sulfite/sulfate crystals from the absorbent liquor which is split into two streams, a portion going to the MgO slurrying tank to produce makeup slurry, and the balance going directly back to the absorber recycle loops. The dewatered crystal cake is conveyed to an oil-fired rotary kiln dryer where it is dried to anhydrous salts. The dry salts are then conveyed to an in-process storage silo. The dryer offgas passes through a cyclone dust collector and a fabric filter before being added to the reheated flue gas stream. A portion of the dryer offgas is recirculated to the dryer for temperature control. The collected magnesium salt fines are conveyed to the storage silo.

MgO Regeneration and SO_2 Recovery — The dried magnesium sulfite/sulfate crystals are fed to an oil-fired fluid bed calciner where they are calcined in the presence of coke to reduce the $MgSO_4$. The calciner offgas passes through a cyclone dust collector, a waste heat boiler, and a fabric filter before entering the sulfuric acid unit. The collected fines are recycled to the calciner. The regenerated MgO is sent to the reagent preparation area.

Sulfuric Acid Production and Storage — The dilute SO_2 stream from the calcination unit is fed to a conventional contact sulfuric acid plant which produces 98% sulfuric acid. Tail gas from the sulfuric acid plant is recycled to the ID fan inlets to minimize overall SO_x emissions from the plant.

GENERAL OPERATING PARAMETERS

Chemical Parameters

The crystals of hexahydrate formed by reaction (1) are usually large (200 μm), and the material is damp and sandlike. The crystals of trihydrate are much smaller and have a mudlike consistency; therefore, they are centrifuged less efficiently and result in a wetter cake which agglomerates in the dryer.

The nature of the hydrate formation depends somewhat on the temperature in the absorber. The hexahydrates are stable at room temperature. As the temperature increases, however, the hexahydrates gradually change to more stable trihydrates. A recent laboratory study shows that (1) the transition temperature is from 60° to 63°C (higher than previously known), (2) the hexahydrates persist even above the transition temperature, and (3) the rate of the transition is very slow.

Magnesium sulfite is formed in the scrubber by the reaction between the SO_2 and the alkaline slurry; this product, however, is not very soluble in water. When the scrubbing slurry is acidic, very soluble $Mg(HSO_3)_2$ is formed. Thus, the fundamental chemical reaction in the scrubber tower results in the conversion of a slightly soluble material to a very soluble material. It is quite possible that this approach could eliminate plugging problems in the scrubber. The pH of the scrubbing slurry is controlled by regulating the rate of addition of $Mg(OH)_2$ slurry.

The equilibrium vapor pressure of SO_2 over a slurry of $MgSO_3$ in a slightly acidic, dilute solution of $Mg(HSO_3)_2$ is higher than that over a slurry of alkaline MgO or $Mg(OH)_2$ in water. It is possible, therefore, that the removal of SO_2 from stack gas may not be as effective with an acidic slurry as with an alkaline slurry.

In practice, however, the scrubbing slurry does not approach the point of saturation or equilibrium, and the SO_2 removal capability of the magnesium oxide scrubbing system is not limited. A recent laboratory study demonstrated no great difference between SO_2 removal at pH's of 6 and 8.

Physical Parameters

An MgO FGD system can achieve SO_2 collection efficiencies of 90% on coal-fired boilers. Key operating parameters include the following:

- Venturi scrubbers operating at pressure drops of 2,500 pascals or greater with an L/G ratio of 5.4 to 6.8 ℓ/m^3 have achieved collection efficiencies in excess of 90%. A TCA scrubber applied on smelter reverberatory furnace off-gas also achieved collection efficiencies of 99% and greater.

- Slurry pH (measured in the absorber discharge tank) should be maintained in the range of 6.0 to 7.5.

- The MgO should be slaked (dissolved) in water in a heated agitation tank. Sometimes it requires preslaking and dilution before it is added as makeup to the absorber.

- Prescrubbing of the flue gases with water for residual fly ash removal and cooling will minimize the formation of the trihydrate of $MgSO_3$ and insure better handling, drying, and calcining of the product.

The flue gas is reheated from 54°C to between 71° and 98°C (maximum 110°C) to prevent condensation and enhance dispersion. Reheating requires approximately 1,100 J/°C of heat input per kilogram of flue gas.

DEVELOPMENT STATUS

Variations of the magnesia slurry process described herein have been tested on three prototype-scale systems with generally satisfactory performance. Data from these units can be used as a basis for design of full-scale systems.

The Philadelphia Electric Company (PECo) process has been tested at the 120 MW scale on a coal-fired utility boiler. Current testing is aimed at resolving problems in the areas of temperature control in the fluid bed calciner to avoid "dead-burning" of regenerated MgO, slaking of regenerated MgO, and control of MgO fugitive losses.

Two other prototypes have been constructed to test the Chemico-Basic Mag-Ox process, one on a 155 MW oil-fired boiler owned by Boston Edison and the other on a 100 MW coal-fired boiler owned by Potomac Electric Power. Both have undergone short-term tests with generally satisfactory performance. They differ from the PECo system primarily in their use of higher pH in the absorbent slurry (8.5 vs 6.3) and rotary rather than fluid bed calciners for regenerating MgO.

Two MgO FGD systems are operating in Japan. One treats smelter gas and the second gases from oil-fired boiler furnaces and Claus furnaces.

The MgO FGD process is a viable system with chemistry well developed. Technology has been shown adequately for oil-fired boilers but insufficiently for coal-fired systems. Since it has no significant solid waste and wastewater problems, it can be located in populous areas with high land costs, suitably in industrial areas with a market for H_2SO_4. It is adaptable to multiple boilers, small or large, each using only one centrally located regeneration facility.

The system requires no raw materials except water and a small amount of MgO to replenish attrition of the initial charge. Although the system still has some mechanical problems, availability of one Japanese plant has been as high as 95%; and SO_2 removal efficiencies are from 99.5% to over 99.8% from a concentrated (2 to 3% SO_2) smelter gas stream.

The major disadvantages of this system are the relatively high energy requirements and insufficient experience on coal-fired boilers, where fly ash could lead to problems.

One U.S. designer of MgO FGD systems guarantees SO_2 and particulate removal efficiencies and opacity compliance on oil- and coal-fired facilities for a period of 12 to 18 months. This company also offers financing, operating, and maintenance services.

RECENT TECHNOLOGICAL DEVELOPMENTS

The design of scrubbers and calciners for MgO FGD equipment has noticeably improved.

In the United States, the first MgO process on a boiler used a single-stage venturi scrubber. No particulate removal equipment was used in this installation, however, because it burned fuel oil. The second MgO installation, the first coal-fired application, involved a two-stage venturi device (Figure 5.2). In this design, particulates were removed in the first stage in a variable-throat venturi and SO_2 was absorbed by magnesium-based slurry in a fixed-throat venturi in the second stage. Particulates were removed in the variable-throat venturi using water and operating at 121°C. The gas was saturated and cooled to 49°C as the fly ash was removed.

Figure 5.2: Two-Stage Venturi Scrubber

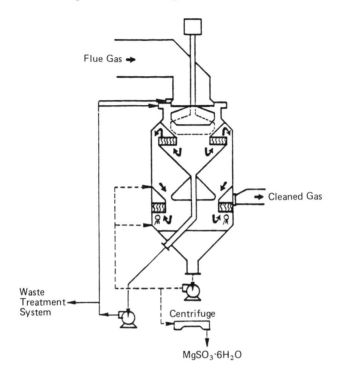

Source: NTIS PB-295010

In a third installation, a rectangular venturi-rod scrubber is used (Figure 5.3). This design essentially has two scrubbing stages in one scrubber. In each stage, a slurry of magnesium sulfite is sprayed upward through openings between cylindrical rods concurrent with the stack gas. This produces a venturi-like effect as the streams flow between the rods, assisting in the contacting between the stack gas and the magnesium sulfite slurry.

When a single-stage venturi scrubber or a venturi-rod scrubber is used for coal-fired utility installations, particulate must be removed before the SO_2 scrubbing step. A venturi prescrubber is commonly used for this purpose as well as to cool and humidify the flue gas. It also removes chloride from the flue gas. When mechanical collectors and/or ESP's are used, a gas cooler and a humidifier are normally required to insure the formation of the hexahydrates in the scrubber. Equipment currently available for fly ash removal had demonstrated efficiencies in excess of 99%. Whatever the type or combination of equipment dictated by the fuel being fired, well-engineered systems are available that can handle the particulate loading.

Figure 5.3: Venturi-Rod Scrubber

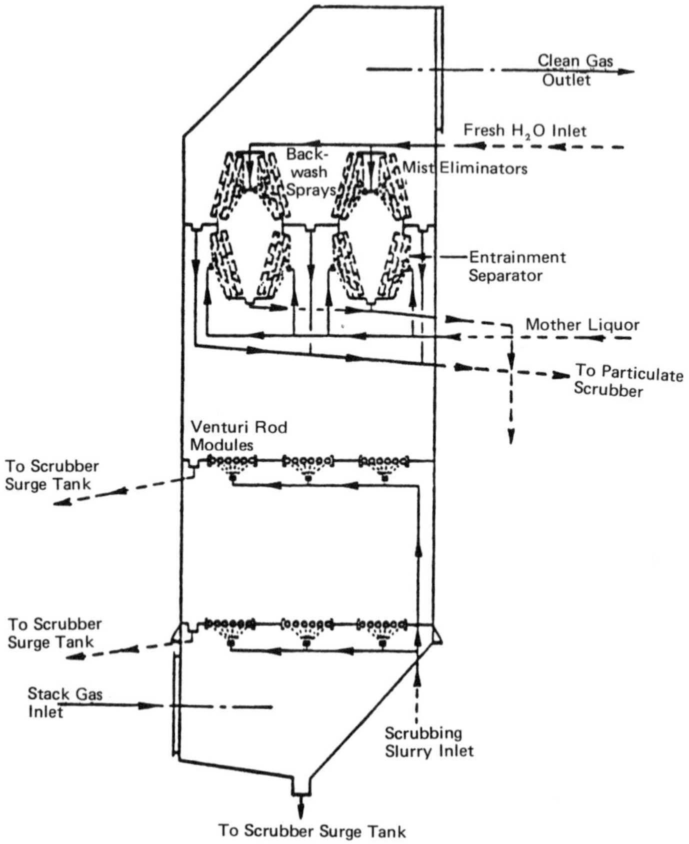

Source: NTIS PB-295010

One Japanese installation uses a turbulent contact absorption tower (TCA) for the SO$_2$ removal step. Little is known about the design. The major difference, however, seems to be the pressure drop across the scrubbers. Well-designed venturi scrubbers often have pressure drops of 4,500 to 5,000 pascals; TCA towers have pressure drops ranging from 2,500 to 3,000 pascals. An SO$_2$ removal efficiency of 99.8% has been claimed for the TCA tower.

As for calcining, it is carried out conventionally in a direct-fired rotary kiln, where coke is added to decompose MgSO$_4$. At the off-site regeneration facility of the Eddystone Generating Station (PECo), an oil-fired, fluidized-bed calciner was investigated. Figure 5.4 presents a flow diagram of this MgO regeneration system.

Figure 5.4: MgO Regeneration Plant

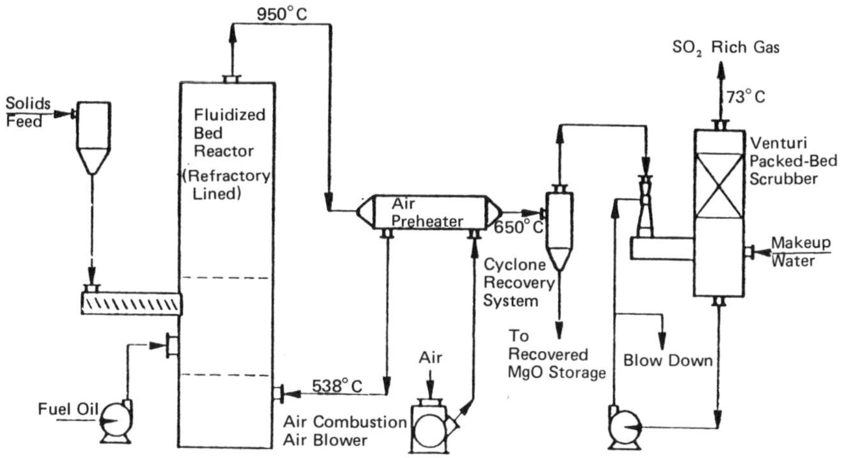

Source: NTIS PB-295010

SOLUTIONS TO PROBLEM AREAS

Problems have been primarily mechanical in nature, but some misunderstanding of magnesium oxide characteristics, such as slaking properties, also occurred. When regenerated MgO was first used, it was found to be less reactive than virgin material. High-speed agitation and heating with steam spargers in a slaker proved to be the most effective way to restore the reactivity of regenerated MgO. Heating the MgO above 82°C must be avoided, however, to prevent scaling problems.

Corrosion and erosion have occurred in both the particulate and the SO_2 scrubber systems. The best solution thus far seems to be the use of rubber-lined pipe, pumps, and valves. Spares also should be installed for critical components.

Inside the scrubber, the use of 316L stainless steel, flaked glass, and corrosion-resistant polyester-resin coatings have given excellent protection; however, some erosion/abrasion occurred. In some cases where 316L stainless steel was attacked, Inconel 625 was substituted and showed no sign of corrosion. With these materials of construction, no problems have occurred involving scaling or buildup inside the scrubbers.

The failure or limited capacity of the solids-handling equipment often caused plant outages. For example, centrifuge failures have occurred because of excessive wear of the internals.

Some considerations for improving centrifuge design are as follows. Moving parts should be given Stellite hard surfacing treatment to lessen/eliminate wear problems. Wash-out connections should be installed that are capable of thoroughly washing the centrifuge internals. The centrifuge should be designed to handle a wide range of solids concentrations and viscosities and provide good separation. The discharge hopper should be designed to minimize product accumulation.

A capacity problem can be solved by building in excess capacity on conveyors and bucket elevators for transporting dried magnesium sulfite to storage, thus eliminating product surges that can cause conveyor overloading.

Flow in the magnesium sulfite dryer should be concurrent, and internal chain sections and/or external rappers should be included to dislodge product buildup. The dryer off-gas system should include a cyclone separator for removal of entrained $MgSO_3$, and the cleaned gas stream should then be piped into the SO_2 scrubber for further removal of $MgSO_3$.

Chlorides in the scrubbing solution occur primarily from the combustion of chloride-bearing fuel and, to a lesser extent, from chloride in the raw water. The chlorides will react with magnesium oxide and build up in the liquid recirculation loop causing potential problems due to corrosion, decreased SO_2 absorption, and increased magnesium oxide makeup. These problems can be reduced by (1) using a prescrubber and keeping that liquid stream separate from the SO_2 absorber streams, (2) designing for chloride corrosion resistance, and (3) incorporating a purge system into the recirculation loop. Depending on wastewater discharge regulations, the purge stream may require further treatment.

WELLMAN-LORD PROCESS

The information in this chapter is based on the NATO-CCMS study, *Flue Gas Desulfurization Pilot Plant Study—Phase I— Survey of Major Installations—Appendix 95-J—Sodium Sulfite Scrubbing Flue Gas Desulfurization Process,* January 1979, (NTIS PB-295 011) prepared by F. Princiotta of U.S. Environmental Protection Agency, R.W. Gerstle and E. Schindler of PEDCo Environmental; *Economic and Design Factors for Flue Gas Desulfurization Technology,* April 1980, EPRI CS-1428, prepared by Bechtel National, Inc. for the Electric Power Research Institute; and "Operating and Status Report, Wellman-Lord SO₂ Removal/Allied Chemical SO₂ Reduction Flue Gas Desulfurization Systems at Northern Indiana Public Service Company and Public Service Company of New Mexico" by D.W. Ross of Davy Powergas, Inc., J. Petrie of Public Service Company of New Mexico and F.W. Link of Northern Indiana Public Service Company, in Symposium Vol 2.

BACKGROUND

The Wellman-Lord process involves the absorption of the SO_2 in flue gas in a solution of sodium sulfite. The sulfite is converted to bisulfite, and the solution is regenerated by thermal decomposition in an evaporator/crystallizer unit. The SO_2 is recovered in a concentrated stream, and it may be used for the production of sulfuric acid, elemental sulfur, or liquid SO_2. This process is applicable to any flue gas containing SO_2. Thirty-one commercial installations are either now in operation or under construction in the United States and Japan.

Davy Powergas developed the Wellman-Lord SO_2 recovery process during the 1960's. Its original purpose was to supply an economical means of recovering sulfur from fertilizer production. In 1965 the original pilot plant work was conducted at Tampa Electric's Gannon Station, using a potassium-based scrubbing solution. Although another pilot plant (at the Crane Station of Baltimore Gas &

Electric) failed because of excessive steam consumption and absorber plugging problems, the experience led to the present sodium sulfite/bisulfite system and rapid commercialization.

The first commercial installation in the United States was on an Olin Chemical sulfuric acid plant in Paulsboro, New Jersey. Most of the early Wellman-Lord systems were smaller units used at sulfuric acid and Claus-type sulfur recovery plants. The process has recently been applied more widely, e.g., on oil-fired boilers, particularly in Japan. Of 26 systems in various parts of the world, 15 are on oil-fired boilers, 7 on Claus plants, 2 on acid plants, and 2 on coal-fired boilers. Treatment capacities of these systems range as high as 739,000 Nm^3/h of flue gas. Sulfur dioxide removal efficiency is typically 90 to 98%. Collected SO_2 is recycled at Claus and acid plants and used for the production of additional sulfuric acid, elemental sulfur, or liquid SO_2.

Three Wellman-Lord systems are under construction in the United States; two are on coal-fired power plants, and one is on a fluid coke-fired steam boiler at a refinery. A particularly noteworthy system was put into operation in June 1971 at the Japan Synthetic Rubber plant, Chiba, Japan. This system operated at 97% availability during its first year and at 100% during its second and third years. This system, which requires only one-half man per shift to operate (including production and disposal), has maintained 90% SO_2 removal efficiency.

The first Wellman-Lord process FGD unit retrofitted to a coal-fired utility boiler is Unit No. 11 (115 MW) at the Dean H. Mitchell Station of Northern Indiana Public Service Company, Gary, Indiana. On September 14, 1977, the acceptance testing for this unit was successfully completed. All the process performance guarantees were met.

The following are the major advantages of the Wellman-Lord FGD system:

(1) No scaling takes place in the system.
(2) High SO_2 removal efficiencies (95% or better) are achievable.
(3) Low liquid-to-gas ratios are required in the scrubber.
(4) The simplicity of the various unit operations enhances the likelihood of reliable performance.
(5) Recovered SO_2 can be used to produce high-grade sulfur, sulfuric acid, or liquid SO_2.
(6) Considerable operating experience has been obtained on oil-fired boilers, sulfuric acid plants, and Claus plants, and some on coal-fired boilers.

The following are the major disadvantages of the Wellman-Lord FGD system:

(1) A small quantity of purge solids (sodium sulfate and other sodium salts) must be sold as a by-product or disposed of. The soluble sodium salts in this purge stream may lead to water pollution problems.
(2) The process uses large quantities of low-pressure steam and the high energy demand results in a derating of the power station, often by as much as 6%.
(3) The evaporator system must be maintained free of solids; therefore, the incoming flue gas must be very low in particulates, the scrubbing solution must be filtered to remove these solids before it is returned to the evaporator, or both.

GENERAL PROCESS DESCRIPTION

Process Chemistry

The Wellman-Lord process is based on absorption of SO_2 in an aqueous sodium sulfite/bisulfite solution. The spent absorbent is regenerated in a steam-heated evaporator/crystallizer which releases SO_2. The SO_2 is sent to a recovery plant, where H_2SO_4, sulfur, or liquid SO_2 are produced. A purge stream removes impurities that build up in the absorbent.

Absorption: After the flue gas from the boiler is cleaned of particulate, it is passed through an absorption tower, where contact with aqueous sodium sulfite promotes absorption as follows:

(1) $$SO_2 + Na_2SO_3 + H_2O \rightarrow 2NaHSO_3$$

Formation of sodium sulfate occurs by oxidation of sodium sulfite.

(2) $$2Na_2SO_3 + O_2 \rightarrow 2Na_2SO_4$$

After absorption, flue gases are demisted, reheated, and vented to the atmosphere. The spent absorber liquor, now rich in sodium bisulfite, is divided into two streams, 90% to the evaporator for regeneration and 10% to the purge processing system.

Regeneration: The regeneration step occurs in the evaporator. Heat is added to the spent absorber liquor to reverse reaction (1) and to promote crystallization of the less soluble sulfite salts.

(3) $$2NaHSO_3 \rightarrow Na_2SO_3\downarrow + H_2O + SO_2\uparrow$$

The water in the evaporator overhead is partially condensed, stripped of SO_2, and recycled to the process to dissolve the slurry discharge from the evaporator. The SO_2/H_2O stream from the evaporator contains about 5% SO_2 by volume and the balance is water.

Makeup Na_2CO_3 or NaOH is added to the dissolving tank, where the stripper and evaporator bottoms are combined to form the new absorber feed. The Na_2CO_3 or NaOH becomes an active absorbent by the following reactions:

(4) $$Na_2CO_3 + SO_2 \rightarrow Na_2SO_3 + CO_2\uparrow$$

(5) $$2NaOH + SO_2 \rightarrow Na_2SO_3 + H_2O$$

Product Stream: The vapor stream from the evaporator is first cooled to condense water vapor and to reach 80 to 90% SO_2 concentration. This concentrated SO_2 can be used to produce several end products. Wellman-Lord systems on sulfuric acid plants or Claus plants recycle the gas, whereas units installed on boilers produce sulfuric acid, elemental sulfur, or liquefied SO_2, depending on available facilities. The production of these products is explained below.

Sulfuric Acid — SO_2 is reacted with oxygen in the presence of a vanadium pentoxide catalyst to form SO_3, which is in turn absorbed into 98% sulfuric acid. This contact process can make sulfuric acid in any strength desired (usually 93 or 98%) or oleum. A single contact plant can be used. The tail gas is recycled to the Wellman-Lord absorber.

Liquid SO_2 — The SO_2 gas is dried with silica gel or sulfuric acid, then compressed and condensed. The resulting liquid SO_2 is collected and stored in a pressurized tank.

Elemental Sulfur — The process for recovery of elemental sulfur with a minimum assay of 99.5% is carried out in two steps. In a reduction section, a portion of the SO_2 reacts with methane and produces a mixture of elemental sulfur, hydrogen sulfide, carbon dioxide, water vapor, and some unreacted SO_2. (Propane or other similar reducing gases can be used as the reductant.)

(6) $2CH_4 + 3SO_2 \rightarrow 2CO_2 + 2H_2O + S + 2H_2S$

In the Claus section of the process, the hydrogen sulfide and remaining SO_2 from the reduction section react to produce additional elemental sulfur and water vapor.

(7) $2H_2S + SO_2 \rightarrow 2H_2O + 3S\downarrow$

The tail gas is incinerated and recycled to the absorber.

Purge Stream: Nonregenerable sodium sulfate is also formed in the absorber by oxidation of sulfite or by absorption of sulfur trioxide:

(8) $Na_2SO_3 + \frac{1}{2}O_2 \rightarrow Na_2SO_4$

(9) $2Na_2SO_3 + SO_3 + H_2O \rightarrow Na_2SO_4 + 2NaHSO_3$

Continuous purge limits the amount of salt cake generated to 5% of the sulfur removed. If the salt cake is assumed to be 100% Na_2SO_4, then 0.11 kg Na_2SO_4 per kg SO_2 is generated. Thiosulfate and dithionate compounds also may be present in quantities. Another source of sodium sulfate and another nonregenerable product, sodium thiosulfate ($Na_2S_2O_3$), occurs in the regeneration section as follows:

(10) $6NaHSO_3 \rightarrow 2Na_2SO_4 + Na_2S_2O_3 + 2SO_2 + 3H_2O$

Sodium sulfate and sodium thiosulfate from the bisulfite-rich absorber stream, are removed in a crystallizer. The solids are dried either for marketing or disposal, or they are acidified, neutralized, and then discharged. Soda ash or caustic soda is added to the dissolving tank to make up for sodium lost in the purge stream. If a 5% loss is assumed, soda ash makeup would be 0.083 kg/kg SO_2 removed.

$$0.05 \times \frac{106 \text{ g/mol Na}_2\text{CO}_3}{64 \text{ g/mol SO}_2} = 0.083 \text{ kg Na}_2\text{CO}_3 \text{ added per kg SO}_2 \text{ removed.}$$

One of the principal process problems is the formation of sodium sulfate from both the absorption of SO_3 and the oxidation of sodium sulfite by oxygen. It is difficult to reverse this action to rejuvenate the inactive sodium; therefore, inactive sodium salts must be purged from the system and their sodium ions replaced by active sodium ions in the form of soda ash or caustic soda. Thus, this purge involves both a chemical makeup requirement and a potential waste disposal problem. The purge stream can remove 5 to 10% of the total sulfur, and treatment is required to prevent water pollution. The amount of the purge stream has increased proportionately with increasing sizes of new Wellman-Lord installations.

Attempts have been made to prevent solution oxidation by the addition of different antioxidants, but this approach has been abandoned because the cost of inhibitors is high and some purge stream would still be required. Selective removal of the oxidized part of the solution by crystallization (by chilling) has been more successful and has resulted in a five- to sixfold decrease in the purge stream and chemical makeup requirements. Concentrated sodium sulfate crystals in the purge stream can be dried for sale or disposal, or can be further treated and discharged into a watercourse. This crystallization process is used at many installations.

Flowsheet and Subsystem Descriptions

Figure 6.1 is a flow diagram of the Wellman-Lord FGD system. It should be noted that the process depicted in Figure 6.1 differs from that previously offered by Davy Powergas, Inc., in the following areas:

- Double-effect evaporation is used in place of single-effect evaporation in the SO_2 regeneration section to reduce overall steam consumption.

- Evaporative crystallization is used for purge treatment in place of thermal crystallization by refrigeration.

A summary of the basic process design parameters for the Wellman-Lord FGD system is presented in Table 6.1. The corresponding raw material and utility requirements are presented in Table 6.2. These data were provided by Davy Powergas, Inc., and are reproduced herein without substantial changes.

Description of Major Process Subsystems: The Wellman-Lord FGD process includes the following major process subsystems.

Raw Material Receiving, Storage, and Preparation — This area includes facilities for receiving and storing soda ash. Makeup soda ash is conveyed pneumatically from self-unloading trucks to a slurry storage tank where it forms a crystal bed. The saturated sodium carbonate solution produced in the tank is drawn off through a floating suction line and pumped to the regenerated sodium sulfite dissolving tank to replace sodium lost in the purge salts cake. Condensate from the SO_2 regeneration section is sparged into the bottom of the slurry storage tank to maintain the liquor level above the crystal bed. A scrubbed vent is employed on the slurry storage tank to prevent loss of soda ash fines during truck unloading.

Prescrubber — Hot flue gas from the ID fans passes through venturi scrubbers where it is cooled and chlorides and residual fly ash are removed. Makeup water is added to the prescrubbers through wash sprays on their chevron mist eliminators. The slurry produced by back flushing the spent absorbent filters is also added to the prescrubbers slurry loop to minimize makeup water requirements. A slip stream of the prescrubber slurry is pumped to the plant ash pond to purge fly ash and chlorides from the loop.

SO_2 Absorption — Saturated gas from the venturi prescrubbers passes through the absorber vessels where it is contacted with a sodium sulfite solution in a series of valve trays. Each of the upper trays has a collector tray which permits recirculation of absorbent to each contacting tray stage. Liquor overflowing from the bottom contacting tray is collected in the absorber bottom. Regenerated absorbent is added to the absorption loop as a makeup stream to the top contacting tray's recirculation loop. The cleaned flue gas passes through chevron mist eliminators which remove entrained liquor droplets, and is then reheated before discharging to the chimney.

Spent absorbent solution is collected from the absorber sumps and pumped through a pressure-leaf filter to remove particulate before flowing to the spent liquor tank. The slurry produced by periodically sluicing the filter is collected and pumped to the prescrubber loop to minimize fresh makeup water requirements.

Purge Treatment — A portion of the spent absorbent liquor is preheated and fed to the sulfate crystallizer. A liquid-solid separation chamber in the crystallizer

Figure 6.1: Wellman-Lord Process Flow Diagram

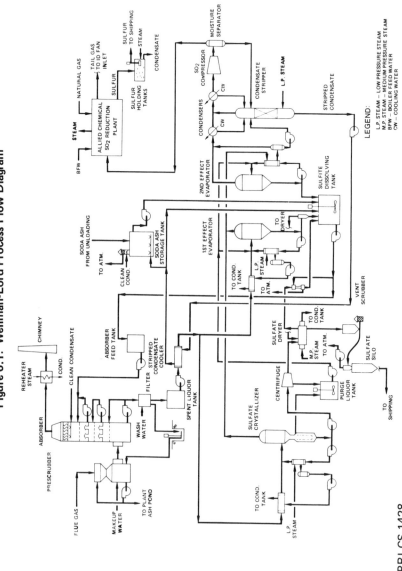

Source: EPRI CS-1428

Table 6.1: Process System Parameters

Data for 1 x 500 MW Unit
(Maximum Load Conditions)

	Units	Eastern Coal	Western Coal
Flue gas flow rate	m^3/sec	855	913
	(10^3 acfm)	(1,812)	(1,935)
Number of absorption trains per unit	each	4	4
Capacity per train	%	33⅓	33⅓
Gas flow per absorber, saturated	m^3/sec	239	261
	(10^3 acfm)	(507)	(552)
Prescrubber type	–	venturi	venturi
Prescrubber L/G	ℓ/m^3	1.07	1.07
	(gal/10^3 acf)	(8)	(8)
Absorber type	–	valve tray	valve tray
Number of contacting stages	each	3	5
Absorber L/G (per tray stage)	ℓ/m^3	0.267	0.267
	(gal/10^3 acf)	(2)	(2)
System total pressure loss	pascals	5,231	7,224
	(inches wg)	(21)	(29)
Mist elimination	–	chevron	chevron
Absorbent oxidation ratio*	–	5	30
Flue gas reheat required	°C	31.1	26.7
	(°F)	(56)	(48)
Number of purge treatment trains per unit	each	1 at 100%	1 at 100%
Number of SO_2 regeneration trains per unit	each	4 at 25%	2 at 50%
Number of SO_2 reduction trains per unit	each	1 at 100%	1 at 100%

*Mols Na_2SO_4/mol SO_2 removed.

Source: EPRI CS-1428

Table 6.2: Raw Materials and Utility Requirements

Data for 1 x 500 MW Unit
(Maximum Load, Average Sulfur)

	Units	Eastern Coal	Western Coal
Raw materials			
Soda ash	10^3 kg/hr	1.23	1.18
	(tons/hr)	(1.36)	(1.30)
Utilities			
Natural gas	10^3 m^3/hr	2.97	0.31
	(10^3 ft^3/hr)	(105)	(11)
Electricity			
Raw material handling and preparation	kW	90	60
Pretreatment and absorption	kW	360	330
Purge treatment	kW	205	185
SO_2 regeneration	kW	3,363	545
SO_2 reduction	kW	80	10
Fans	kW	5,480	8,080
Total	kW	9,578	9,210

(continued)

Table 6.2: (continued)

Data for 1 x 500 MW Unit
(Maximum Load, Average Sulfur)

	Units	Eastern Coal	Western Coal
Steam			
Purge treatment	10^6 Btu/hr	43.9	39.9
Regeneration	10^6 Btu/hr	192.8	21.4
SO_2 reduction	10^6 Btu/hr	26.0	3.0
		credit	credit
Reheat	10^6 Btu/hr	96.6	89.4
Sulfur storage	10^6 Btu/hr	0.3	0.1
Total	10^6 Btu/hr	307.6	147.8
Water			
Process makeup	10^3 ℓ/hr	107.5	110.1
	(10^3 gal/hr)	(28.4)	(29.1)
Cooling water	10^3 ℓ/hr	3,774	856.2
	(10^3 gal /hr)	(997.2)	(226.2)
Boiler feedwater to SO_2 reduction	10^3 ℓ/hr	11.8	1.4
	(10^3 gal/hr)	(3.12)	(0.36)
Products			
Sulfur	10^3 kg/hr	7.26	0.80
	(tons/hr)	(8.0)	(0.9)
Purge salts (at 65 wt % Na_2SO_4)	10^3 kg/hr	1.60	1.52
	(tons/hr)	(1.76)	(1.68)
Incremental Heat Rate Penalties			
Steam (basis gross heat rate)	%	6.82	3.16
Electricity (basis unit net rating)	%	1.92	1.84
Corrected net heat rate	Btu/kWh	10,403	10,395
Variation from base	%	+4.18	+0.99

Source: EPRI CS-1428

produces a clear liquor overflow stream and a purge salt slurry stream. The slurry is centrifuged to separate the crystals from the remaining liquor. The centrate is combined with the crystallizer overflow liquor and pumped to the regeneration area. The overhead vapors from the sulfate crystallizer are also sent to the regeneration area.

The wet cake from the centrifuge and a portion of the mother liquor from the first-effect evaporator are fed to a steam heated dryer. The dried crystals are discharged into a sulfate surge hopper from which they are conveyed pneumatically to a sulfate storage bin. The dryer off-gas passes through a vent gas scrubber for removal of SO_2, and is discharged to the atmosphere.

SO_2 Regeneration — The balance of the spent absorbent liquor is combined with the purge liquor stream and fed to the double-effect evaporator. A purge stream of the mother liquor is separated from the first-effect evaporator slurry stream and pumped to the sulfate dryer to control the level of unreactive sodium thiosulfate ($Na_2S_2O_3$) in the regenerated absorbent. The remaining slurry flows to the dissolving tank.

Overhead vapors from the first-effect evaporator and the sulfate crystallizer are partially condensed in the second-effect heater. The sour condensate produced is steam stripped in the condensate stripper to remove dissolved SO_2. Overhead vapor from the second-effect evaporator is combined with the uncondensed vapor from the second-effect heater and sent to the primary condenser, where most of the water is removed. The sour condensate from the primary condenser is also stripped to remove dissolved SO_2.

The vapor from the primary condenser is combined with the stripper overhead vapor and sent to the secondary condenser. The sour condensate from this condenser also flows to the stripper for removal of dissolved SO_2. Stripped condensate is cooled by exchanging heat with spent absorbent liquor before it is returned to the dissolving tank.

The concentrated SO_2 stream from the secondary condenser is compressed and passes through a moisture separator before it is sent to the Allied Chemical SO_2 reduction plant. The moisture removed from the compressed SO_2 stream is sent to the condensate stripper for removal of dissolved SO_2.

Condensate is added to the slurry in the dissolving tank to redissolve the sulfite crystals. Soda ash solution from the soda ash storage tank is added to replace the sodium lost in the purge stream. A portion of the regenerated absorbent solution is pumped to the vent gas scrubber to remove SO_2 from the dryer vent gas. The regenerated absorbent solution is pumped from the dissolving tank to the absorber feed tank for storage.

SO₂ Reduction — The dried, compressed SO_2 gas is fed to an Allied Chemical SO_2 reduction unit where it is mixed with natural gas, preheated, and partially reduced in a catalytic reactor to a mixture of sulfur, H_2S, and SO_2. The gasses from the reduction stage are partially cooled to separate the sulfur, and the residual SO_2 and H_2S are converted to sulfur in a modified Claus catalytic reaction system.

The product sulfur is stored molten in concrete-lined steam-heated pits for shipment in heated tank cars. The tail gas from the reduction unit is incinerated with natural gas and air to oxidize residual H_2S to SO_2, after which it is returned to the inlet of the scrubbing system.

GENERAL OPERATING PARAMETERS

Chemical Parameters

The pH of the Wellman-Lord absorption systems is maintained between 6 and 7. A pH of 6.5 is the normal control point. A higher pH increases carbon dioxide absorption, whereas a lower pH increases SO_2 absorption. A system with a pH of 6.5 is highly buffered and can adapt rapidly to changes in flue gas inlet conditions. A pH of approximately 5.5 is normally found at the absorber discharge.

Physical Parameters

The primary scrubber in the system can be a particulate scrubber or just a gas cooler and humidifier. Flue gas temperature is normally lowered to 50° to 55°C. Liquid-to-gas ratios for representative systems are 0.5 to 1 ℓ/Nm^3, flue gas velocities across the throat are about 2 m/sec, and the pressure drop across the throat is from 1 to 3 kPa.

The SO_2 absorber has an overall L/G ratio of 0.067 ℓ/Nm^3, whereas the L/G for each tray is about 0.5 ℓ/Nm^3. Internal scrubber flue gas velocities are about 3 m/sec. Pressure drops in the absorber vary from 1 to 4 kPa, depending on the number of trays.

Flue gas velocities in a vertical demister average 6 m/sec. The optimum flue gas velocity in horizontal demisters is about 3 m/sec. Increasing the velocity increases the water removal rate, but it also increases reentrainment of water before it can leave the demister. Pressure drop across the demister ranges from 50 to 150 Pa.

Flue gas reheating temperatures at representative boilers range from 28° to 45°C and energy expenditure approximates 1,100 J/kg of flue gas per °C. Pressure drop across the reheater is 250 to 700 Pa.

The purge rate of dry Na_2SO_4 cake is about 62 kg/hr at a 100-MW station for each 1% of O_2 in the flue gas. In a system where the Na_2SO_4 is purged as a neutralized wastewater, about 1 kg of water is needed as makeup for each mol of SO_2 removed.

DEVELOPMENT STATUS

The Wellman-Lord process as employed for SO_2 emission control on oil-fired boilers is well-developed technology. The same technology is applicable to coal-fired systems providing equivalent control of inlet gas quality is employed. Two such systems are now in operation on coal-fired utility boilers in the United States.

Further development of the process for large-scale coal-fired application is directed toward reducing the oxidation of the absorbent and/or reprocessing the oxidized purge of sodium sulfate to carbonate for subsequent reduction to the sulfite form required by the system. Improvements in system economics by reduction of system pressure drop and absorbent regeneration heat consumption may also be possible.

A major factor affecting large-scale use of the Wellman-Lord process is the peripheral conversion of the liberated SO_2 to marketable products. The current U.S. coal-fired application employs natural gas for reduction of SO_2 to sulfur. A coal-based conversion process (such as RESOX) could markedly expand the opportunities for using this process.

RECENT TECHNOLOGICAL DEVELOPMENTS AND SOLUTIONS TO PROBLEM AREAS

Sodium-based scrubbing processes are not subject to the inherent plugging and scaling problems and limited removal capabilities associated with direct lime- and limestone-based scrubbing processes because the high solubility of sodium alkalis and their sulfur-containing reaction products tend to minimize these problems.

pH Control

The scrubbing solution pH range provides the key to optimal SO_2-removal efficiency. A control value of 6.5 permits maximum SO_2 removal efficiency without chemical-related problems. If the scrubbing solution pH value exceeds 7, carbon dioxide absorption becomes significant. A scrubbing solution pH level below 5 must also be avoided; it causes the vapor pressure of SO_2 to increase on sulfite-bisulfite systems and can lead to equilibrium-limited scrubbing conditions where outlet concentrations at or below 200 ppm are desired. Graphical evidence for this chemical phenomenon is provided in Figure 6.2, in which the sulfur dioxide vapor pressure is plotted as a function of pH for a solution temperature of 54°C, a typical saturation temperature for a boiler flue gas stream. For an active alkali system, the vapor pressure of SO_2 as a function of pH is significant. This is evident from the line in the figure, which was calculated from Henry's constant for SO_2 and water and from the first and second ionization constants for sulfurous acid (H_2SO_3) for a solution with a total oxidizable sulfur (TOS) level of 0.03 gram mol/liter and an ionic strength of 0.3 gram mol of Na^+/liter. The SO_2 vapor pressure at a pH of 6 is approximately 100 ppm and increases rapidly as the pH drops.

Figure 6.2: Concentrated Active Alkali System—Sulfur Dioxide Vapor Pressure versus pH at 54°C

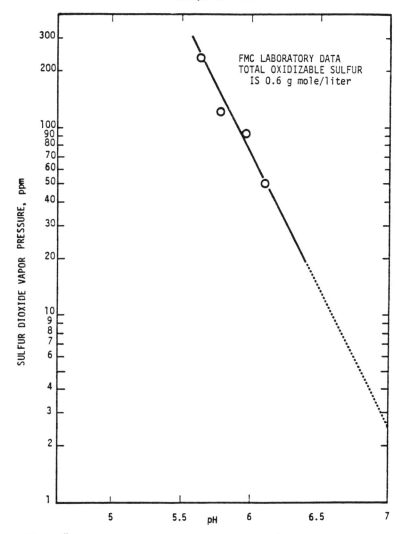

(From: "The FMC Concentrated Double-Alkali Process," by L.K. Legatski, K.E. Johnson, and L.Y. Lee, FMC Corp., Glen Ellyn, Illinois)

Source: NTIS PB-295 011

For these reasons, Wellman-Lord systems tend to operate in the pH range of 6 to 7, preferably at a set point of 6.5, where the sulfite-bisulfite system is highly buffered. The highly buffered system is preferred because the scrubbing solution is able to adapt to rapid changes in flue gas inlet conditions. In contrast, scrubbing systems operating at a pH of 6 or less can be very sensitive to and adversely affected by rapid changes in flue gas inlet concentrations.

Based on the relatively low SO_2 vapor pressure in solution at the optimum pH control value of 6.5, SO_2 concentration levels can be maintained at about 50 ppm in the exit gas stream.

All inactive Na salts must be purged from the scrubber solution and their Na ions replaced by active Na ions from soda ash or caustic soda. Methods are being researched for eliminating the purge stream by reactivating the Na ion. Several processes have been investigated in both laboratory and full-scale applications, but few details have been disclosed. In one process, the purge stream would be incinerated to liberate the SO_2 and recover the Na salts, a concept similar to the recovery boiler used in the pulp and paper industry. Two processes are under development to reduce the Na_2SO_4 chemically. One uses natural gas and the other uses coal as the reductant. Pilot plant testing has been successful. A method employing high-temperature separation of the sulfate also has been tested successfully on a bench scale.

The recovery of elemental sulfur could be limited by natural gas supplies. A 500 MW plant would require about 1,360 Nm^3/hr of natural gas, depending upon the sulfur content of the coal and the efficiency of the sulfur recovery plant. High-sulfur residual oils, coal gasification, and alternative hydrocarbon fuels such as propane are being investigated as alternate fuels.

The presence of fly ash and particulate matter in the flue gas from a coal-fired boiler presents a potential operating difficulty for a Wellman-Lord installation. Particulate matter must be kept away from the scrubbing solution because large quantities of particulate will affect the system's operating efficiency.

Chlorides

Chlorides in the scrubbing solution occur primarily from the combustion of chloride bearing fuel and to lesser extent from chloride in the raw water and limestone. Depending on the amount of these chlorides, problems in the scrubber system may occur due to corrosion, decreased SO_2 absorption, and wastewater discharge. Chloride concentrations in the scrubber liquor also increase with the amount of liquor recycled. These problems can be reduced by (1) using a pre-scrubber and keeping that liquid stream separate from the SO_2 absorber streams, (2) designing for chloride corrosion resistance, and (3) increasing the system purge rate. Depending on wastewater discharge regulations, the purge stream may require further treatment.

NORTHERN INDIANA PUBLIC SERVICE COMPANY'S SYSTEM

This section deals with the Unit No. 11 Boiler FGD System Northern Indiana Public Service Company's Dean H. Mitchell Station located in Gary, Indiana.

The FGD system will be referred to as the NIPSCO FGD Plant. By definition the NIPSCO FGD Plant includes the flue gas booster fan, flue gas isolation damper and the flue gas louver by-pass damper, all of which are outside battery limits as well as the primary battery limits portion of the FGD plant consisting of the prescrubber, the Wellman-Lord absorption, regeneration and purge treatment units and the Allied Chemical SO_2 Reduction Unit. By-product storage and loading facilities are included within battery limits.

NIPSCO Project Background

NIPSCO and the U.S. Environmental Protection Agency entered into a cost shared contract in June of 1972 for the design, construction and operation of a regenerable flue gas desulfurization (FGD) demonstration plant. The system selected for the project was a combination of the Wellman-Lord SO_2 Recovery Process and the Allied Chemical SO_2 Reduction Process. The FGD plant was to be retrofitted to NIPSCO's 115 MW pulverized coal-fired Unit No. 11 at the Dean H. Mitchell Station in Gary, Indiana. NIPSCO entered into contracts with Davy Powergas Inc. for the design and construction of the FGD plant and with Allied Chemical Corporation for operation of the plant.

A successful performance test was run from August 29, 1977, through September 16, 1977.

A one-year demonstration test period began on September 16, 1977 and was extended to March 15, 1979 (1).

TRW under contract to the U.S. Environmental Protection Agency, has continued to monitor and report the performance of the boiler and the FGD plant during the demonstration test period.

This section will discuss the FGD plant operating experience from the beginning of the demonstration test period on September 16, 1977 through December 31, 1978.

NIPSCO is continuing to assess the Wellman-Lord/Allied Chemical option for SO_2 emission control.

NIPSCO is still optimistic that longer periods of continuous operation will be achieved during the extended demonstration test period. Through continued operations at NIPSCO it will be possible to make a more accurate evaluation of the economics and plant reliability.

NIPSCO FGD Performance Design Criteria and Results (2)(3)

During the plant acceptance test, from August 29, 1977, through Sept. 14, 1977, the FGD system performance criteria and obtained results were:

92 MW Equivalent Test

SO₂ Removal

Required: Minimum SO_2 removal of 90%, measured continuously and averaged every 2 hours for a period of 83 hours.

Results: SO_2 removal averaged 91% over the 12 day test period. In only two 2 hour periods (out of 144) was the SO_2 removal less than 90%, and for those periods, it averaged 88 and 89%.

Particulate Removal

Required: Particulate emission measured once daily will not exceed the Federal NSPS for fossil fuel fired steam generators of 0.1 lb/million Btu heat input.

Results: Particulate emission averaged 0.04 lb/million Btu, or 40% of the maximum allowable. Of the 12 days, tests could not be run on 4 days due to inclement weather. On one day, the test data was not valid.

Soda Ash Consumption

>   Required: Average over the 12 day test period not to exceed 6.6
>   STPD.

>   Results: Soda ash consumption determined by daily inventory
>   obtained from storage bin measurement (official result) averaged
>   6.2 STPD, or 94% of the maximum allowable. Consumption
>   determined by manual weighing the feeder output every 2 hours
>   throughout the 12 day test period averaged 5.7 STPD, or 86% of
>   the maximum allowable.

Sulfur Purity

>   Required: Minimum sulfur purity 99.5%, suitable for conversion
>   to quality sulfuric acid by standard production practice.

>   Results: Sulfur purity determined from a composite sample
>   collected over the 12 day test period was 99.9%, easily exceed-
>   ing the required purity.

110 MW Equivalent Test

SO_2 Removal

>   Required: Minimum SO_2 removal of 90%, measured continuously
>   and averaged every 2 hours for a period of 83 hours.

>   Results: SO_2 removal averaged 91% over the 3½ day test period.
>   In only one 2 hour period (out of 42) was the SO_2 removal less
>   than 90%, and for that period, it averaged 89%.

Particulate Removal

>   Required: Particulate emission measured once daily will not ex-
>   ceed the Federal NSPS for fossil fuel fired steam generators of
>   0.1 lb/million Btu heat input.

>   Results: Particulate emission averaged 0.04 lb/million Btu, or 40%
>   of the maximum allowable. Of the 3½ days, a test could not be
>   run on one day due to inclement weather.

Viability of the NIPSCO System (2)(3): During the 12 day performance
period at the 92 MW equivalent rate, there was a total of 26 hours in interruptions
in the fully integrated operation of the FGD system. Of the 26 hours, 18 were
related to boiler problems and 8 were related to problems in the FGD system. In
addition, there was a 4 hour period in which the SO_2 removal averaged "only"
88.5%; this 4 hour period was added to the performance test at the end of the
12 day test period. It should be mentioned that outages in the FGD system did
not interrupt SO_2 removal. Furthermore, an SO_2 removal of 88.5% at the NIPSCO
site results in an emission well below the NSPS of 1.2 lb/million Btu heat input.
During the acceptance test the parameters used by EPA for judging the viability
of a FGD system were: availability, 94%; reliability, 100%, operability, 100%;
utilization factor, 94%.

Problems Causing FGD Outages

Figure 6.3 illustrates the FGD Plant Operating Factor based on hours of
FGD operation divided by hours of generating unit operation. Table 6.3 summarizes the FGD outages and indicates whether the outage was attributable

to the boiler (B), the FGD plant (FGD) or a combination of both (C). The four categories of major FGD outages are discussed below.

Booster Fan: *Problem* — Imbalance of the air foil flue gas booster fan has been a continuous problem since early in the demonstration period. Operating conditions at or below the acid dew point of the flue gas (and below the specified design temperature of 300°F) plus an ineffective guillotine isolation damper upstream of the booster fan contributed to an accumulation of fly ash, water and ice in the fan housing and on the fan blades. Over a period of several months during which FGD system (and the booster fan) was idle for substantial periods because of high silica problems in Unit No. 11 make-up water, these conditions resulted in corrosion and erosion of the air foil blades which finally required a complete reblading of the fan. Turbine governor malfunctions have apparently been caused by exposure of the governor to outdoor weather and dust conditions.

Figure 6.3: FGD Plant Operating Factor (O.F.)

$$O.F. = \frac{HRS. \ FGD \ INTEGRATED \ OPERATION}{HRS. \ GENERATOR \ ON \ LINE} \times 100$$

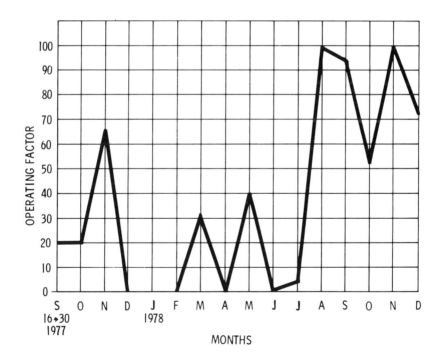

Source: Symposium — Volume 2

Solution — The following revisions and additions were made to resolve the booster fan problems: (a) The intermediate layer of Unit No. 11 air preheater elements was removed to raise the flue gas exit temperatures; (b) the booster fan inlet and outlet duct and the booster fan housing were insulated; (c) the booster fan was rebladed and Inconel shields were installed on the blade leading edges; (d) a steam soot blower was installed in the booster fan for on line blade cleaning; and (e) a booster fan drive turbine enclosure is presently under construction.

Results — Preliminary results are very encouraging. The air preheater element removal has resulted in an increase of approximately 30°F for the exit flue gas temperature with no apparent adverse effects on the boiler other than a slight decrease in boiler efficiency. The flue gas temperature is now consistently at or above 300°F and condensation has not occurred in the ductwork and booster fan as it did in the past.

High Silica Levels in Boiler: *Problem* — In October 1977, routine Unit No. 11 boiler water chemistry tests indicated high levels of silica. The first attempts to solve the problem focused on inspection of Unit No. 11 boiler and condenser for leaks. Frequent or continuous blowdown of the boiler and reduced steam pressures made the steam supply to the FGD plant so unreliable that it precluded integrated FGD operation. After extensive investigation including ultrasonic condenser tube testing, the condenser was found not to be the cause of the contamination. The causes of the silica build-up appear to have been a boiler chemistry upset due to high condensate make-up requirements imposed on the boiler by the FGD plant requirements. This was compounded by the continuous boiler blowdown and lack of continuous silica monitoring on a portable demineralizer being used to supplement the station demineralizer to supply Unit No. 11 and FGD condensate make-up demand.

The need for using a portable unit resulted from failure of a reverse osmosis unit, installed by NIPSCO, to operate successfully.

The design of the FGD system was based on the use of filtered lake water for use as flushing water to the packing gland on the process pumps. During plant startup it was found that the quantity of fine silt in the lake water made it impossible to adequately filter the water for packing gland flushing. As an expedient the use of condensate as a flushing fluid was implemented and the resulting consumption of condensate in the FGD system became substantially greater than initial design.

Since the water treating facilities (station demineralizer and reverse osmosis unit) available at the NIPSCO plant were expected to have adequate capacity, this increase in condensate consumption was not seen as a problem at the time except from the cost standpoint. The unit cost for treating water with the portable demineralizer equipment is very high.

Solution — A continuous silica analyzer was installed on the portable demineralizer outlet and the silica content of the treated make-up water was limited to 10 parts per billion. Condensate from the FGD system is diverted to waste at a low pH reading of 6.5 or a high reading of 9.5 and at a conductivity reading of 5.5 micromohs/cm and higher. The FGD plant condensate return diverter system was changed from manual to an automatic reset mode. If the condensate is diverted to waste because of high conductivity, low or high pH, it will automatically return to the Unit No. 11 condenser when the pH and conductivity readings are again within acceptable limits. As part of the solution to this problem, a comprehensive program was undertaken in the FGD plant to reduce condensate consumption and losses.

Results — The above modifications have been effective in controlling Unit No. 11 boiler silica. Combined Unit No. 11 and FGD condensate make-up is now generally in the 40 to 50 gpm range with an occasional excursion to higher volumes if the return condensate has been diverted to waste.

Guillotine Isolation Damper: *Problem* — A top entry guillotine damper was installed between the outlet of the existing Unit No. 11 boiler induced draft fans and the FGD plant booster fan inlet. During operation with the damper open a build-up of fly ash occurred in the bottom and lower side channels of the damper frame. When the damper was closed it would strike the fly ash build-up and the bottom seals of the damper sustained mechanical damage. Considerable corrosion damage has also been experienced with the damper.

Solution — Considerable work was done on maintenance and modification of the damper. A bottom channel purge air system was installed and modified. Damper seals were replaced with seals made of a more corrosion resistant alloy.

Results — The fly ash build-up continued to occur at the bottom of the damper. A decision was made not to invest any additional money in the damper. Manual slide gate dampers were installed in each of the booster fan inlet housings and will be used to isolate the fan when necessary for maintenance.

Note: During October through December 1978, the guillotine isolation damper operated successfully but did not provide total shut-off capability, probably caused by fly ash damaged seals. The modifications to the Unit No. 11 air preheaters and installation of thermal insulation on the flue gas duct work allowed the flue gas temperatures to be maintained above the dew point and alleviated some of the problems with the guillotine isolation damper as well as the flue gas booster fan.

Coal Quality: *Problem* — Wet, poor quality coal caused erratic Unit No. 11 boiler operation. As a result, steam pressure and flow to the FGD plant fluctuated and prevented sustained integrated FGD operation.

Solution — NIPSCO's coal procurement department negotiated with suppliers to insure a consistent supply of a more acceptable quality coal.

Results — The coal quality has improved and is no longer considered a problem.

Table 6.3: Summary of FGD Outage Causes

Cause	Attributable to (B), (FGD) or (C)	Total Number of Days in Outage
Booster fan cleaning balancing and reblading	FGD	67
High boiler silica levels	C*	53
Booster fan guillotine isolation damper	FGD	32
Booster fan bearing oil leak, booster fan turbine governor repairs, and turbine bearing repair	FGD	26
Scheduled boiler maintenance	B	22
Wet, poor quality coal—erratic boiler operation	B	20
FGD plant main steam pressure reducing valve malfunction	B	13

(continued)

Table 6.3: (continued)

Cause	Attributable to (B), (FGD) or (C)	Total Number of Days in Outage
Unit No. 11 boiler tube leaks	B	12
Unit No. 11 electrostatic precipitator malfunction	B	7
Evaporator circulating pump repacking	FGD	7
Boiler baseline test	C	6
Evaporator heat exchanger tube sealing and tube leaks	FGD	6
Heat balance and flue gas flow rate tests	B	6
Bearing failure—evaporator circulating pump turbine drive	FGD	5
Induced draft fan imbalance	B	5
Coalescer pluggage	FGD	5
Gasket failure—evaporator solution line	FGD	4
Turbine-generator stop valve repair	B	4
Air preheater cleaning	B	3
Various instrument problems	FGD	3
Tail gas incinerator malfunction	FGD	2
Sulfur condenser tube leaks	FGD	2
SO_2 compressor gasket leak	FGD	2
Process water booster pump failure	FGD	2
Steam leak at drum pressure transmitter line.	B	1

Recapitulation

Total days in period September 16, 1977 through December 31, 1978	472
Total days integrated FGD operation	157
Total days FGD outage	315
Total days FGD attributable outage	163
Total days boiler-turbine attributable outage	80
Total days combination attributable outage	59
Total days boiler and turbine operated (on line)	424
Total days boiler outage (off line)	48
Longest continuous integrated FGD operation period, days	43**

*The cause of high silica in the boiler feedwater was a silica breakthrough in the portable demineralizer unit that was being used to supplement the increased water demand. The additional usage was compounded by increased boiler blowdown. This was corrected by setting more stringent silica limits and closer monitoring equipment of the portable demineralizer unit. It was also followed up by a program to reduce condensate consumption and losses including installation of automatic reset controls on the return condensate diversion system.

**Disregarding 2-2hour interruptions in reduction section.

Problems Not Causing FGD Outages

There were some problems experienced during the early plant operations which are mentioned here. These situations did not cause FGD Plant outages.

Absorber Tray Leakage: Sodium balances and analyses of the fly ash sump discharge indicated that there were solution losses occurring in the absorber. Inspection of the absorber lower collector tray showed that solution was splashing over the chimney free-boards and leakage was also occurring between the tray perimeter and the absorber wall. The chimney free-board height has been raised and the tray perimeter recaulked and sealed with rubber.

Evaporator Circulating Pump Driver — The evaporator slurry circulating pump was originally equipped with a steam turbine drive, utilizing main steam supplied by Unit No. 11 boiler to the FDG plant, at 550 psig. Exhaust steam from the drive turbine was utilized as part of the heating steam requirements for the evaporator.

During the demonstration period the main steam supply to the FGD plant was partially or totally interrupted on many occasions resulting in slowdown or stopping of the circulating pump. To avoid settling of solids and plugging of the evaporator it was necessary to dilute the solution and dump the evaporator slurry into a holding tank. After the steam supply was reestablished, two days would be required to refill and reestablish solids content before evaporator operation could be resumed.

To avoid these lengthy interruptions to the FGD operation an electric motor drive was installed on the pump in September 1978.

Purge Dryer Capacity — System sulfate inventories and dry sodium salt production indicated that the purge dryer possibly was not operating at rated capacity. A test program was undertaken by Davy Powergas. Mechanical modifications were aimed at increasing the purge dryer capacity.

References

(1) "Sulfur Recovered From Flue Gas at Large Coal Fired Power Plants": Roberto I. Pedroso, Davy Powergas Inc., Lakeland, Florida. Paper presented to the Third Symposium on Sulphur and other Airborne Emissions, Salford, England — April 1979.

(2) The Operating History and Performance Experience of Unit Number 11 and the FGD System During the Pre-Acceptance Test and Acceptance Test Periods. F.W. William Link of NIPSCO and Wade Ponder of EPA, presented in November 1977 at the FGD Symposium, Hollywood, Florida.

(3) Sulphur Recovered From SO_2 Emissions at NIPSCO's Dean H. Mitchell Station: Howard A. Boyer, Allied Chemical Corporation, Morristown, New Jersey and Roberto I. Pedroso, Davy Powergas Inc., Lakeland, Florida.

PUBLIC SERVICE OF NEW MEXICO'S SYSTEM

This section pertains to the FGD facilities of the Public Service Company of New Mexico's San Juan Station located in Waterflow, New Mexico (near Farmington, New Mexico). The FGD facilities will be referred to as the PNM plant.

PNM Project Background

Public Service Company of New Mexico (PNM) selected the Wellman-Lord/Allied Chemical FGD processes early in 1974 from 4 different systems that were being considered. The process compared favorably both economically and technically with the lime-limestone, double alkali, and with a dry char adsorber.

The Wellman-Lord/Allied Chemical process was also considered advantageous because the end product, elemental sulfur, could be marketed. The initial concept was to regenerate the purge salt, sodium sulfate. In the meantime, a contract had been obtained to sell this material. The elimination of the sulfate regeneration process reduced the capital requirements for Units No. 1 and 2 by $4,000,000.

A major advantage of the Wellman-Lord process is using a clear scrubbing solution that prevents absorber pluggage. A more soluble substance is produced in the scrubbing liquor as SO_2 is absorbed. The materials handling is also much simpler since relatively small volumes of materials enter and exit the system compared to the calcium based system.

The main disadvantage of the Wellman-Lord process, is that it is somewhat more complex than the lime-limestone systems because the solutions are regenerated.

PNM Design Criteria

The Wellman-Lord system for Units No. 1 and No. 2 was designed to remove 90% of the SO_2 from the flue gas when firing coal ranging in sulfur from 0.59 to 1.3% by weight with an average of 0.8%. Relatively high pressure drop prescrubbers were specified for the system because of New Mexico's stringent regulation of a fine particulate emission and to provide some backup for the electrostatic precipitators. The purge from the prescrubber is sent to the plant wastewater system where it is treated and then recycled.

At PNM's request four scrubber absorber modules were installed on each power plant unit each sized to handle one-third of the total gas flow. Therefore, the plant has one complete spare scrubber-absorber available. This permits a maintenance program to be established whereby absorbers can be rotated in and out of service for routine and preventative maintenance purposes.

The chemical plant has two double effect evaporators which provide steam conservation (the overhead vapors from the first effect are utilized as the heat source in the second effect). For reliability each evaporator is connected to steam and offgas compressor manifolds so any one evaporator can be taken out of service for maintenance without affecting the operation of the three absorber units.

The purge treatment plant consists of three low temperature crystallizers where sodium sulfate is precipitated in the decahydrate form. The crystallizers are followed by a melt tank and evaporator. The evaporator is similar to, but smaller than, the main evaporators. Water is driven off and the sulfate purge is centrifuged, then dried in a flash dryer. The entire purge treatment plant was specified to obtain a concentration of 70% sodium sulfate and 30% sodium sulfite in the dried purge salt. The actual sulfate content has been very high; purities have been achieved in the area of 90% sodium sulfate. Residual moisture in the dried purge salt has been less than 1.0%.

Two identical Allied Chemical SO_2 reduction trains were installed as part of Units 1 and 2 FGD systems. Each of the trains has a design capacity of more than 50% of the total FGD system capacity based on the use of low grade coal (1.3% sulfur) in the boilers. A requirement for two SO_2 reduction trains was specified by PNM with the objective of achieving an FGD system with essentially a 100% on-stream reliability.

The system being designed for the Unit Number 3 and 4 power plants will be somewhat similar with the following exceptions. The prescrubbers on Unit

Number 3 and 4 systems will have a lower pressure drop for energy considerations. The five stage tray absorbers also function quite well for residual particulate removal; sulfate purge quality will not be degraded since the fly ash is filtered out of the solution.

In the interest of economy the absorbers have been designed for 4 units in operation per boiler when burning low grade coal; i.e., this removes the one module spare as on Units 1 and 2. The coal being used at San Juan Station has rarely exceeded 0.95% sulfur for long durations of time hence this should not decrease the plant operability and maintenance capability.

A sulfuric acid unit will be installed in the FGD system for Units 3 and 4. The sulfuric acid plant design capacity will be based on low grade coal (470 tons per day as 100% sulfuric acid).

Plant Operation Personnel

General: The plant scrubber operation is controlled by two separate PNM groups; the power generation plant personnel are responsible for the booster blowers, scrubbers, absorbers, fly ash filters, and wastewater treatment. The chemical plant operations personnel are responsible for supplying solutions for the absorbers, regeneration of SO_2 from the absorbing solutions, operation of the sulfur dioxide reduction unit, purge treatment, and loading and shipping of the products.

Personnel: There are approximately 100 people involved in support of the FGD process.

Scrubber operators are in their fourth year apprenticeship and are near the end of their formal training as Journeymen Operators for the power generation plant. All operators are rotated through this position as part of their training. First and second year apprentices are assigned the field activities for the scrubbers and the fly ash filters areas. All of these operators are responsible to the power generation plant supervision for direction of their activities. The formal training given to the operators included lectures by Davy. In-plant training during start-up was directed by Davy start-up engineers, while some of the "on the job training" was by PNM supervision. The chemical plant operators were given more extensive training initially because of the chemical process involved. This included training during initial equipment testing for acceptance by the engineering contractor, as well as the plant commissioning and subsequent plant operations.

Approximately one-third of the FGD personnel were selected from the power plant operations group and the remainder from the chemical industry. The creation of an operationally experienced crew for the chemical plant, with the necessary levels of experience and ability presented numerous problems of finding and relocating personnel. Recruiting outside of the local plant area was necessary because of local competition for experienced people. Past experience of operators and supervisors included ammonia plants, power generation plants, sulfur plants, chemical plants, refineries, and PNM's own power plant.

How the Plant Operates

Scrubber-adsorber operations do not affect the power plant operation because the FGD system can be by-passed. During a normal unit startup, scrubbers and absorbers are put on line after the electrostatic precipitators are functioning. Power for the electrostatic precipitators, booster blowers, and absorber circulating pumps is supplied from PNM's power plant.

Each scrubber-absorber module is operated independently (except the control data transmission to the control panel board), and put on line separately. Each unit has a reheat system that is needed to protect the stack from corrosive products resulting from condensation in the flue gas when three modules are in operation simultaneously.

Two 750,000 gallon tanks for absorber product and feed solution provide surge in the system to prevent the chemical plant operation from being affected by normal operating fluctuations in the scrubber-absorber area. The ideal operating situation is to have the feed solution tank full and the product tank level very low.

The chemical plant's ability to operate depends on the power generation unit operations. To date this has been the major operating problem. Steam and water to the chemical plant originates in the power plant area; clean condensate and process offgas are returned from the chemical plant. The sumps and wastewater streams are collected and neutralized in a water treatment plant.

SO_2 is recovered from the absorber product solution in a double effect evaporator. Precise control of the rate of SO_2 recovery from the slurry in the evaporators is possible. The oxygen level in the recovered SO_2 stream is controlled at less than 0.35%, as per the design which facilitates operation of the SO_2 reduction units at maximum natural gas utilization. Evaporator slurry concentrations up to 70% by volume have been achieved with little or no solids accumulation in any of the vessels. Condensed water from the overhead SO_2-H_2O stream is recycled and used to dissolve the solids from the evaporators. Both evaporator trains have operated for many consecutive days without interruptions (other than steam availability) over a wide range of operating conditions. Critical parameters for SO_2 recovery rate are: steaming rates, slurry solids percent, and chemical composition. Enough versatility has been provided so that all four evaporators may be operated simultaneously or independently.

High speed, dry compressors are used to lower the absolute pressure in the evaporators and pressurize the SO_2 gas stream through the reduction process. Offgas blowers downstream of the SO_2 reduction units serve to return incinerated tail gas and vessel vent gases back to the scrubber-absorbers.

Each of the Allied Chemical SO_2 Reduction units includes a primary reactor system where a portion of the SO_2 is reduced to sulfur and H_2S using natural gas as a reductant. Exothermic heat of reaction is stored in combination reactor generator vessels for subsequent use in preheating the feed gas. The flow through the primary reactor system is reversed on a periodic basis. The sulfur formed is condensed and the cooled gas, containing proper proportions of SO_2 and H_2S, is processed through a Claus conversion system for recovery of additional sulfur. The Claus system off-gas is incinerated and recycled to the Wellman-Lord absorbers.

The raw natural gas available at the San Juan Station contains quantities of C_5 and higher hydrocarbons which make it unacceptable for use as a reductant in the SO_2 reduction process. The heavy ends are removed in a gas treating unit and utilized as fuel in the tail gas incinerators.

Sulfate formed by the oxidation of sulfite in the absorber solution must be purged from the solution since sulfate is inert in the system and will not absorb or release SO_2. Sulfate is separated from the absorber product solution by low temperature crystallization.

The sulfate is recovered as a decahydrate, melted and sent to an evaporator for removal of the water of hydration, then separated and dried to less than 1% water for ease of storage and shipment.

Separate plant utilities are dedicated to the chemical plant for cooling towers, air compressors, steam reducing stations, water supply and recovery systems. Steam, high quality water, and electricity as stated previously are provided from the PNM generating plant. Steam condensate is returned to the generation plant from the Chemical Plant.

Substantially Trouble-Free Plant Operations

The following units have operated as designed and with little or no start-up problems.

Computer control of the scrubbers from the remote control room has worked well from the beginning. As with most computer-controlled systems, the success or failure of the control system depends on the sensor elements in the field.

Absorbers do a good job of removing SO_2 from the flue gases. With nearly design solutions and normal amounts of soda ash being added to the system, Unit Number 2 has made compliance with State and Federal regulations as demonstrated on November 29, 1978.

Evaporators and SO_2 compressors have been operated for long periods of time. The only problem has been some erosion of the SO_2 compressor impellers probably caused by a condensing condition at the inlet of the compressors. Corrections and revisions have been made to prevent this condensation. These units have been operated, since the corrections, over a wide range of flow rates with no difficulty.

Soda ash addition to the process has been relatively easy and very dependable. Soda ash is stored dry and dissolved when needed for use at each absorber or at the evaporator slurry dissolving tank.

Plant Operating Problems

As in most new plants there were various mechanical problems experienced during the initial start-up. At times these problems created operating difficulties. Most of these problems have been resolved with a few to be completely cleared. The following is a brief summary of the major problems that have occurred.

Steam and Water Availability: *Problem* — The steam and water supply to the FGD plant has frequently been interrupted or curtailed thereby greatly reducing or stopping the SO_2 chemical plant operations. This resulted from operating difficulties in power generating units.

Solution — The original design parameter was that relatively unlimited steam and water supplies would be available from the power plant. This has not been true. Studies and evaluations of the steam and water systems are underway to assure more continuous supply, or to provide standby quantities of steam and boiler quality water solely dedicated to the chemical plant.

Solution Losses: *Problem* — There have been solution losses occurring which have not been readily identifiable, especially in the winter months. This solution loss appears as though it may be coming from the bottom tray of the absorber into the scrubber sump when less than minimum design gas flow rate is being processed.

The design turndown is 50% for each absorber. The solution losses occur when the absorbers are operated below the design turndown in order to limit the amount of incoming fly ash during periods when the electrostatic precipitators malfunctioned. Additional solution losses have occurred at the fly ash filter whenever the automatic cycling fails.

Solution — (a) Improved operator attention, (b) correction of the electrostatic precipitator problems and (c) operate the absorbers at or above the minimum design gas flow rate.

Electrostatic Precipitator: *Problem* — The electrostatic precipitators have failed or malfunctioned leading to a condition which overloads the scrubber solution with fly ash. Design specifications are that the scrubber should be capable of accepting a complete precipitator failure for a period of two hours. The scrubbers have demonstrated the capability of accepting a complete precipitator failure for time periods up to 36 hours. Furthermore, the scrubbers have operated for extended periods of time (several days) when the precipitators operated at less than 60% efficiency. Solution density measurement devices have proved inadequate to warn of pending recirculation problems. At one time significant amounts of fly ash were collected in the recirculating solution to increase the density where recirculation was reduced to allow hot flue gases to bypass. The hot gases contacted the downstream heat-sensitive chevron mist eliminators that separate the scrubber and absorber. Some of the chevrons were warped and required replacement in four modules.

Solution — (a) Correction of problems in the electrostatic precipitators, (b) improved operator attention and manual sampling and (c) establishment of operating techniques to handle sudden and unexpected fly ash, such as setting a 15% by volume fly ash concentration limit for shut down.

Purge System: *Problem* — The purge system evaporation of crystallized and remelted sulfate to a dry solid has not operated sufficiently to provide evaluation. This is the result of power plant shutdowns and steam and water shortages. There has been some indication that solids being formed in the purge evaporator are too small to be separated by a screen type centrifuge. A problem has also been experienced in the method of feeding the wet centrifuge cake into the flash dryer.

Solution — (a) Improved utilities, such as steam, water and power, (b) sieve sized changes in the centrifuge, (c) improved operating techniques for improved crystal growth and (d) improvement in conveying chutes, conveyor and ducts.

Claus Catalyst Beds: *Problems* — Portion of catalyst in the SO_2 reduction area are found to have shifted past the Claus converter catalyst support screens into the bottom of the vessels. Investigation revealed that support screens had been improperly installed during plant construction with substantial gaps between sections of screen and adjacent to the vessel walls.

Solution — Correct installation of the screens in the Claus converter has been completed.

Results — The problem has not reoccurred since correction was made.

Future Plans and Chemical Sales Program

The sales of FGD process products, the main concern in the immediate future, has several effects on the operating and the administration costs of the scrubber operations. Since beginning the chemical sales contracts, PNM has received numerous contacts concerning contracts and logistics involved in selling and shipping these materials. The following is a brief discussion that attempts to answer the typical inquiries received on this program. The sales of these commodities have been executed through a broker.

The most obvious cost reduction effected by the sales program is the elimination of disposal costs. On an average basis the four units at San Juan Station will produce about 300 tons/day of SO_2. This translates into 150 tons/day of elemental sulfur or 1,340 tons/day of 60% calcium sulfate and water sludge had limestone scrubbers been used. Whereas the sulfur produced in a regenerable system is a salable product which results in a significant monetary credit to help offset the operating cost of the FGD system, use of a limestone scrubber FGD system would necessitate the disposal of some 45 truck loads of sludge per day. Gypsum disposal costs for the plant would have exceeded $1,000,000 per year for hauling, mine preparation costs excluded.

PNM plans to install a sulfuric acid plant on the FGD Units Number 3 and 4. The advantage of producing sulfuric acid instead of sulfur is the elimination of natural gas consumption. At $1.70/MCF Units Number 1 and 2 consume about $450,000 per year of natural gas. With sulfur plants on Units Number 3 and 4, San Juan Station would require $1,007,000 per year of natural gas. A sulfuric acid plant costs less in capital than a comparable capacity SO_2 reduction unit. Although the natural gas is classed as chemical plant feed stock, it is still subject to curtailment during the winter months, which is another reason to eliminate this requirement.

When the San Juan complex is complete the following materials will be produced from the scrubber systems. (On an average coal basis):

Table 6.4: Anticipated PNM Material Produced

Item	Approximate Short Tons per Day	Truck Loads per Day
Salt cake	60	2
Sulfur	60	2
Sulfuric acid	250	10

Source: Symposium – Volume 2

While it would seem inappropriate to discuss unit prices here, it can be stated that the revenues PNM will receive for these materials will reduce the overall FGD system operating costs by about 10%, or roughly $1,000,000 per year, in addition to the savings discussed earlier.

Revenues received by others for product chemicals depend on site location and the availability of transportation. The net return on chemicals is freight sensitive and distances of 100 miles can easily double the selling price if one expects to recover the freight. Easy access to rail or barge networks would allow cheaper freight, hence the radius encompassing final customers is extended somewhat if those options are available. At times these commodities may have to be sold on a freight equalized basis; that is, the seller absorbs part of the freight in order to offer a competitive price for the material. The extent of freight equalization depends upon market conditions, the proximity of other production plants, and upon what modes of transportation are available at the plant site. In any case, it is possible that freight costs can reduce the possible return on the product sales.

Since the market prices for sulfur and sulfur products have varied considerably over the years, the contract between a broker and a producer should be structured to protect both parties. One form is the standard "evergreen" contract with a floor price for the producer with an additional increment for the broker to cover his costs. Anything beyond the increment could be split on a percentage basis that is mutually agreeable to the contracting parties.

One aspect of scrubber operations that is affected through sales of products is that the quality of the end product must be considered in daily operations. For salt cake this means color, particle size, and composition. For sulfur, quality is based on color, purity, and the absence of remelt sulfur. The sulfuric acid quality will be clear (free of particulates) and have a low iron content. The Wellman-Lord scrubber system is expected to produce electrolytic grade acid because the SO_2 from the evaporators is very clean.

DRY SYSTEMS

The information in this chapter is based on *Survey of Dry SO₂ Control Systems,* February 1980, EPA-600/7-80-030, by G.M. Blythe, J.C. Dickerman and M.E. Kelly of Radian Corporation for the U.S. Environmental Protection Agency.

INTRODUCTION

The purpose of this chapter is to summarize the status of dry flue gas desulfurization (FGD) processes in the United States, for both utility and industrial application. Throughout, dry FGD will be defined as any process which involves contacting a sulfur-containing flue gas with an alkaline material and which results in a dry waste product for disposal. This includes (1) systems which use spray dryers for a contactor, with subsequent baghouse or electrostatic precipitator (ESP) collection of waste products; (2) systems which involve dry injection of alkaline material into contact with flue gas, and subsequent baghouse or ESP collection; and (3) other varied dry systems which include concepts such as addition of alkaline material to a fuel prior to combustion or contacting flue gas with a fixed bed of alkaline material.

This definition of dry systems excludes several dry adsorption or "acceptance" processes, such as the Shell/UOP copper oxide process, or the Bergbau-Forschung adsorptive char process.

Also excluded is the regenerable Rockwell Aqueous Carbonate Process (ACP) which, although it does use a spray dryer for a flue gas contactor, does not fit the limitation of being a "throwaway" system. However, the open loop, spray dryer contactor portion of the Rockwell process has been adapted for a "throwaway" system and as such has been included here.

Technical Glossary: Definitions for several terms that are used frequently throughout this report to describe the operation of dry FGD systems are defined as follows:

Stoichiometry for dry scrubbing is defined as the mols of fresh sorbent introduced to the system divided by the mols theoretically required for complete reaction with all of the SO_2 entering the system whether or not it is all removed. This is opposed to wet scrubbing where stoichiometry is generally based on mols of SO_2 removed by the system.

Sorbent utilization is defined as the percent SO_2 removal by the system divided by the stoichiometry:

$$\frac{\left[\dfrac{\text{mols } SO_2 \text{ removed}}{\text{mols } SO_2 \text{ entering system}}\right] \times 100}{\left[\dfrac{\text{mols sorbent entering system}}{\text{mols sorbent required to react with } SO_2 \text{ entering}}\right]} = \text{percent utilization}$$

If one defines the sorbent for a calcium-based system as CaO and the sorbent for a sodium-based system as Na_2O, one mol of sorbent reacts with one mol of SO_x. Consequently, the above expression reduces to:

% utilization = mols SO_2 removed/mols sorbent entering system x 100

Since the mols of sorbent do not include alkalinity from other sources such as recycled fly ash, it is possible to see apparent utilizations of greater than 100 percent. That is, the alkalinity in recycled fly ash can react to remove SO_2 so that there are more mols of SO_2 removed than mols of fresh sorbent feed.

A spray dryer is defined as any apparatus in which flue gas is contacted with a slurry or solution such that the flue gas is adiabatically humidified and the slurry or solution is evaporated to apparent dryness. For FGD applications the material dried is often a calcium-based slurry or a sodium solution which reacts with flue gas sulfur during and following the drying process. The spray dryer can use rotary, two-fluid or nozzle atomization, and the vessel can be anything from the back-mix reactor typically used in spray dryer technology to a large horizontal duct.

Dry injection is defined as the process of introducing a dry sorbent into a flue gas stream. This can take the form of pneumatically injecting sorbent into a flue gas duct, precoating or continuously feeding sorbent into a flue gas duct, precoating or continuously feeding sorbent onto a fabric filter surface, or any similar form of mechanically introducing a dry alkaline sorbent into a flue gas stream.

Coal/limestone combustion is defined as the process of burning a mixture of coal and limestone whereby the SO_2 released from the coal reacts with the limestone to form solid calcium salts that are collected with the ash. Two specific combustion processes are discussed: one involves burning a coal/limestone pellet in a stoker-fired boiler, and the other involves burning a pulverized coal/limestone mixture in a low NO_x burner.

PROCESS ASSESSMENT

Spray Dryer-Based Systems

In these systems, flue gas at air preheater outlet temperatures (generally 250° to 400°F) is contacted with a solution or slurry of alkaline material in a vessel of relatively long residence time (5 to 10 seconds). The flue gas is adiabatically humidified to within 50°F of its saturation temperature by the water evaporated from the solution or slurry. As the slurry or solution is evaporated, liquid-phase salts are precipitated and remaining solids are dried to generally less than one percent free moisture. These solids, along with fly ash, are entrained in the flue gas and carried out of the dryer to a particulate collection device. Reaction between the alkaline material and flue gas SO_2 proceeds both during and following the drying process. The mechanisms of the SO_2 removal reactions are not well understood, so it has not been determined whether SO_2 removal occurs predominantly in the liquid phase, by adsorption into the finely atomized droplets being dried, or by reaction between gas phase SO_2 and the slightly moist spray-dried solids.

Sodium carbonate solutions and lime slurries are common sorbents. A sodium carbonate solution will generally achieve a higher level of SO_2 removal than a lime slurry at similar conditions of inlet and outlet flue gas temperatures, SO_2 level, sorbent stoichiometry, etc. Lime, however, has become the sorbent of choice in many circumstances because of the cost advantage it enjoys over sodium carbonate and because the reaction products are not as water-soluble. Through the use of performance-enhancing process modifications, such as sorbent recycle and hot or warm gas bypass, lime sorbent has been demonstrated at the pilot scale to achieve high levels of removal (85 percent and greater) at sorbent utilization near 100 percent.

Using a spray dryer for a flue gas contactor involves adiabatically humidifying the flue gas to within some approach to saturation. With set conditions for inlet flue gas temperature and humidity and for a specified approach to saturation temperature, the amount of water which can be evaporated into this flue gas is set by heat balance considerations. Liquid to gas ratios are generally in the range of 0.2 to 0.3 gal/Mcf. The sorbent stoichiometry is varied by raising or lowering the concentration of a solution or weight percent solids of a slurry containing this set amount of water. While holding other parameters such as temperature constant, the obvious way to increase SO_2 removal is to increase sorbent stoichiometry.

However, as sorbent stoichiometry is increased to raise the level of SO_2 removal, two limiting factors are approached: (1) sorbent utilization decreases, raising sorbent and disposal costs on the basis of SO_2 removed; and (2) an upper limit is reached on the solubility of the sorbent in the solution, or on the weight percent of sorbent solids in a slurry.

There are at least two methods of circumventing these limitations. One method is to initiate sorbent recycle, either from solids dropped out in the spray dryer or from the particulate collection device catch. This has the advantage of increasing the sorbent utilization, plus it can increase the opportunity for utilization of any alkalinity in the fly ash.

The second method of avoiding the above limitations on SO_2 removal is to operate the spray dryer at a lower outlet temperature, that is, a closer approach to saturation. Operating the spray dryer outlet at a closer approach to saturation has the effect of both increasing the residence time of the liquid droplets and increasing the residual moisture level in the dried solids. As the approach to saturation is narrowed, SO_2 removal rates and sorbent utilization generally increase dramatically. Since the mechanisms for SO_2 removal do not appear to be well understood, it is not obvious whether it is the increase in liquid phase (droplet) residence time, or the increase in residual moisture in the solids, or both which accounts for the increased removal.

Unfortunately, the approach to saturation at the spray dryer outlet is set by either the requirement for a margin of safety to avoid condensation in downstream equipment or restrictions on stack temperatures. The spray dryer outlet can be operated at temperatures lower than these restrictions would otherwise allow if some warm or hot gas is bypassed around the spray dryer and used to reheat the dryer outlet. Warm gas (downstream of the boiler air heater) can be used at no energy penalty, but the amount of untreated gas involved in reheating begins to limit overall SO_2 removal efficiencies. Significantly less hot gas (upstream of the air heater) is required to heat, but an energy penalty associated with the decrease in heat load to the air heater comes with bypassing the plant air heater. Figure 7.1, a general flow diagram of a spray dryer-based system, illustrates these two "reheat" options.

The spray dryer design can be affected by the choice of particulate collection device. Bag collectors have an inherent advantage in that unreacted alkalinity in the collected waste on the bag surface can react with remaining SO_2 in the flue gas. Some process developers have reported SO_2 removal on bag surfaces on the order of 10 percent (1). A disadvantage of using a bag collector is that since the fabric is somewhat sensitive to wetting, a margin above saturation temperature (on the order of 25° to 35°F) must be maintained for bag protection. ESP collectors have not been demonstrated to achieve significant SO_2 removal. However, some vendors claim that the ESP is less sensitive to condensation and hence can be operated closer to saturation (less than a 25°F approach) with the associated increase in spray dryer performance.

The choice between sorbent types, use of recycle, use of warm or hot gas bypass, and types of particulate collection device tends to be rather site specific. Vendor and customer preferences, system performance requirements, and site-specific economic factors tend to dictate the system design for each individual application.

Dry Injection Process

Dry injection schemes generally involve pneumatically introducing a dry, powdery alkaline material into a flue gas stream with subsequent particulate collection. A generalized flow diagram of this process is shown in Figure 7.2. The injection point has been varied from the boiler furnace area all the way to the flue gas entrance to an ESP or bag collector. Most dry injection schemes use a sodium-based sorbent. Lime has been tested but has not been demonstrated with much success. Many dry injection programs have used nahcolite as a sorbent. Nahcolite is a naturally occurring mineral, associated with western oil shale reserves, and is about 80 percent sodium bicarbonate. Sodium bicarbonate appears to be more reactive than sodium carbonate, because it loses two mols of CO_2 and one mol of water in reaction, while sodium carbonate loses only one mol of CO_2 in reaction with SO_2. The following overall reactions illustrate this point:

$$2NaHCO_3 + SO_2 \rightarrow Na_2SO_3 + H_2O + 2CO_2$$

$$Na_2CO_3 + SO_2 \rightarrow Na_2SO_3 + CO_2$$

Since bicarbonate loses three mols for every mol of SO_2 removal, bicarbonate particles tend to have larger pore volumes and are apparently less susceptible to blinding on reaction than are sodium carbonate particles. Unfortunately, the availability of raw nahcolite in commercial quantities in the near future is questionable due to the substantial investment necessary before commercial-scale mining can begin. Since the potentially favorable economics of dry injection are based to some extent on the use of inexpensive sorbents, the use of commercially refined sodium bicarbonate is prohibitively expensive.

Figure 7.1: Typical Spray Dryer/Particulate Collection Flow Diagram

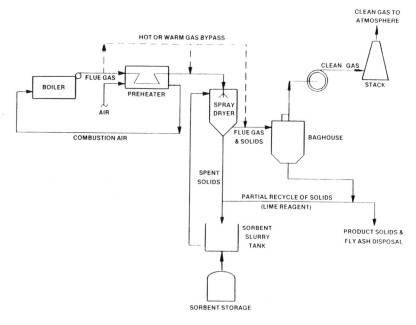

Source: EPA-600/7-80-030

Figure 7.2: Nahcolite Dry Injection Flow Diagram

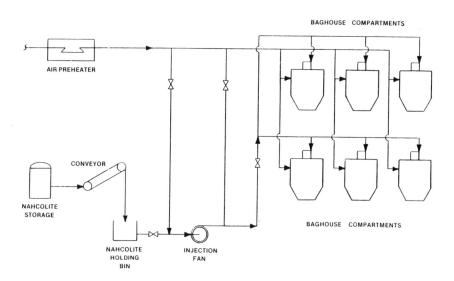

Source: EPA-600/7-80-030

Recent research has been aimed at studying the use of raw trona ore, which is currently mined in large quantities both in the Green River, Wyoming area and the Owens Lake, California area. The mineral trona contains one mol of sodium carbonate, one mol of sodium bicarbonate and two waters of hydration ($Na_2CO_3 \cdot NaHCO_3 \cdot 2H_2O$). Trona has the potential for providing a good compromise between reactivity, cost, and availability for use in dry injection schemes.

An unresolved problem with this technology is disposal of the sodium-based waste materials in an environmentally acceptable manner. Sodium waste materials are highly soluble and can result in contamination of aqueous streams. Disposal of sodium compounds is an area requiring further investigation.

Both baghouse and ESP collection devices have been tested with dry injection processes. However, the effect of the reaction between unspent sorbent on collecting bag surfaces and SO_2 remaining in the flue gas seems to overwhelmingly favor the bag collector (2). Since a major portion of the SO_2 removal reaction appears to take place on the bag surface, various methods of feeding have been tested:

(1)   Continuous. After the bag is cleaned, sorbent entrained with the flue gas is added to the bag surface continuously from injection points located upstream of the baghouse.

(2)   Batch. After the bag is cleaned, all sorbent is added to the bag as a precoat before flue gas flow is resumed.

(3)   Semibatch. This feeding method is a compromise between types (1) and (2). After bag cleaning some sorbent is added initially as a precoat and the remainder is added continuously throughout the bag cycle from an upstream injection point.

Sorbent stoichiometry, sorbent particle size, point and temperature of injection, baghouse air-to-cloth ratio, and bag cleaning frequency are also varied in dry injection programs.

Combustion of Coal/Limestone Fuel Mixture

The current research on combustion of a coal/limestone fuel mixture has taken two forms:

(1)   Combustion of a coal/limestone pellet in an industrial spreader-stoker boiler.

(2)   Combustion of a pulverized coal/limestone mixture in a low-NO_x burner system.

Preliminary results of bench-scale test work on both processes have indicated that up to 80% of the available sulfur in the fuel can be retained by the limestone. The ratio of calcium to sulfur in the coal/limestone fuel mixture is important in determining how much sulfur is retained.

A spreader-stoker boiler (20 bhp) has been used in testing the combustion and sulfur retention characteristics of the coal/limestone pellet. A Ca:S mol ratio of 7:1 has been used so far, but further work with a 3:1 Ca:S pellet is planned. The emissions generated are dependent upon burner design, coal properties and combustion operating parameters. The inherent staged combustion of the stoker-fired boiler (accomplished by supplying the total combustion air as primary air through the grates and secondary air through over-fire jets above the bed) results in lower NO_x emissions relative to conventional pulverized coal-fired boilers.

The two-staged combustion concept was employed by Babcock & Wilcox (B&W) to design an advanced low-NO_x burner system. EPA has funded test work to develop a concept of firing a coal/limestone fuel mixture in B&W low-NO_x burners to reduce SO_2 emissions. Tests conducted on a 12×10^6 Btu/hr unit by the Energy and Environmental Research Corporation (EERC), with a Utah low-sulfur coal, have demonstrated 88 percent SO_2 removal with a 3:1 Ca:S mol ratio. This high SO_2 removal has been attributed to the lower flame temperature found in the low-NO_x burner which may help maintain limestone reactivity. EERC has reported that SO_2 removal increased substantially when the reagent was passed through the pulverizer with the coal.

Further research on a larger scale for both systems is needed to determine the effects of combustion of a coal/limestone fuel mixture on boiler operation and maintenance. Collection of the increased ash loading and investigation of the properties and disposal of the waste products must also be studied.

COMPARISON OF DRY AND WET SCRUBBING

Comparisons between dry and wet scrubbing systems can be drawn in five major areas: waste disposal, reagent requirements, operation and maintenance, energy requirements, and economics. These comparisons will focus on general aspects of dry FGD systems as compared to conventional lime/limestone wet scrubbing systems.

With regard to waste disposal, dry FGD systems have an inherent advantage over wet lime/limestone systems in that they produce a dry, solid waste product that can be handled by conventional fly ash handling systems, eliminating requirements for a sludge handling system. However, the waste solids from sodium-based dry FGD systems are quite water-soluble and can lead to leachability and waste stability problems. Waste solids from lime spray drying systems and coal/limestone fuel systems should have similar environmental impacts as waste from lime/limestone wet systems, for which waste disposal technology is better defined.

In general, dry FGD systems require a higher stoichiometric ratio of sorbent to entering SO_2 to achieve the desired removal efficiency than do conventional limestone wet scrubbing systems. In addition, the reagents employed in spray drying and dry injection systems (soda ash, lime, commercial and naturally occurring sodium carbonates and bicarbonates, such as nahcolite and trona) are significantly more expensive than limestone. Consequently, limestone wet scrubbing systems will have an advantage with regard to reagent utilization and sorbent-related operating costs.

It has been claimed that dry systems will have lower maintenance requirements than comparable wet systems. Dry systems require less equipment than wet systems as the thickeners, centrifuges, vacuum filters and mixers required to handle the wet sludge waste product from wet systems are eliminated. In addition, slurry pumping requirements are much lower for spray drying and are eliminated in dry injection and combustion of coal/limestone fuel systems. This is important because wet systems have reported high maintenance requirements associated with large slurry circulation equipment. Finally, the scaling potential in wet limestone systems requires extra effort to maintain proper scrubber operation and possibly makes dry systems somewhat more flexible as far as their ability to adjust process operations to respond to variations in inlet SO_2 concentrations and flue gas flow rates.

With regard to energy requirements, dry FGD systems appear to have a significant advantage over wet systems due to savings in reheat and pumping requirements. Many wet FGD systems reheat the flue gas before it enters any downstream equipment to prevent corrosion. This reheat requirement is eliminated in most dry system configurations and results in considerable energy savings. Spray dryer systems are usually designed with a 30° to 50°F approach to the adiabatic saturation temperature of the flue gas at the outlet of the spray dryer. Energy savings from reduced pumping requirements result from the fact that wet scrubbers may require liquid to gas (L/G) pumping rates of up to 100 gallons per 1,000 acfm of gas whereas spray dryers only require an L/G rate of 0.2 to 0.3.

One of the major driving forces for development of dry SO_2 removal systems is the opportunity for reduction in both capital and operating costs. Although costs are quite site specific, the three types of dry FGD technologies considered here offer several potential possibilities for cost savings. This is due to the reduction in equipment and operation and maintenance requirements relative to conventional wet lime/limestone systems, especially in utility applications. Basin Electric evaluated the costs of the two spray drying systems they have purchased (Antelope Valley and Laramie River Stations) to be about 20 to 30 percent less costly over the 35-year life of the plants than comparable wet systems. However, it should also be noted that these economics are based on pilot-scale data and should be better determined after the operation of commercial systems has begun. Reagent costs for sodium-based spray dryer systems will be considerably higher than for lime-based systems although vendors claim the capital costs, excluding waste disposal, will be lower for sodium-based systems.

The minimal equipment and operating requirements for dry injection systems make the process economically attractive as far as capital costs are concerned, but high sorbent requirements and uncertainties in sorbent availability and cost are slowing further development of the technology on a commercial scale. Capital costs for both the pellet and low-NO_x burner coal/limestone fuel mixture systems should also be low since they will consist mainly of the equipment needed to produce the mixtures. However, since these systems have the potential for impacting the design and/or operation of the boiler, more information on the overall operability of these systems is needed before overall operating costs can be estimated.

In summary, dry systems do offer potential advantages over wet systems, especially in the areas of energy savings and costs. However, crucial issues such as waste disposal and demonstration of commercial-scale systems, which may continue to limit the overall acceptability of this technology, remain to be answered.

DRY FGD RESEARCH REVIEW

This section provides a summary of research prior to April 1979 conducted on dry FGD methods. The research activities have been divided into three areas: (1) dry injection/particulate collection systems, (2) spray drying/particulate collection systems, and (3) other research including combustion of a coal/limestone fuel mixture and fixed and fluidized bed reactors.

The research activities in each area are listed in chronological order as reported. The results of previous test work have generally been published and are available in great detail. Results from current research activities are often not readily available because test work may not be completed, conclusions may not have been finalized, etc. Since it was the intent to report as much information regarding dry FGD research as was available, the results of older, completed studies presented here are generally more complete and detailed than those of current or recent studies. It was not intended, however, to bias the content of the report toward earlier studies. Unfortunately, although the more recent studies are discussed in less detail, these are probably most indicative of the state of the art of dry FGD.

Dry Injection/Particulate Collection

Owens-Corning Fiberglas, Laboratory Tests, August 1970: *Scope* — In an effort to develop fabric filters that would withstand high temperatures, Owens-Corning Fiberglas (OCF) developed an S-glass cloth treated with a proprietary inorganic finish. Under contract with the National Air Pollution Control Administration, now EPA, a stainless steel test facility to assess the filtration and contacting characteristics of the new fabric was constructed and operated during 1970. The emphasis was on investigating various sorbents for SO_2 removal from flue gas at high temperatures.

Process Description — The facility was a single-compartment stainless steel baghouse with six bags. Two methods of sorbent feeding were tested: (1) continuous injection of the dry sorbent into the simulated flue gas, upstream of the baghouse, and (2) precoating of the bag by injecting the sorbent into SO_2-free gas to build up a dust cake on the bag before the regular gas stream (containing SO_2) was started.

The synthetic flue gas was supplied by a natural gas boiler-heat exchanger and spiked with SO_2. Hot and warm flue gas streams were controlled to give a gas temperature between 250° and 1000°F at varying flow rates and constant composition.

Bag cleaning was accomplished by conventional reverse-air shaking methods. Inlet and outlet SO_2 concentrations, pressure drop across the bags, and baghouse temperatures were continuously monitored.

Parameters Investigated — The tests were divided into four series: an additive study, a flue gas flow rate study, a precoated bag study, and a fly ash study. Parameters investigated within these series included: type of sorbent, gas stream temperature, flue gas flow rate, stoichiometric ratio, and additive feeding method.

Test Conditions — Inlet SO_2 concentration was maintained at 2,800 ppm. The test cycle was usually 60 minutes. The average flue gas composition was 3.5% H_2O, 81.0% N_2, 5.6% O_2, 1.0% Ar, and 9.0% CO_2.

Results — The percentage SO_2 removal was calculated using conventional breakthrough curves.

Slaked lime and dolomite achieved maximum SO_2 removals at 800°F. However, additive utilization was low, with a value of only 40% at 81% SO_2 removal.

Promoted slaked limes did not perform significantly better than slaked limes. Both MnO_2 and alkalized alumina gave good SO_2 removals (70% for MnO_2 and 70 to 90% for alkalized alumina depending on test conditions). Again, additive utilization was low and a regenerative process would definitely be required for these expensive sorbents.

Increasing turbulence by increasing the flue gas flow rate improved SO_2 removal and additive utilization when lime sorbents were used. This indicates that lime sorption of SO_2 may be gas-phase mass transfer limited at the high temperatures used in these tests.

In the tests using nahcolite sorbent, an increase in flow rate did not enhance SO_2 removal, but it did improve additive utilization. Overall, nahcolite utilization was greater than lime utilization, 75% versus 40% at 290 cfm.

Precoating of the bags was not an effective method of removing SO_2 when lime sorbents were used. This was attributed to the absorption of the CO_2 present in the SO_2-free flue gas used in coating the bags, thus reducing the amount of lime available to react with the SO_2.

Fly ash addition did not affect SO_2 removals.

Conclusions and Comments — Slaked lime and dolomite achieved satisfactory SO_2 removals at 800°F but with low reagent utilization. These high temperatures are not likely to be encountered in commercial dry injection FGD systems, as the hot flue gas passes through an air preheater upstream of the injection point and/or fabric filter.

Removal of SO_2 using nahcolite was about 60% with 75 to 85% additive utilization at 300°F (the inlet SO_2 concentration was 2,800 ppm). Also, the dry product formed is quite water-soluble and waste disposal problems may occur.

Information Sources — Veazie, F.M. and Kielmeyer, W.H., "Feasibility of Fabric Filter as Gas-Solid Contactor to Control Gaseous Pollutants," Report No. APTD-0595 for the National Air Pollution Control Administration, HEW Contract No. Ph-22-68-64, Owens-Corning Fiberglas Corporation, Granville, OH, August 1970.

Air Preheater at Mercer Station, March 1971: *Scope* — The Air Preheater Corporation was contracted by the National Air Pollution Control Administration (now EPA) to conduct tests to investigate the potential of the top inlet fabric filterhouse as a chemical contactor for SO_2 removal from flue gas. The study was carried out at a pilot test facility at the Mercer Station of Public Service Electric and Gas Company in Trenton, New Jersey during 1971.

Process Description — The filterhouse (baghouse) test facility of top entry design had four compartments, each containing nine Fiberglas filter bags. The total filter area was 4,300 ft². The bags were cleaned by the reverse-air deflation method. In addition, two different types of baghouse operation, cyclic and parallel, and both continuous and batch sorbent feeding methods were tested. Sorbent could be added upstream of the baghouse, just before the baghouse or precoated onto the filter bags.

Flue gas was available at temperatures from 270°F (leaving the air preheater) to 680°F (just ahead of the air preheater) and was mixed to give a variety of operating temperatures. Inlet and outlet SO_2 concentrations, as well as temperatures and pressure drops, were continuously monitored.

Parameters Investigated — The study was performed in two phases. The first phase evaluated the feasibility of the process using sodium bicarbonate, $NaHCO_3$. The second phase investigated the process variables at higher temperatures using different sorbents (nahcolite, hydrated dolomite, promoted hydrated dolomite, and hydrated lime). The process variables investigated were: stoichiometric ratio, operating temperature, filter air-cloth ratio (flue gas flow), method of sorbent feeding, and mode of baghouse operation.

Test Conditions — Flue gas flow rates of 7,500, 10,000 or 15,000 acfm were used to give corresponding air-to-cloth filter ratios of 1.75, 2.3, or 3.5., respectively. The flue gas temperatures ranged from 270° to 650°F. No SO_2 inlet concentrations were reported. The moisture level was given as 5%. The tests were usually conducted in 40-minute cycles.

Results — About half of the total 62 tests were run with sodium bicarbonate at various operating conditions. Increasing the stoichiometric ratio increased the SO_2 removal with accompanying decreased sorbent utilization. An increase in operating temperature also increased SO_2 removal.

Nahcolite tests results showed trends similar to those in sodium bicarbonate tests. Nahcolite gave a somewhat higher SO_2 removal than sodium bicarbonate at the same temperature and stoichiometric ratio. The investigators claim this may have been due to the smaller particle size of the nahcolite. Sulfur dioxide removal efficiencies were low in tests with lime at 350°F.

As to the variation with mode of baghouse operation, parallel operation resulted in higher removals for continuous additive feeding. This seems to be a result of the lower air-to-cloth ratio of parallel operation, where all the compartments are on-stream during the cycle, as compared to the higher ratio resulting when one of the compartments is always off-stream for cleaning during cyclic operation. This is in agreement with results of tests with increased air-to-cloth ratios that show a decrease in SO_2 removal and additive utilization for both $NaHCO_3$ and nahcolite at temperatures from 280° to 350°F. When lime was used, the removal efficiency and additive utilization increased with higher flue gas flow rates.

Data from the chemical analyses of fallout and shakedown samples were used to approximate the conversion distribution in the filterhouse. Although only approximations could be made, it appeared that most of the SO_2 removal, up to 80%, occurred in the filter cake for tests with sodium bicarbonate and nahcolite at 350°F. However, for lime the bulk of the SO_2 removal was in the gas suspension. These calculations were based on the assumption that no chemical reaction took place in the fallout collected from the hopper.

Conclusions and Comments — The use of a fabric filter as a chemical contactor to remove SO_2 through reaction with a dry sorbent was proven feasible. The additive utilization was fairly low at flue gas temperatures below 350°F. Higher temperatures (650°F) are required for lime to be an effective sorbent.

Stoichiometric ratio and operating temperature appear to have the strongest effect on SO_2 removal. The inlet SO_2 concentration was not specified for any of the tests. The moisture content (5%) of the flue gas was somewhat lower than most of the other similar studies on the dry injection/baghouse process.

In tests with nahcolite and sodium bicarbonate at flue gas temperatures of about 300°F, the effect of increasing flue gas flow was small. This suggests that the reaction is not gas-phase mass transfer limited at these temperatures. However, in tests with lime at 650°F, the increase in flue gas flow rate substantially increased the SO_2 removal efficiency, indicating that this reaction may be gas-phase mass transfer limited at higher temperatures.

Information Source — Liu, Han and R. Chafee, "Evaluation of Fabric Filter as a Chemical Contactor for Control of SO_2 in Flue Gas," presented at the Air Pollution Control Office Fabric Filter Symposium, Charleston, SC, March 1971.

Wheelabrator-Frye at Nucla Station, July 1974: *Scope* — In July 1974 Wheelabrator-Frye conducted tests at the Nucla Station of the Colorado Ute Electric Association. The tests were performed to demonstrate the applicability of dry nahcolite injection into a commercially available baghouse to remove SO_2 from the flue gas.

Process Description — The tests were conducted on a unit comprised of an 11 MW spreader-stoker-fired boiler burning low-sulfur coal and a Wheelabrator-Frye continuous automatic fabric filter collector. Figure 7.3 shows the schematic process flow diagram. Fifteen of sixteen tests were conducted in a batch feeding mode. The most frequently employed procedure was to inject all the nahcolite at the beginning of the test cycle by blowing it into the bags through a common inlet manifold using the injection fan. In some runs the nahcolite was injected into a single compartment, in which case the injection fan was used to overcome the added pressure drop that resulted across that compartment.

The filter bags were of silicone-graphite-treated fiber glass and were cleaned by reverse air-shaking methods.

The SO_2 inlet and outlet gas concentrations were monitored continuously.

Parameters Investigated — The primary parameters investigated for their effects on SO_2 removal and additive utilization are as follows: SO_2 inlet concentration, method of additive feeding, stoichiometric ratio, air-to-cloth ratio, and cycle time between bag cleaning.

Test Conditions — Most of the tests were conducted with 0.8% S coal, resulting in a 480 ppm inlet SO_2 concentration. When a higher-sulfur coal (1.1% S) was burned, the SO_2 inlet concentration was around 900 ppm. The flue gas flow to the baghouse was 32,000 scfm at a temperature of 285°F (44,000 acfm). The stoichiometric ratios tested ranged from 0.2 to 3.4. The nahcolite was 60% $NaHCO_3$. Flue gas composition was not given.

Results — Sulfur dioxide removal efficiencies for 90-minute cycles varied from 50 to 70%. The additive utilization was 56% at 70% SO_2 removal. Pressure drop over the fabric filter ranged from 2 to 5 inches H_2O. A change in air-to-cloth ratio did not affect SO_2 removal.

Single compartment injection resulted in a slightly lower overall removal efficiency in a continuously operating system. This is due to the fact that the lower cake resistance encountered after bag cleaning resulted in lowered SO_2 removal.

Conclusions and Comments — This series of tests conducted by Wheelabrator-Frye demonstrated the feasibility of SO_2 removal by nahcolite injection into the baghouse. The fact that 70% SO_2 removal, with reasonable additive utilization, was achieved on a system with commercial hardware not specifically designed for optimum sorbent utilization is significant. The success of the tests at Nucla Station led to more demonstration work by Wheelabrator-Frye at Basin Electric's Leland Olds Station in early 1977.

Information Source — Bechtel Corp., *Evaluation of Dry Alkalis for Removing SO_2 from Boiler Flue Gases.* EPRI Final Report FP-207, October 1976.

American Air Filter, Laboratory Tests, 1976: *Scope* — In 1972 American Air Filter, with substantial funding from Arizona Public Service, conducted bench-scale studies to investigate the use of fabric filters as chemical contactors for SO_2 removal from flue gas.

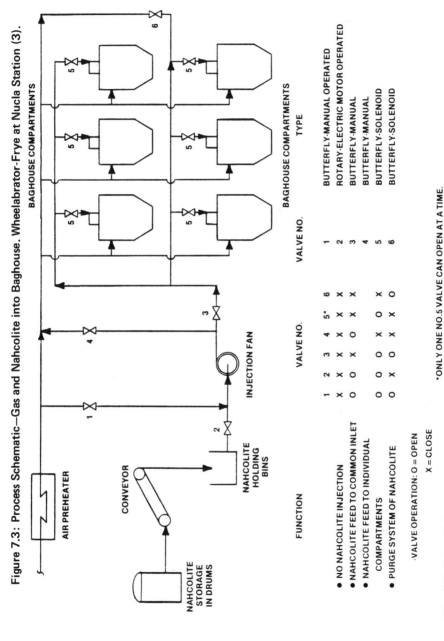

Figure 7.3: Process Schematic—Gas and Nahcolite into Baghouse. Wheelabrator-Frye at Nucla Station (3).

FUNCTION	VALVE NO.						BAGHOUSE COMPARTMENTS
	1	2	3	4	5*	6	

VALVE NO.	TYPE
1	BUTTERFLY-MANUAL OPERATED
2	ROTARY-ELECTRIC MOTOR OPERATED
3	BUTTERFLY-MANUAL
4	BUTTERFLY-MANUAL
5	BUTTERFLY-SOLENOID
6	BUTTERFLY-SOLENOID

FUNCTION	1	2	3	4	5*	6
● NO NAHCOLITE INJECTION	X	X	X	X	X	X
● NAHCOLITE FEED TO COMMON INLET	O	O	X	O	X	X
● NAHCOLITE FEED TO INDIVIDUAL COMPARTMENTS	O	O	O	X	O	X
● PURGE SYSTEM OF NAHCOLITE	O	X	O	X	X	O

·VALVE OPERATION: O = OPEN
X = CLOSE

*ONLY ONE NO.5 VALVE CAN OPEN AT A TIME.

Source: EPA-600/7-80-030

Process Description — Reactant and fly ash were introduced into synthetic flue gas flowing in a vertical tube 8.3 inches in diameter and 29.5 inches long. A filter made of commercial fiber glass bag fabric was clamped and sealed across the bottom of the tube. After investigating both batch and continuous modes of reactant/fly ash feeding and observing no difference in performance, most of the remaining tests were conducted using the easier-to-control continuous mode.

Parameters Investigated — Tests were conducted to evaluate the following parameters over the given ranges: sorbent type, flue gas velocity (2.0 to 4.4 ft/min), sorbent stoichiometric ratio (0.75 to 4.5), temperature (200° to 250°F), moisture content of flue gas (1.1 to 7.7%), and fly ash concentration (0.18 to 15.6 gr/ft^3). The stoichiometric ratio was defined as mols of Na_2CO_3 fed per mol of S in the inlet flue gas stream.

Test Conditions — The primary sorbents investigated were nahcolite and trona ($Na_2CO_3 \cdot NaHCO_3 \cdot 2H_2O$). The SO_2 concentration ranged from 450 to 600 ppm. The flue gas averaged 1.5% O_2, 14 to 15% CO_2, and 76 to 82% N_2, depending on the moisture content.

Results — Removal efficiencies of up to 90% were achieved when the moisture content was greater than 5%. Removal decreased significantly below this moisture level, but little effect was observed upon increasing the moisture content above 5%. Increasing the stoichiometric ratio increased SO_2 removal. Higher operating temperatures resulted in better additive utilization as well as greater SO_2 removal. Fly ash content had no effect on performance.

In a test conducted without a filter, essentially no SO_2 was removed. This indicated that the reaction occurred primarily on the filter cake and not in the gas-solid suspension.

Conclusions and Comments — Both nahcolite and trona were shown to be effective SO_2 removal agents. Trona was found to be less reactive than nahcolite, but the authors suggested that reactivity could be improved by such means as particle size optimization.

A moisture content of at least 5% was required for effective SO_2 removal. Increases in moisture content above 5% did not significantly increase removal.

The observation that no reaction occurred in the gas-solid suspension may be explained in part by the laminar nature of flow in the test apparatus. Because the relative gas/particle velocity is small, a stagnant film of gas surrounds the particle and mass transfer can only take place by molecular diffusion (no convective diffusion by turbulence). This requires high bulk SO_2 concentrations to overcome film resistance.

Information Source — Rivers, R.D., et al, "The Role of Fabric Collectors in Removing SO_2," presented at the 1st National Fabric Alternatives Forum, Denver, CO, July 1976.

Wheelabrator-Frye at Leland Olds, March 1977: *Scope* — Wheelabrator-Frye conducted pilot testing of a dry injection process using nahcolite for SO_2 removal with a baghouse collection device. The tests were carried out at Basin Electric's Leland Olds Station in Stanton, ND during the period of January through March 1977. Bechtel Power Corporation, acting for the Otter Tail Power Company, coordinated the overall effort. The tests were performed to demonstrate the viability of the dry injection process for use at the full-scale Coyote Station being planned by Otter Tail Power Company. Process conditions at Leland Olds were similar to those expected at the new facility.

Process Description — Leland Olds Unit 2 is a 440 MW facility with a North Dakota lignite-fired boiler. The Wheelabrator-Frye process, based on test work at the Nucla Station in 1974, involved injecting dry nahcolite and collecting the solids with a baghouse filter. The SO_2 is removed through chemical reaction with the nahcolite. Several possible feeding procedures were considered: (1) feeding all of the nahcolite at the start of the cycle in the batch mode, (2) feeding nahcolite continuously throughout the cycle, and (3) combinations and variations of the two methods. Provisions were made for SO_2 spiking of the flue gas and for high-temperature testing by injecting the nahcolite into hot flue gas upstream of the air preheater. The use of a booster fan to blow the gas that, depending on the feeding method, may or may not have contained SO_2 helped to disperse the nahcolite if it was present.

The Wheelabrator-Frye twelve-compartment baghouse contained fiber glass bags with a silicone-graphite finish that comprised a total filter area of 1,080 ft^2. The cleaning of the filters was accomplished with conventional deflation-shaker methods.

Temperatures and SO_2 inlet and outlet concentrations were continuously monitored.

Parameters Investigated — The primary parameters investigated were nahcolite feeding method, stoichiometric ratio, and SO_2 inlet concentration. The effect of pretreating the nahcolite by heating to decompose $NaHCO_3$ to more porous Na_2CO_3 was also examined.

Test Conditions — The air-to-cloth ratio in the baghouse was 3:1 with the normal flue gas flow of 3,100 acfm. The flue gas composition ranges were given as 10.5 to 15% CO_2, 77 to 83% N_2, and 5.5 to 8% O_2 on a dry basis with a moisture content of 13 to 16%. The stoichiometric ratio varied between 0.8 and 1.7. The SO_2 inlet concentration was usually between 850 and 1,000 ppm with extremes of 830 and 2,700 ppm SO_2. Baghouse temperatures ranged from 286° to 301°F.

Results — It was found that 90% SO_2 removal could be achieved at a stoichiometric ratio of 1.6 under optimum conditions. Variations in SO_2 inlet concentration appeared to have only a small effect on nahcolite performance with removal at higher concentrations being slightly better. Collection of filterable particulates was 99+ percent in all cases, and baghouse performance remained satisfactory even when large quantities of nahcolite were injected. It was also reported that feeding procedures had a significant effect on additive utilization. However, these effects were not detailed. No results of high-temperature testing were given.

Conclusions and Comments — The pilot plant test work at Leland Olds by Wheelabrator-Frye appeared to show that substantial SO_2 removal by nahcolite injection is possible under optimum conditions. These "optimum" conditions were not specified in the available literature. However, additive utilization at 90 percent SO_2 removal is only about 56 percent, even under optimum conditions. The resulting large nahcolite requirements pose a serious problem due to great uncertainties in nahcolite availability. (See Bechtel report listed below.)

Although it was noted that nahcolite feeding procedures had a significant effect on additive utilization, these effects were not discussed nor was an optimum method of feeding given.

Information Sources — Bechtel Corp., *Evaluation of Dry Alkalis for Removing SO_2 from Boiler Flue Gases,* EPRI Final Report FP-207, October 1976.

Wheelabrator-Frye, *Nahcolite Pilot Baghouse Study — Leland Olds Station, Non-Confidential Test Data,* unpublished, March 1977.

Grand Forks Energy Development Center, DOE; Bench-Scale Dry Injection/ESP or Baghouse Collection, 1975 to Present: *Scope* — DOE-Grand Forks dry FGD work began in 1975 during bench-scale combustor/ESP testing of North Dakota Lignite when sulfur retention was observed in fly ash alkalinity. The current dry FGD program was begun around the first of 1978, with the testing of nahcolite injection and ESP collection. The scope of the project was broadened to include Green River area trona as a sorbent. Later, when it was apparent that high SO_2 removal efficiencies were not demonstrable with this configuration, the ESP collector was replaced by a bag collector and testing was resumed.

Process Description — The Grand Forks bench-scale coal combustor produces a nominal 200 scfm flue gas flow rate. Ductwork downstream of the combustor is equipped with water cooling so that the time/temperature profile of the flue gas can be varied. Injection points are also variable.

Parameters Investigated — The test program with a bag collector has so far included over 30 tests, where injection and bag temperatures, sequencing of sorbent addition and bag cleaning, bag materials, sorbent (nahcolite and trona), and bag air-to-cloth ratios were varied.

Results — The dry injection/ESP system was not effective for SO_2 removal. The dry injection/bag collector was much more suitable. SO_2 removal efficiencies on the order of 90 percent at 50 to 60 percent utilization were demonstrated at an air-to-cloth ratio of 3 ft/min. At higher air-to-cloth ratios (up to 6 ft/min), lower sorbent utilization resulted, as low as 10 to 30 percent. Plans were to operate the dry injection/baghouse collection system another six months at various selected "optimum" conditions with detailed results available when the final report was published.

Grand Forks has also identified upcoming dry FGD related work. After the startup of the Coyote plant, in spring 1981, Grand Forks will sample the FGD system for particulate and SO_2 removal efficiencies.

Another upcoming program has actually begun, but very little progress has been made. This program involves column work and beaker tests with samples of spent sorbent and fly ash products from the Leland Olds dry FGD pilot operations. Initial studies involve leaching and solubility studies and toxicant extraction of the untreated wastes. Future studies may include similar testing of chemically fixed wastes.

Information Sources — Blythe, Gary, telephone conversation with Stanley J. Selle, DOE, June 7, 1979.

Blythe, Gary, telephone conversation with Harvey Ness, DOE, July 3, 1979.

KVB Incorporated, Bench-Scale Tests, Late 1977 to Present: *Scope* — This bench-scale study was funded by the Electric Power Research Institute, Inc. (EPRI) for the purpose of obtaining basic process data for a dry injection/baghouse collection FGD system. The test work was performed on KVB's bench-scale coal-fired combustor in Tustin, California. Six sodium-based sorbents were tested over a wide range of operating conditions.

Process Description — The source of flue gas for the test work was the coal-fired KVB bench-scale combustor which produced a flue gas flow of approximately 725 scfm. A baghouse was installed downstream of the combustor for collecting fly ash and spent sorbent. A heat exchanger was installed in the duct between the combustor and baghouse in order to control the flue gas temperature at the baghouse inlet. The duct was designed for sorbent injection at several points between the boiler and baghouse, so that sorbent injection temperature and residence time in the duct could be varied.

Parameters Investigated — The sodium-based sorbents investigated in this extensive test program were: commercial bicarbonate, nahcolite, Green River (Wyoming) trona, Owens Lake (California) trona, soda ash solution (sprayed into duct for evaporation), and predecomposed nahcolite (heated to release CO_2 and water).

The following data summarizes the parameters and ranges investigated in this test program: stoichiometric ratio, 0 to 4:1; injection temperature, 550° to 800°F upstream of heat exchanger, 230° to 320°F at baghouse inlet; particle size, 35 to 400 mesh; inlet SO_2 concentration, 350 to 750 ppm; baghouse air-to-cloth ratio, 1 to 4 ft^3-min/ft^2; sorbent feed method, continuous, batch, semibatch (bag precoat).

Results — The final report of these results was due to be completed in late 1979. Since the reporting of the results was in draft form, EPRI was reluctant to discuss results in detail. However, the results were described as being generally very encouraging as to the future of dry injection as an FGD system.

Information Sources — Blythe, Gary, telephone conversation with Navin Shah, EPRI Project Director, July 17, 1979.

Shah, N.D., et al, "Application of Dry Sorbent Injection for SO_2 and Particulate Removal," paper presented at the EPA Symposium on Flue Gas Desulfurization, Hollywood, Florida, November 11, 1977.

Carborundum, Dry Injection/Baghouse Collection Pilot on Stoker-Fired Boiler, 1976 to Present: *Scope* — Carborundum (now a division of Kennecott Development Company) has tested a dry injection/baghouse collection system using 100 acfm slipstream from a small stoker-fired boiler near their Knoxville, Tennessee offices. A larger, 1,000 acfm baghouse has been installed, and dry injection testing is continuing at this scale. The 100 acfm unit was also equipped for spray dryer/baghouse testing, and plans were to equip the 1,000 acfm unit with a spray dryer as well.

Process Description — Dry injection/baghouse collection studies have been completed with sodium bicarbonate, nahcolite, and ammonia sorbents. The smaller (100 acfm) baghouse is an industrial size unit with small bags. The larger (1,000 acfm) baghouse has 4 to 6 full-size (11½-inch by 32-foot) bags. A small Bowen spray dryer employing rotary atomization was used for spray dryer testing on the 100 acfm scale.

Information Sources — Blythe, Gary, telephone conversation with Don Boyd, Carborundum, May 16, 1979.

Blythe, Gary, telephone conversation with Hank Majdeski, Carborundum, July 6, 1979.

Majdeski, H.M., personal communication with Gary Blythe, July 17, 1979.

Spray Drying/Particulate Collection

Atomics International (Rockwell) at Mohave Station, 1972: *Scope* — Under an agreement with Southern California Edison (SCE), Atomics International conducted pilot plant test work of their (AI's) aqueous carbonate process (ACP). The test work was conducted at SCE's Mohave Generating Station in the first half of 1972. Funding was provided by the WEST (Western Energy Supply and Transmission) Association of utilities.

The major objective of the test program was to determine optimum operating conditions for the spray dryer scrubber using aqueous sodium carbonate solutions for removal of SO_2 from flue gases. The operating results were also to be used in designing a full-scale open-loop ACP process.

Process Description — In the open-loop (no sorbent regeneration) ACP process, the SO_2 is removed by contacting the hot flue gas with an atomized solution of sodium carbonate (Na_2CO_3). The dry product formed is a combination of sodium sulfite, sodium sulfate, and unreacted reagent.

The Mohave test installation included a 23-year-old modified spray dryer, 5 feet in diameter, a sodium carbonate feed system, and a multicyclone solids collection system. A slipstream of flue gas, obtained downstream of the station ESPs, was fed to the spray dryer. Provisions were made for SO_2 spiking and for testing at various temperatures. SO_2 concentrations were continuously monitored.

Parameters Investigated — The test work was designed to investigate the following parameters with respect to their effect on SO_2 removal and system performance: sorbent concentration, stoichiometry, inlet flue gas temperature, flue gas flow rate, concentration of SO_2 in inlet gas, and recycle of products to feed. One series of tests was conducted with the objective of demonstrating effective scrubber performance under optimum design conditions.

Test Conditions — The flue gas flow rate varied from 1,150 to 1,375 scfm at an inlet temperature ranging from 250° to 340°F. The inlet SO_2 concentration was usually about 400 ppm, except for testing to determine the effects of higher SO_2 concentrations in which case concentrations of up to 1,465 ppm SO_2 were used. Temperature drop across the spray dryer was usually 120°–160°F, but when operation in the "wet" mode (>25% moisture in the product) was required to meet particulate removal requirements in the downstream cyclone, the temperature drop was 180°F (although the exit gas was always maintained at least 25°F above its dew point). Inlet gas moisture content was about 10% by volume. A liquid-to-gas-ratio of 0.3 gal/1,000 scf was employed in most tests. The weight percent of Na_2CO_3 in the feed solution varied between 4 and 5.5, except in the tests investigating the effects of sorbent concentrations where it varied between 20 and 32 percent.

Results — The results of sorbent concentration testing with 400 ppm SO_2 flue gas indicated that equivalent SO_2 removal could be achieved more efficiently with low weight percent solutions. The most efficient SO_2 removal from 400 ppm SO_2 flue gas at 300°F was obtained with a 4 percent weight Na_2CO_3 solution. This solution concentration gave a stoichiometric ratio (Na_2CO_3 to SO_2) of about 1. A lower temperature flue gas would require a more concentrated solution to avoid saturating the exit gas. Higher inlet SO_2 concentrations would also require a more concentrated feed solution for the same degree of removal.

A larger temperature drop over the spray dryer resulted in increased removal efficiency. At a fixed feed rate, lowering the inlet gas temperature also increased removal efficiency. In both cases, the increased wet contact time enhanced removal.

At a constant feed rate of a 5.4 percent weight Na_2CO_3 solution, increasing the gas flow rate from 1,150 to 1,300 scfm resulted in a linear increase in removal efficiency.

Sorbent utilization was found to increase as the inlet SO_2 concentration increased. In some cases, utilization exceeded 100 percent. The authors suggest that this may be the result of sodium bisulfite ($NaHSO_3$) formation.

Recycled sulfate product salts did not inhibit removal of SO_2. Some results indicated that sorbent utilization increased as the recycle amount increased.

Reasonably consistent results showed 2 to 3 percent NO_x removal under most conditions, although the accuracy of the analytical methods available was suspect.

In the series of tests designed to demonstrate system performance under optimum conditions, 90 percent SO_2 removal efficiency was regularly achieved using a 5.3 percent weight Na_2CO_3 solution at an L/G of 0.3 gal/1,000 scf and a stoichiometric ratio of 1.15. The total system pressure drop was between 9 and 11 inches H_2O. During these tests it was established that higher atomizer wheel speeds increased removal efficiencies. No mechanical or maintenance problems were encountered during the testing. Additive utilization approached 90 percent.

Conclusions and Comments — The spray dryer was shown to be an efficient method of contacting a dilute Na_2CO_3 solution with hot flue gas for SO_2 removal. For a fixed feed rate, maximum removal efficiency occurred at lower inlet gas temperatures. Both additive utilization and removal efficiency were near 90 percent.

The Mohave facility was equipped with a multicyclone collector which sometimes required the spray dryer to operate in the "wet" mode (>25 percent H_2O in the product) to achieve the required particulate removal requirements.

Information Sources — Gehri, D.C. and Gylfe, J.D. *Pilot Test of Atomics International Aqueous Carbonate Process at Mohave Generating Station.* Final Report AI-72-51, Atomics International Division/Rockwell International, Canoga Park, CA, September 1972.

Koyo Iron Works, Pilot Unit, 1973: *Scope* — Performance characteristics of a spray drying process with NaOH and Na_2CO_3 aerosols were studied by Koyo Iron Works in Japan. The tests were carried out on a pilot-scale facility during 1973.

Process Description — Flue gas from a 3,000 lb steam/hr oil-fired boiler was passed through a spray dryer for removal of SO_2 by reaction with NaOH or Na_2CO_3 droplets. A two-fluid atomizer was used to disperse the liquid from the top of the dryer into the flue gas. The gas, containing dry product solids, then passed through a multicyclone and an electrostatic precipitator to remove the solids. The dry powder collected was a mixture of Na_2SO_3, Na_2CO_3 and Na_2SO_4.

Parameters Investigated — The parameters investigated in the tests included: inlet sulfur dioxide concentration, temperature drop across the dryer, and stoichiometric ratio (ratio of NaOH fed to theoretical amount needed to react with all incoming SO_2).

Test Conditions — The flue gas flow rate was about 850 scfm in all cases. The temperature of the gas varied from 320° to 428°F. The fuel oil for the boiler contained 1 to 3 percent sulfur, resulting in inlet SO_2 concentrations of from 600 to 1,700 ppm.

Results — Sulfur dioxide removal was reported to increase as the stoichiometric ratio or inlet SO_2 concentration increased. An SO_2 removal of 89 percent was reported for flue gas containing 1,300 ppm SO_2. The stoichiometric ratio required for this removal was 1.2. At the same inlet SO_2 level, but with a stoichiometric ratio of 1.0, removal decreased to 84 percent. The outlet flue gas was 278°F in both cases, resulting in an 80°F temperature drop over the dryer.

At constant inlet gas temperature, it was found that sulfur dioxide removal increased as the temperature drop across the dryer increased. This reflects the increase in stoichiometric ratio that occurs when feed rate is stepped up to obtain a lower outlet gas temperature (greater ΔT). It was also observed that for a constant temperature drop over the dryer, a lower inlet gas temperature resulted in greater removal efficiency. This is attributed to the lower water evaporation rate at the lower temperature, thus allowing more reaction time in the liquid droplet where the bulk of the sorption reaction occurs.

The product composition was given as 70 to 80 percent Na_2SO_3, 10 to 20 percent Na_2CO_3 and 5 to 15 percent Na_2SO_4. It was suggested that the collection particles could be used as chemicals in a kraft or sulfite pulping process.

The authors reported 99 percent collection efficiency for the ESP downstream from the multicyclone.

Conclusions and Comments — The sulfur dioxide removals and utilizations reported show spray drying to be an efficient method of sorbent-gas contacting. The report states that either sodium hydroxide or sodium carbonate may be used as the sorbent, but it is not absolutely clear which was used for the data reported.

Information Sources — Isahaya, E.F., "A New FGD Process by a Spray Drying Method Using NaOH Aerosols as the Absorbing Chemical," *Staub Reinhaltung der Luft* (in English), 33(4), April 1973.

Rockwell/Wheelabrator-Frye at Leland Olds Station, 1977–1978: *Scope* — The primary objective of the Leland Olds pilot plant was to demonstrate the applicability of Rockwell's open-loop Aqueous Carbonate Process (ACP) on an existing lignite-fired boiler. The tests were conducted on Unit 2 at Basin Electric's Leland Olds Station in Stanton, ND during 1977 and 1978. The results of the test work were used in designing a full-scale FGD system for the proposed Coyote Station of the consortium headed by Otter Tail Power Company and to be located in Beulah, ND.

Sodium carbonate and trona were used as the scrubbing media for demonstration tests. Another series of tests was conducted to explore the feasibility of less costly sorbents such as lime or limestone. Also, a week-long endurance test (using trona) was conducted to demonstrate long-term reliability.

Process Description — The open-loop ACP includes a spray dryer for contacting the atomized sorbent solution with hot SO_2-laden flue gas and a downstream solids collection device to remove fly ash and entrained product solids. At Leland Olds the collection device was a Wheelabrator-Frye two-compartment fabric filter with a total filter area of 1,098 ft². The bags were cleaned by reverse air-shaking methods. The spray dryer was a 7-ft diameter Bowen model equipped with a rotary atomizer. The clean flue gas vented to the stack from the filter did not require reheat, nor did the gas exiting the spray dryer as it was usually maintained about 40°F above the dew point.

Tests were conducted on a 1 MW (6,000 acfm) sidestream of flue gas from a 440 MW cyclone boiler burning North Dakota lignite.

Scrubber feed was prepared by dissolving or slurrying sorbent in a makeup tank from which it was pumped to the scrubber. In the open-loop process the solids are discarded or sometimes used for recycle.

Parameters Investigated — Test subprograms included sodium carbonate tests, trona tests, hydrated lime and limestone tests, and trona endurance tests. The parameters investigated in each series are listed below.

Sodium carbonate, trona, and hydrated lime and limestone tests included: sorbent concentration, inlet SO_2 concentration, inlet temperature of flue gas, temperature drop over the dryer, and flue gas flow rate (except for lime/limestone tests).

Endurance tests with trona included: inlet SO_2 concentration, and inlet temperature of flue gas.

Test Conditions — Table 7.1 lists test conditions for the various series.

Table 7.1: Open-Loop ACP Pilot-Plant Test Conditions — Rockwell at Leland Olds (4)

Series	Sorbent Concentration (wt %)	Inlet SO_2 Concentration (ppm)	Flue Gas Flow (scfm)	Dryer Inlet (°F)	Dryer ΔT* (°F)
Sodium carbonate	4-22	900-2,300	1,500, 2,500, 3,100	260-350	90-170
Trona	7-12	800-1,400	1,500-2,500	250-350	100-170
Hydrated lime and limestone	3, 5, 10	600-2,000	2,500	310-350	120-170
Trona endurance tests	8-18	700-1,400	2,400-2,500	293-329**	100-130

*The difference in flue gas temperature between the spray dryer inlet and outlet.
**With boiler operation.

Source: EPA-600/7-80-030

Results — Although specific findings of this study are considered proprietary information, overall the open-loop ACP was demonstrated to be an effective SO_2 removal method. Trona and sodium carbonate were equally effective in removing SO_2 with up to 85 percent of the SO_2 being removed in the spray dryer and an additional 10 to 20 percent removal occurring on the fabric filter cake. Sorbent utilizations ranged from 80 to 100 percent. In all cases the product was a dry, free-flowing powder.

Hydrated lime was not as effective an absorbent as the sodium alkalis on a once-through basis. Removals were in the 45 to 60 percent range and were accompanied by low additive utilization. The single test in which limestone was used gave poor SO_2 removal despite a high stoichiometric ratio.

The parameters that seemed to influence the reaction of SO_2 with material collected on the filters were filter temperature, pressure drop, gas flow rate, and sorbent utilization in the spray dryer. An increased temperature drop across the spray dryer, caused by increasing liquid rate, increased removal.

The endurance test with trona was successful. No degradation of SO_2 removal or serious equipment malfunction was observed.

Conclusions and Comments — The open-loop ACP was shown to be a viable SO_2 removal process. The additional removal occurring in the filter cake did not affect baghouse performance.

As a result of this and subsequent test work at the Leland Olds pilot-plant facility, Rockwell/Wheelabrator-Frye was awarded a contract to design and build a full-scale open-loop ACP system at the proposed Coyote Station. The system was to be completed in spring, 1981.

Information Sources — Dustin, D.F., Report of Coyote Pilot Plant Test Program, Test Report, Rockwell International, Atomics International Division Canoga Park, CA, November 1977.

Joy/Niro at Hoot Lake Station, 1977-1978: *Scope* — In late 1977 Joy Manufacturing and its Western Precipitation Division entered into an exclusive agreement with Niro Atomizer to design and market dry FGD systems. Niro had begun test work using a spray dryer for FGD applications in 1974, on the 1,000 to 3,000 acfm scale, at their Copenhagen facility. The tests investigated various alkaline sorbents such as lime, limestone, and sodium carbonate. During the six-month period of November 1977 to April 1978 Joy/Niro designed, constructed, and operated a pilot plant test facility at Otter Tail Power Company's Hoot Lake Station in Fergus Falls, Minnesota. The purpose of this pilot unit was to demonstrate the viability of the Joy/Niro process and to provide the design basis for a bid on a commercial unit to be constructed at Basin Electric's Antelope Valley Station. Additional test work was conducted during the period of September to December 1978 to acquire data for preparation of a bid for a dry FGD system for the Basin Electric Laramie River Station.

Process Description — Flue gas from a 53 MW CE boiler was supplied to the spray dryer at an average rate of 20,000 acfm. The gas temperature was 310°F. The Joy/Niro dry FGD process consisted of a spray dryer absorber equipped with a Niro atomizer and a Joy baghouse that was used for collection of solids from the existing flue gas.

The process flow diagram for the Hoot Lake test facility is shown in Figure 7.4. Hot flue gas entered through a roof gas disperser and contacted the atomized slurry. Some of the dry product was collected at the conical bottom of the spray dryer, the rest being swept out with the exiting flue gas and collected, along with the fly ash present, in the downstream baghouse or electrostatic precipitator (ESP). The quantity of slurry was controlled using variable-speed Moyno pumps to maintain the flue gas outlet temperature at 30° to 40°F above saturation.

The spray dryer used was 11.25 ft in diameter and was equipped with a Niro atomizer coupled to a variable-speed motor. The four-compartment baghouse had a 9,000 acfm capacity with bag cleaning by both reverse-air and shaking methods. Both acrylic and fiber glass bags were tested with no difference in performance observed. The ESP employed in some tests was a single field unit capable of handling up to 5,000 acfm of flue gas.

Parameters Investigated — The testing was conducted in two phases. First, a series of short (10-to 12-hour) tests covering the expected operating ranges was performed. The second phase consisted of a 100-hour endurance test conducted to determine the ability of the system to meet maximum SO_2 removal requirements during continuous operation.

The first phase of tests was conducted using various types of lime and lime-slaking methods. Optimum hydration/lime-slaking methods were determined for use in the 100-hour endurance test. The other parameters investigated during the first phase were spray dryer inlet temperature, flue gas temperature drop across the spray dryer, SO_2 inlet concentration, flue gas flow rate, sorbent concentration, atomizer speed and wheel configuration.

Additional short tests were conducted to investigate the Joy/Niro solids recycle concept. Figure 7.5 depicts the recirculation flow scheme.

Figure 7.4: Hoot Lake Pilot Plant Flow Sheet (5)

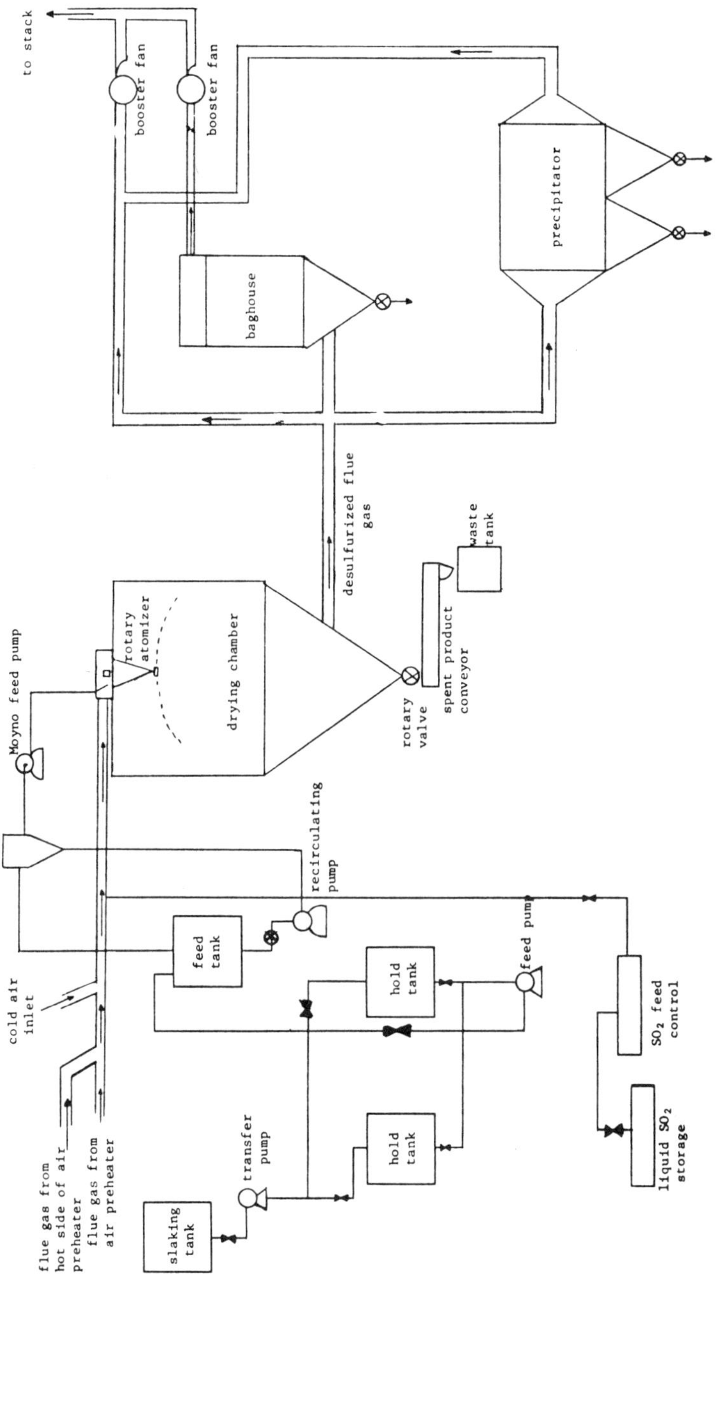

Source: EPA-600/7-80-030

Figure 7.5: Flowsheet of Pilot Plant Operation with Partial Solids Recycle (5)

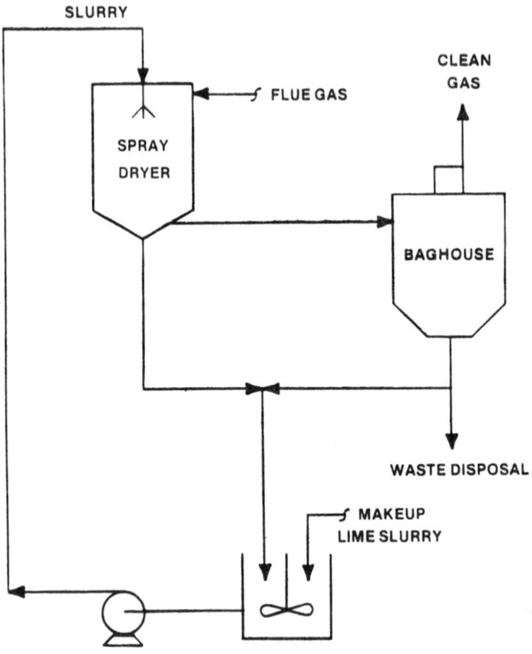

Source: EPA-600/7-80-030

Tests using soda ash as the sorbent were also conducted. As a final verification of process reliability, two types of upset condition testing were performed to check (1) the control system and (2) the effects of flow spray dryer exit gas temperature on baghouse performance.

Test Conditions — The primary variables used in the first phase of test work are as follows: SO_2 inlet concentration, 800, 1,200, and 1,600 ppmv; lime slurry concentration, 5, 10, 15, 20, and highest possible weight percent; temperature difference across the spray dryer, 70°, 105°, 140°, and 175°F. These test conditions were specifically related to the proposed Antelope Valley dry scrubbing system. Additional test work was conducted over a wide range of conditions. SO_2 inlet concentration varied from 200 to 4,000 ppm and the stoichiometric ratio was varied from 0.9 to 3.0.

Results — Specific correlations between SO_2 removal and stoichiometric ratio at a given SO_2 level were not available. The results presented in the literature covered wide ranges of SO_2 levels but gave only general removal figures. For a "once-through" system, the SO_2 removal was reported to be 90 percent at an inlet SO_2 concentration from 200 to 2,000 ppm with a stoichiometric ratio of 2.0 to 3.0, depending on the gas temperature drop across the dryer. Removal versus stoichiometric ratio curves were presented for "typical" cases, but they were not accompanied by SO_2 concentration levels.

Because of the limited data available, only general trends can be presented:

(1) Additive utilization was significantly improved in tests where partial recycle of solids was employed. Ninety percent removal was obtained with stoichiometric ratios between 1.3 and 1.7.

(2) Soda ash was found to be much more reactive than lime, with stoichiometric ratios of 1.0 to 1.2 resulting in 90 percent SO_2 removal at inlet SO_2 concentrations ranging from 800 to 3,000 ppm.

(3) Increasing the temperature drop over the dryer at a constant stoichiometric ratio was found to increase the SO_2 removal. If the outlet temperature of the gas is taken to be 30°F above the adiabatic saturation temperature of 130°F, a temperature drop of 155°F corresponds to a dryer inlet flue gas temperature of 315°F, very near normal operating flue gas temperature at both the Hoot Lake Station and the proposed Antelope Valley Station.

The control system responded well to upset conditions, such as a large temperature drop in the inlet gas to the dryer or operation at temperatures low enough to result in a wet product. Baghouse performance was not affected during periods of "wet" operation as the exiting gas remained unsaturated.

Particulate removal exceeded 99+ percent in all cases with both baghouse and ESP collection.

During the mid-September to mid-December 1978 test period, coal from the source to be used at the proposed Laramie River Station was shipped to the Hoot Lake Station for use in several days of test runs. The Laramie River ash was reported to be quite cementitious, but no operating problems were reported. Some tests were conducted adding water treatment sludge to the atomizer feed. This was found to be a suitable method of handling for the sludge.

In limited test work with commercially available ground limestone, the best SO_2 removal obtained was 50 to 60 percent.

Conclusions and Comments — The Joy/Niro dry FGD system performed well during pilot plant parametric and endurance testing. No major problems were encountered with the equipment or process chemistry.

Though no specific removal data were given, it appears that 85 percent SO_2 removal was achieved with near theoretical amounts of lime when partial recycle of solids was employed. The exact removal is, of course, dependent on the inlet SO_2 concentration, among other factors.

The literature also states that the "initial" stoichiometric ratios were considered conservative due to large air leakages. It is not clear at what point during the test period this problem was corrected.

It was also concluded that the method of lime slaking is an important factor in determining the effectiveness of lime as a sorbent.

Sulfur dioxide removal is limited by, among other factors, the amount of slurry that can be supplied to the dryer. To avoid moisture condensation in downstream particulate removal equipment or stack, the slurry delivery rate must be maintained below levels that would cool the flue gas below its adiabatic saturation temperature.

After the Hoot Lake Station test work was completed, Joy/Niro bid on and was awarded a contract for a 440 MW commercial dry FGD system to be built at Basin Electric's proposed Antelope Valley Station. The system to be constructed would be a large-scale version of the pilot plant facility described above with the only major change being the method of gas dispersion into the spray dryer.

Information Sources — Blythe, Gary, meeting notes, meeting at Joy Manufacturing, June 14, 1979.

Davis, R.A., et al, "Dry SO_2 Scrubbing at Antelope Valley Station," presented at the American Power Conference, April 25, 1979.

Felsvang, Karsten, "Results of Pilot Plant Operations for SO_2 Absorption," presented at the Joy Western Precipitation Division Seminar, Durango, CO, May 21, 1979.

Kaplan, Steven, "The Niro-Joy Spray Absorber Development Program — Pilot Plant Description and Test Results," presented during Joy/Niro-sponsored tour for executives of U.S. power industry, Copenhagen, September 23–30, 1978.

Babcock and Wilcox, Pilot-Scale Spray Dryer/ESP at Velva, 1978: *Scope* — B&W was one of four vendors which piloted a spray dryer-based dry FGD system at Basin electric. B&W began testing at Basin Electric's William J. Neal Station, near Velva, N.D., using technology developed by Hitachi of Japan. This pilot plant eventually employed two-fluid nozzles for atomization and a horizontal reactor chamber, whereas the Hitachi process employed nozzle atomization in a vertical reactor. Gas flow out of the Hitachi reactor was through a duct in the bottom of the vessel, which tended to plug during testing. This led B&W to abandon the Hitachi configuration and instead adapt the steam-atomized oil burner ("Y-jet" nozzle) technology in a horizontal reactor. Figure 7.6 illustrates the Y-jet nozzle configuration.

Process Description — The reactor at Velva was 30.5 ft long by 5.75 ft square, with one atomizing nozzle and three hoppers located in the floor to handle solids drop out. The gas capacity of the unit was 8,000 acfm. While the Hitachi reactor was tested with both a 16-bag Mikropul pulse-jet baghouse and an ESP collector, the Y-jet configuration was tested only with an ESP collector. Reagents tested included soda ash, pebble lime, hydrated lime, ammonia addition with lime, and precipitated limestone. A ball mill slaker was used for lime slaking. Reactor inlet and outlet gas temperatures, inlet SO_2 concentration, gas flow, and sorbent stoichiometry were also varied during the test program.

The Y-jet configuration pilot unit began operation in June 1978. Future studies include full parametric studies on a 1,500 acfm pilot plant at the company's research facilities in Ohio and a 120,000 acfm demonstration unit treating flue gas from a 500 MW unit at a large western utility.

Results — Because the results of this privately funded research effort are considered proprietary, no specific performance results are available. However, it was found that due to reagent costs, lime was the preferred reagent. Also, B&W found that they could operate more comfortably near adiabatic saturation at the reactor outlet when using an ESP rather than a bag collector for downstream particulate collection. The increase in reagent utilization due to closer approach to saturation was found to be greater than that resulting from SO_2 removal on a bag collector surface. Apparent utilizations approaching or even exceeding 100% were found to be possible during lime tests when the reactor outlet temperature closely approached saturation. Utilizations greater than 100% may be due to the alkaline species in the recycled fly ash reacting with the flue gas SO_2.

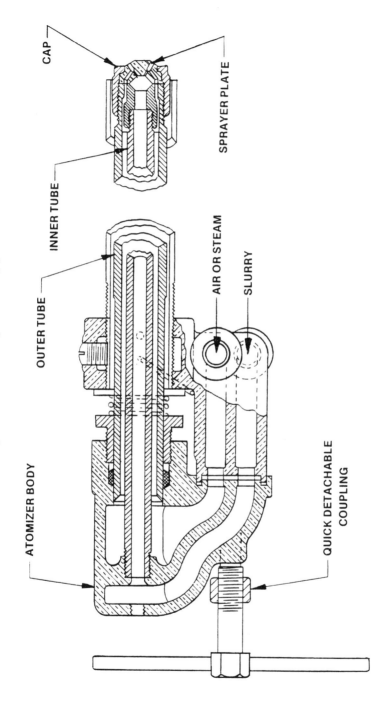

Figure 7.6: Y-Jet Slurry Atomizer (6)

CAP

SPRAYER PLATE

INNER TUBE

OUTER TUBE

AIR OR STEAM

SLURRY

ATOMIZER BODY

QUICK DETACHABLE
COUPLING

Source: EPA-600/7-80-030

In test work using only recycled fly ash as a reactant, it was found that fly ash reactivity was increased by recycling through the ball mill slaker rather than by recycle as a slurry. Presumably, this increase was at least partially caused by the reduction in particle size resulting from treatment of the fly ash in the ball mill.

Information Sources — Blythe, Gary, meeting notes, meeting at Babcock and Wilcox, June 28, 1979.

Janssen, Kent E. and Eriksen, Robert L., "Basin Electric's Involvement With Dry Flue Gas Desulfurization," paper presented at the Fifth EPA Symposium for Flue Gas Desulfurization, Las Vegas, Nevada, March 5–8, 1979.

Slack, A.V., "A.V. Slack FGD Report No. 62," January 1979, pp. 15–30.

Carborundum/DeLaval Spray Dryer Pilot Plant at Leland Olds, 1978: *Scope* — As one of the four process vendors invited to pilot a spray dryer-based FGD system for bidding on the Coyote station, Carborundum participated in pilot testing with a 15,000 acfm unit at Basin Electric's Leland Olds Station. The pilot unit employed a DeLaval spray dryer as a flue gas contactor. A Carborundum baghouse was used as a particulate collection device.

Process Description — The pilot unit was a spray dryer baghouse system. Lime, soda ash, and ammonia were tested as sorbents.

Results — The lime tests by Carborundum at Leland Olds are significant because a wide range of spray dryer outlet temperatures were tested. Outlet temperatures ranged from around 122°F (a 4°F approach to saturation) to over 200°F (90°F approach to saturation). SO_2 removal varied from 55 to 91 percent (stoichiometric ratio of about 1, inlet SO_2 of 700 ppm). Higher removals occurred at the lower dryer outlet temperatures.

Information Sources — Blythe, Gary, telephone conversation with Don Boyd, Carborundum, May 16, 1979.

Blythe, Gary, telephone conversation with Hank Majdeski, Carborundum, July 6, 1979.

Majdeski, H.M., personal communication with Gary Blythe, July 17, 1979.

Bechtel Power Company, Conceptual Spray Dryer/Baghouse and Dry Injection/Baghouse Study, August 1978 to Present: *Scope* — Bechtel is being funded by EPRI in an economic study comparing various spray dryer-based and dry injection-based dry FGD systems with a wet limestone scrubber. Initial work was begun in August 1978; a final report was expected by the end of 1979. Most of the economic cases were for a hypothetical 500 MW, 85 percent SO_2 removal unit, although two 200 MW cases were calculated. Four coals were considered, ranging from 0.5 to 4.0 percent sulfur. Trona and nahcolite sorbents are being considered for the dry injection cases; lime and soda ash sorbents are being considered for the spray dryer cases.

The study compares capital and operating costs of the dry systems to a wet limestone scrubber. Also, the study considers the engineering aspects of applying dry FGD, such as integration of the unit with the utility boiler.

Information Source — Blythe, Gary, telephone conversation with Navin Shah, EPRI, July 20, 1979.

Other Research

This area includes test work with bench- or pilot-scale bed reactors and combustion of a coal/limestone fuel.

FMC Corporation, Bench-Scale Fixed Bed, June 1970: *Scope* — During the period of June 1969 to July 1970, FMC Corporation undertook a series of bench-scale screening tests designed to evaluate potential SO_2 sorbent. The study was funded by the National Air Pollution Control Administration (now EPA).

Process Description — The experimental apparatus was designed to evaluate the sorption characteristics of a given material in less than an hour. The sample holder was a vertically mounted stainless steel tube 4 inches long and ½ inch in diameter. The sorbent was supported on a sintered steel plate near the bottom of the tube. Synthetic flue gas, prepared by blending bottled gases, was passed through the fixed bed of sorbent.

Parameters Investigated — Various types of sorbents were tested: sodium carbonates, impregnated silica gels, impregnated fly ash, hydroxides, and sulfides.

Test Conditions — All experiments were carried out with one-gram samples of sorbent. The gas flow rate was about 0.6 ft^3/sec with an inlet SO_2 concentration of 3,500 ppm at 257°F. The "typical" dry gas composition was 82.7 percent N_2, 14.9% CO_2 and 3.0% O_2. The moisture content was 4 or 6 mol percent.

Results — The sodium carbonates were considerably more effective than any of the other materials tested.

Conclusions and Comments — The results of this study indicate significant sorption of SO_2 by "fixed-bed" samples of sodium carbonate and sodium sesquicarbonate. Natural soda ash and sesquicarbonates were superior to the commercial carbonates, both in amount of SO_2 absorbed after a given time and in utilization of the sorbent.

Information Source — Friedman, L.D., *Applicability of Inorganic Solids Other Than Oxides to the Development of New Processes for Removing SO_2 from Flue Gas,* FMC Corporation for the National Air Pollution Control Administration, Contract No. CAA 22-69-02, Princeton, NJ, December 1970.

Nagoya Institute of Technology, Multistage Bed, September 1973: *Scope* — In 1973, bench-scale studies were conducted to evaluate the use of soda ash in removing SO_2 from air in a multistage bed. Also, the rate of SO_2 sorption was investigated in bench-scale equipment.

Process Description — The active soda ash was added to the top stage of the bed and moved to lower levels with rakes. The dimensions of the apparatus were not available. The test gas was air with controlled amounts of SO_2 and water vapor added to achieve the desired composition.

Parameters Investigated — The tests were conducted to investigate the effects of the following on SO_2 removal: concentration of water vapor, particle size, temperature, and mol ratio of soda ash to inlet SO_2.

Test Conditions — Inlet SO_2 concentration varied from 1,050 to 2,460 ppm.

Results — Removals of 97+ percent were reported for all conditions.

In other tests with a thermobalance, active soda ash was found to absorb SO_2 at the rate of 1 mg/cm^2-min below 170°C (338°F). Soda ash containing up to 50 percent sodium sulfite absorbed SO_2 as fast as pure soda ash. When more than 50 percent sulfite was present, the absorption rate fell off slowly. Sulfite was oxidized to sulfate at temperatures above 428°F.

Conclusions and Comments — The study concluded that the multistage bed removal method was suitable for small plants. Using the scale-up factors proposed by the researchers, a 330 MW plant with 2,000 ppm SO_2 flue gas would require about 14 desulfurizers, each 30 feet in diameter.

Information Source — Yamada, Tamotsu, et al, *Desulfurization of Combustion Exhaust by Active Soda Ash,* Nagoya Institute of Technology, Japan, 25 395-403, 1973.

Stearns-Roger/Superior Oil, Fixed-Bed and Countercurrent Reactors, 1974: *Scope* — In December 1973, Superior Oil contracted with Stearns-Roger for the design, construction, and management of a pilot facility to investigate the feasibility of using nahcolite as a stack gas SO_2 removal agent. Bench-scale tests using a fixed bed reactor were conducted to investigate reaction kinetics. A pilot-scale countercurrent reactor was used to verify bench-scale results and to provide data for a conceptual design and a hypothetical 500 MW power plant.

Bench-Scale Test Work — Process Description: After passing through a filter to remove particulates, the hot flue gas flowed downward through a thin bed (1-inch deep) of screened and sized nahcolite or sodium bicarbonate. The flue gas flow was large enough with respect to the thin bed so that SO_2 concentration was essentially constant across the bed. Provisions were made for spiking the gas with SO_2 and for heating it to the desired temperature. Portions of the nahcolite bed were withdrawn periodically to yield conversion versus time results.

Parameters Investigated — The main parameters investigated were granule size and reaction time. The grade of sorbent (commercial sodium bicarbonate and nahcolite) and SO_2 concentration were also varied. The prime objective was to correlate experimental data with mathematical models to determine the rate controlling step of the reaction.

Test Conditions — The nahcolite and sodium bicarbonate granules ranged in size from 10 mesh to ½ inch. SO_2 concentration was 450 to 10,000 ppm.

Results — Comparison of conversion versus time experimental data and predicted results of mathematical models of reaction kinetics led to the observation that the nahcolite-SO_2 reaction is controlled by the diffusion of SO_2 through the layer of sodium sulfate that builds up on the outer surface of the particle. The experimental data correlated with the ash diffusion model, whereas the gas-film diffusion and chemical reaction controlled models were not confirmed by data. Other results suggested that pretreatment of the nahcolite by heating to form Na_2CO_3 may improve SO_2 removal.

Conclusions and Comments — Bench-scale study found the $NaHCO_3/SO_2$ reaction is controlled by diffusion of SO_2 through the sodium sulfate layer that builds up on the particle. This observation only applies to the fairly large particle sizes; the reaction may be controlled by gas-film diffusion for smaller particles.

Pilot-Scale Test Work — Process Description: The pilot-scale countercurrent reactor was 42 inches in diameter. Hot flue gas flowed up through a slowly descending bed of nahcolite or sodium bicarbonate. Since the reactor was insulated, temperature drop was negligible. The solids residence time ranged from 100–200 hours. Nahcolite was added to maintain a constant stoichiometric ratio, and spent solids were removed to maintain a constant bed level. Temperature, pressure drop across the reactor, and inlet/outlet SO_2 concentrations were continuously monitored.

Parameters Investigated — The test variables were: temperature, inlet SO_2 concentration, bed height, and flue gas flow rate.

Results — Sulfur dioxide removals of 80 to 84 percent were achieved with less than stoichiometric amounts of sodium bicarbonate. Additive utilization was about 97 percent. The SO_2 removal efficiency was lower in the nahcolite test, 67 percent with 75 percent additive utilization. The investigators pointed out that the purpose of the nahcolite test was to verify the predictive ability of their mathematical model, not to optimize SO_2 removal or additive utilization.

Pressure drops over the reactor ranged from 16 to 22 inches of H_2O. No results were presented as to the effect of varying temperature, bed height, or flue gas flow, although the temperature was 20°F lower in the nahcolite run. Nitrogen oxides removal was reported to be 42 percent in the test with nahcolite, the only run for which such a determination was made.

Conclusions and Comments — Although high SO_2 removals with good additive utilization were achieved in the countercurrent reactor, there remain certain disadvantages to the process. The countercurrent reactor is a somewhat mechanically complicated piece of equipment and might be prone to operating and maintenance problems on a full-scale size. Also, the pressure drop over the reactor is much greater than that for dry injection or spray dryer systems.

Information Source — Stearns-Roger, *Nahcolite Granule Scrubbing System Feasibility Study,* vol. 1, for Superior Oil Company, November 1974.

R.W.E. Tests in Germany: *Scope* — R.W.E., a German utility company, has evaluated various calcium-based alkali materials for removing SO_2 from flue gas by injecting them through low-NO_x burners in an existing 60 MWe lignite-fired boiler. R.W.E. has been using limestone for about a year in this process to achieve compliance with local air pollution regulations.

Process Description — This is a very simple process that takes limestone from storage, pulverizes it along with coal, and injects the coal/limestone mixture into the boiler through low-NO_x burners. This is a retrofit installation that uses the boiler's existing burners. Process control is simple and straightforward; an instrument is used to monitor outlet SO_2 concentration which in turn regulates the flow of limestone to the coal pulverizer. The coal fired by the utility is a German brown coal which is similar to a low-grade U.S. lignite. Its heating value is reported to vary from 4,300 to 5,000 Btu/lb with a sulfur content of 0.4 to 0.7 percent.

Results — R.W.E. has apparently examined the use of three sorbent materials: limestone, lime, and calcium sludge from water treating. Their experience has shown that limestone is the best sorbent for their use due to its superior handling characteristics over the other sorbents.

Results of their testing indicate that SO_2 removals of 60 to 90 percent can be achieved with stoichiometric calcium to sulfur ratio of up to 3. No time intervals were, however, given for these high removal rates so it may not be known how effective this process will be in achieving high SO_2 removals over long periods of time.

Over the last year, the system has operated well achieving compliance with local air pollution regulations. However, due to the low-sulfur coal being burned, coupled with nonstringent SO_2 control requirements, the system has only had to achieve 25 to 50 percent SO_2 removal during this time. The stoichiometric ratio required to achieve this level was reported to be about 1. Capital investment costs (including modifications to the particulate control equipment) were reported to be about $150,000 for the 60 MWe system.

Conclusions and Recommendations — Results of the test work described are promising and illustrate the ability of this technique to remove SO_2 without plugging of boiler tube spaces. The major unanswered question is the ability of this technique to achieve the stringent removal levels required by U.S. air pollution regulations under sustained periods of operation. Although it appears that the application of this technique in the U.S. may be limited to only low sulfur Western coals, this technique, if proven, appears to provide a very economical SO_2 removal alternative.

Also, it should be noted that boiler tube spacings for German lignite-fueled boilers are larger than those used in conventional coal-fired boilers. In addition, the flow arrangement is such that the hottest portion of the boiler is not at the "top" of the boiler. Consequently, coal-fired boilers may need to be redesigned to avoid plugging if coal-limestone mixes are to be used successfully.

Information Source — Dickerman, J.C., telephone conversation with R.M. Statnick, U.S. EPA, October 4, 1979.

ONGOING ACTIVITIES

The discussions of current and ongoing development activities are presented in alphabetical order by company name. For companies which have been in the dry FGD business for several years, the information here includes description of substantial research efforts and details of commercial sales. For companies which have only recently entered this market, only a brief discussion of their intentions may be given here. The results of most current activities reported in this section are very preliminary and are unpublished. Consequently, most data were obtained through personal contact, either by telephone or meetings held with process vendors.

Babcock and Wilcox

Background: B&W was one of four companies which piloted spray dryer-based dry FGD systems at Basin Electric in late 1977. B&W originally tested Japan's Hitachi two-fluid nozzle atomization technology in a Hitachi vertical spray dryer/reactor. When this concept proved inadequate for utility FGD application, B&W developed a horizontal reactor using "Y-jet" steam-atomized nozzles which were adapted from their standard oil burner technology. In this aspect B&W has a rather novel approach to spray drying technology. Other spray dryer-based dry FGD system vendors tend to use more standard spray drying technology with a back-mix type reactor and two-fluid or rotary atomization. B&W also takes a different approach in that they favor ESP collection of the spent sorbent/fly ash mixture rather than baghouse collection which appears to be favored by other vendors.

Research: As mentioned above, initial test work by B&W at Basin Electric's William J. Neal station near Velva, North Dakota was with a Hitachi two-fluid nozzle in a vertical reactor. When this configuration proved inadequate, B&W went to a modified "Y-jet" steam-atomized oil burner for slurry atomization. Flue gas enters the reactor through registers or vanes, which impart a spinning motion to the flue gas around the nozzle area. The reactor itself is a box designed to give proper retention time. Some dried sorbent drops into hoppers at the bottom of the horizontal reactor, while the remainder is collected with fly ash in a precipitator or baghouse.

The pilot unit at Velva was rated at a nominal 8,000 acfm gas flow. Only ESP collection was tested with the "Y-jet" unit. The Hitachi reactor had been tested with both ESP and baghouse collection, but B&W felt they could safely operate with lower dryer outlet temperatures using an ESP collector than with a baghouse collector. The increase in sorbent utilization resulting from operating the spray dryer outlet nearer to saturation was apparently greater than that corresponding to reaction on a bag surface. Reagents tested included soda ash, pebble lime, hydrated lime, ammonia addition with lime reagent, and precipitated limestone.

The pilot plant in Velva has been shut down and is currently being dismantled. Research work is being continued with a 5×10^6 Btu/hr (1,500 acfm) combustor using a single reactor with one nozzle. As well as doing full parametric studies, B&W hopes to achieve an increased understanding of the SO_2 removal reaction mechanisms in the spray dryer with this unit.

Commercial Status: Following the work at Neal Station, B&W successfully bid on the FGD system for Basin Electric's Laramie River Station Unit 3, a 500-MW unit to be built near Wheatland, Wyoming. Engineering activities for the Laramie River Station are on schedule with construction slated to begin soon. The unit is scheduled to go into commercial operation in April, 1982.

The system at Laramie River will consist of four reactors each followed by an electrostatic precipitator. Normal plant operation will call for the use of three reactors, with one as a spare. Double-louver dampers (1 set isolation, 1 set control) regulate flow to the modules, with one set on each reactor inlet and a downstream set on each ESP outlet. Air flow control to individual "burners" in each reactor is set by vanes in the distribution box. A first pass on the vane design for Laramie River has been model-tested to give equal flows (within 1%) to the burners.

Each reactor will be equipped with 12 "Y-jet" nozzles in three rows of four nozzles. Reactor size was chosen to correspond to the size of the ESP used. B&W precipitators of the same size and design as those on the existing Laramie River Units 1 and 2 will be used. Units 1 and 2 use wet limestone FGD systems. The turndown ratio for individual "burners" or "Y-jet" nozzles on the Laramie River design is about 2:1 with the fixed registers and relatively low nozzle ΔP employed. By using variable registers and higher design ΔP, a greater individual turndown ratio could have been achieved. The nozzles to be used will weigh about 64 pounds and take one to two men from 2 to 5 minutes to change.

The Laramie River design employs no recycle, even though the cost contains a theoretical 4.5:1 Ca/S ratio. The design calls for 90% SO_2 removal at maximum fuel sulfur and flow rate, with a maximum stoichiometry of 1.12 based on entering sulfur. The gases leaving the spray dryer will be 10°F above saturation; 3% hot gas bypass will raise the dryer outlet temperature 15°F before gases enter an ESP. The Laramie River design calls for 30,000 lb/hr of 150 psig steam with 50°F superheat for atomization. Figure 7.7 is a flow diagram of the Laramie River dry FGD system.

Although a demonstration and not strictly a commercial FGD system, a 120,000-acfm reactor (~40 MWe) is scheduled for startup at Pacific Power and Light's Jim Bridger Station in late 1979. Tests have been delayed due to a boiler outage. The reactor will treat a slipstream of flue gas from a 600-MW boiler fired with a low-sulfur Wyoming coal (~450 ppm SO_2) and return the treated gas to one of six ESPs. Slipstreams from this reactor will be treated by a nominal 4,000 to 5,000 acfm baghouse and a 5,000 to 6,000 pilot ESP. This demonstration will use one six-nozzle reactor with the same automatic control system proposed in the Laramie River design. It will provide a check on the control logic for Laramie River. The system will have no hot gas bypass, as Laramie River will have. Baseline parametric studies will be with lime; the utility wants to test ammonia addition, soda ash, and an available waste sodium-based liquor. Basin Electric hopes to compare results of pilot ESP testing on the spray dryer outlet with original pilot ESP data used to design full-scale precipitators for the Laramie River Station. B&W will operate the spray dryer and pilot ESP for 4 to 6 months for their baseline testing but can extend operation if the utility funds additional testing. During baseline testing, target SO_2 removal is the 85% required at Laramie River.

Testing is proceeding on a 1,500-acfm pilot unit at B&W's Alliance, Ohio research facility. The unit, intended for parametric studies, uses a single Y-jet

Figure 7.7: Laramie River Station Flow Diagram

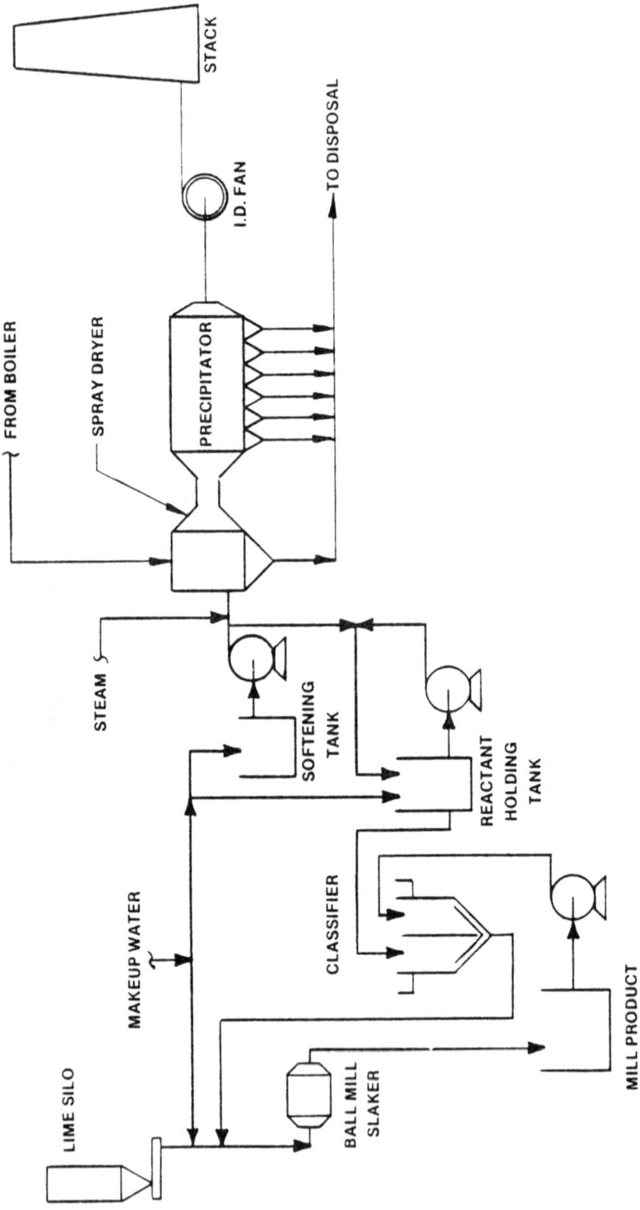

Source: EPA-600/7-80-030

nozzle in the reactor, with a cyclone for particulate collection. Plans call for the addition of an ESP and a baghouse. Initial plans call for several coals to be tested and flue gas SO_2 concentrations of 500 to 2,000 or perhaps 3,000 ppm.

Lime reagent will be used; test parameters include inlet and outlet gas temperatures, sorbent stoichiometry, and recycle schemes. A paste slaker will be used for fresh limes. Tests on low-sulfur lignite (0.5% S), low sulfur subbituminous (0.9% S) and low-sulfur bituminous (2.2% S) coals have been completed. Earlier research indicated recycled fly ash/spent sorbent mixtures were more reactive when mechanically ground to smaller particle size than when reslurried directly. Thus, some recycle work will be done with the ball mill slaker from the now-disassembled Velva pilot plant used to reduce particle size of recycled material.

One additional anticipated result of this test program is that B&W expects to derive a better understanding of the SO_2 reaction mechanisms in the spray dryer. B&W feels that neither gas-liquid reaction where absorbed gas-phase SO_2 reacts with alkaline droplets nor gas-solid reaction where flue gas SO_2 adsorbs onto dried alkaline solids can account for all of the SO_2 removal. They expect to find that chemisorption of SO_2 across several molecular layers of residual moisture on spray-dried alkaline material will account for a major portion of the SO_2 removal.

Continuing work will focus on validating the preliminary reaction mechanisms that have been formulated to date. Immediate plans are to burn oil and spike the flue gas with SO_2, with a possibility of testing high-sulfur fuels in the near future. B&W also plans to add a baghouse downstream of the spray dryer to compare SO_2 and particulate removal efficiencies to those obtained with the spray dryer/cyclone configuration. Longer range plans include burning different fuels and the investigation of the SO_2 removal capabilities of virgin fly ash in a spray dryer system.

Information Sources: Blythe, Gary, meeting notes at B&W, Barberton, Ohio, June 28, 1979.

Janssen, K.E. and Eriksen, R.L., "Basin Electric's Involvement with Dry Flue Gas Desulfurization," paper presented at Fifth EPA Symposium on FGD, Las Vegas, Nevada, March 5-8, 1979.

Kelly, M.E., telephone conversation with John Doyle, B&W Alliance Labs, October 11, 1979.

Kelly, M.E., telephone conversation with Tom Hurst, B&W, October 18, 1979.

Slack, A.V., "Lime Scrubbing by 'Dry Processes'," A.V. Slack Report No. 62, January 1979, pp. 15–30.

Buell/Anhydro and EPA/Buell

Background: Buell, a division of Envirotech Corporation, has two dry FGD efforts under way. One is a totally in-house development of a dry injection/baghouse collection system which would use a Buell bag collector. In the other, Buell is working with Anhydro, a Copenhagen-based spray dryer company which began as a spin-off from Nitro Atomizer, to develop a spray dryer/baghouse system. Test work on both systems will be conducted at the City of Colorado Springs' Martin Drake Station.

Research: EPA-funded test work on a 3,000 acfm dry injection/baghouse system at the City of Colorado Springs' Martin Drake Station was due to start in late October. The test work will investigate three sodium-based sorbents: nahcolite, raw trona ore, and upgraded trona. The parametric testing, to be conducted through December 1979, will include investigation of various stoichiometric ratios (0.7, 1.0 and 1.5), inlet gas temperatures (325°, 425° and 500°F), and inlet

SO_2 concentrations (400 to 600 ppm). The fuel used in the Martin Drake unit is a Colorado bituminous coal with a higher heating value of about 12,000 Btu/lb and a sulfur content of from 0.3 to 0.7 percent. The nahcolite to be used in the tests has been obtained from a Bureau of Mines pilot shaft sunk in a Colorado nahcolite/oil shale deposit near Denver. The nahcolite from this mine will be made available for government and industry dry FGD test work. This pilot shaft is currently one of the few available sources of nahcolite.

Buell is also conducting EPA-funded waste disposal studies. The major process to be studied is sintering of the dry waste product, based on the Sinterna process (patented by Industrial Resources, Inc.). The waste disposal studies are expected to last for about nine months and are being performed by Battelle Memorial Institute (Columbus Laboratories) under a subcontract from Buell.

In a parallel privately funded research program, Buell and Anhydro are developing a dry FGD system with a spray dryer and baghouse. Initial work with both lime and soda ash sorbents was done at Anhydro's Copenhagen facility with a 3,000 acfm spray dryer. Future plans are to conduct EPA-funded demonstration studies on a 20,000 acfm spray dryer/baghouse system at the Martin Drake Station. The spray dryer, supplied by Anhydro under an exclusive technological agreement, is a 13-ft diameter tower equipped with a rotary atomizer. Sorbents to be tested include lime and the three sodium-based sorbents used in the dry injection studies. Recycling of the spent sorbent/fly ash products will also be investigated. The spray dryer/baghouse system was expected to start up in December 1979, with tests run for 6 months to a year. The primary purpose of the unit was to obtain design data for development of a full-scale spray dryer system.

Commercial Status: At this point Buell's top priority is the completion of test work on the spray dryer and dry injection systems at Martin Drake.

The Buell/Anhydro joint venture submitted five budgetary prices to utilities for spray drying systems. They have also submitted a budgetary price for one industrial application and bid on another industrial application.

Information Sources: Blythe, Gary, meeting notes at Buell, Lebanon, PA, June 19, 1979.

Kelly, Mary E., telephone conversation with Dale Furlong, Buell Envirotech R&D, October 11, 1979.

Kelly, Mary E., telephone conversation with Lloyd Hemenway, Buell Envirotech, October 11, 1979.

Combustion Engineering

Background: Combustion Engineering has conducted several in-house studies related to dry FGD. In 1972 they studied sludge drying as a dewatering method, and they have an ongoing program studying char-ash drying in the C-E coal gasification process. C-E has also studied an ammonium sulfate dry SO_2 scrubbing process available from a licensor.

Regarding a lime-based dry FGD system, C-E has gone through a literature survey and a study of available pilot data. They have tested several atomizing devices, presumably for use in a spray dryer, and have completed a conceptual design of a lime-based dry FGD system.

Research: As a result of the above, C-E has begun test work on a pilot unit consisting of a 20,000 acfm C-E spray dryer equipped with a two-fluid nozzle atomizer using air as the atomizing fluid. The spray dryer is followed by a fabric filter and ESP in parallel. The pilot unit will be installed at the Northern States Power Sherburne County Unit No. 1, which burns Sarpy Creek (Montana) coal (1.0 percent sulfur, 10 percent ash, 8,000 Btu/lb). The pilot unit will use lime as a sorbent. Test work will involve parametric studies of temperatures, air-to-cloth ratios, L/G ratios, SO_2 levels, scrubber velocity, and recycle of spent sorbent/fly ash mixtures. In conjunction with this pilot program, C-E is also planning a dry scrubbing waste disposal study.

In the future, C-E plans to continue pilot plant testing of their spray dryer system although no definite plans were disclosed at this time. C-E is currently preparing bids for two dry FGD systems for utility applications.

Information Sources: Blythe, Gary, telephone conversation with K.W. Malki, C-E, June 6, 1979.

Malki, K.W., C-E System Design, personal communication with Gary Blythe, July 11, 1979.

Kelly, M.E., telephone conversation with Kal Malki, C-E System Design, October 11, 1979.

DOE/Grand Forks Energy Technology Center

Background: Grand Forks Energy Technology Center (GFETC) has been conducting research on dry injection systems on the 200 scfm scale, comparing nahcolite and trona sorbents. Parameters being investigated include inlet SO_2 concentration, inlet gas temperature, bag materials, air-to-cloth ratios, and the sequencing of sorbent addition and bag cleaning cycles. SO_2 removal efficiencies of up to 90 percent have been achieved with nahcolite at 50 to 60 percent sorbent utilization, with the lower sorbent utilization observed at a high air-to-cloth ratio.

Research: Dry injection test work on the 200 scfm scale was expected to be completed in early 1980. Tests to optimize performance with trona have been completed. Test work to optimize performance with nahcolite is proceeding after some delay in obtaining nahcolite. (A Bureau of Mines pilot shaft near Denver has provided a new source of nahcolite.) SO_2 levels representative of low-sulfur Western coals, up to 1,500 ppm SO_2 on a dry basis, are being investigated in the current test work.

Research conducted at GFETC has indicated that trona is somewhat less reactive than nahcolite as an SO_2 sorbent. This diminished reactivity can be explained in terms of specific surface area (m^2/g). At equivalent conditions of temperature and time, GFETC studies have shown that nahcolite has a significantly larger surface area (7.0 m^2/g vs 5.0 m^2/g at 500°F and 30 minutes activation time). The difference is greatest at high temperatures and long activation times. The GFETC investigators suggest that it is possible that the use of a sorbent with a smaller particle size, injection at elevated temperatures, and operation of the baghouse at the highest practical temperatures would overcome the apparent lower reactivity of trona enough to make it "a practical alternative to nahcolite" for use in dry injection/baghouse systems.

GFETC is planning to expand their current dry injection program in the near future. They are presently designing a 150 scfm baghouse that will be dedicated to dry injection tests (the current baghouse is also used for particulate characterization studies). The new baghouse will be a pulse-jet type. Plans are to test both nahcolite and trona sorbents while varying parameters such as temperature, residence time (air-to-cloth ratio), and stoichiometry. Higher SO_2 levels, above 1,500 ppm, may also be investigated. Bench-scale waste disposal studies have been completed and a final report is being prepared.

Future Research Plans: In addition to dry injection work, Grand Forks will be involved with the Rockwell/Wheelabrator-Frye spray dryer/baghouse FGD system being constructed at Otter Tail Power's Coyote Station. GFETC will sample for particulate and SO_2 removal efficiencies. In addition, they are currently negotiating with Otter Tail Power to determine the ranges in which GFETC will be allowed to vary such parameters as temperature and sorbent feed rate to characterize the FGD system. The Coyote Station FGD system was scheduled to start up in the spring of 1981. As a supplement to the proposed test work at Coyote Station, Grand Forks is also conducting conceptual and small-scale studies to investigate large-scale recovery of sodium from the spent sorbent/fly ash mixture for reinjection into the system. The advantage of sodium recovery is seen as two-fold: (1) it will help to stabilize the waste products by reducing their soluble sodium content, and (2) reinjection of recovered sodium will reduce fresh sorbent consumption, which would provide considerable savings in operating costs.

As a final note on dry injection, there remain uncertainties in both trona and nahcolite availability. In the case of trona, depletion allowance complications may keep trona from being available in the quantities needed for full-scale dry injection systems. The new Bureau of Mines pilot shaft is currently the only available source of nahcolite.

Information Sources: Ness, H.M., Selle, Stanley, DOE/GFETC and Oscar Manz, University of North Dakota. *Power Plant Flue Gas Desulfurization for Low-Rank Western Coals,* presented at the 1979 Lignite Symposium, Grand Forks, ND, May 30–31, 1979.

Blythe, Gary, telephone conversation with Stanley J. Selle, DOE, June 7, 1979.

Blythe, Gary, telephone conversation with Harvey Ness, DOE, July 3, 1979.

Kelly, M.E., telephone conversation with Harvey Ness, DOE, October 16, 1979.

DOE/Morgantown Energy Technology Center

Background: The use of powdered dry lime or limestone in dry injection/baghouse systems has resulted in much lower removal efficiencies than in systems where sodium-based sorbents, such as nahcolite and trona, have been used. However, the waste solids from sodium alkali-based processes are water-soluble and pose potential disposal problems, whereas calcium-based product solids are considerably more stable. Studies have been under way at Morgantown Energy Technology Center (METC) to develop a "dry" limestone FGD process.

Research: A patented technique, involving the addition of water vapor to hot flue gas (300°F) to increase the saturation temperature of the gas above a critical minimum before it is passed over a bed of limestone chips to remove the SO_2, was developed by Shale and Cross in 1976. Both laboratory kinetic studies and bulk evaluation studies have been conducted. Figure 7.8 is a flowsheet for the bulk evaluation tests for this "modified dry" limestone process (MDLP).

Results of the kinetic studies show reaction rates equivalent to those found in high-temperature fluid bed processes involving calcined limestone. Other results of the kinetic studies showed that reaction occurred on the limestone chip surface and that calcium sulfate was the major product.

Figure 7.8: Flow Sheet for Bulk Evaluation Studies of Modified Dry Limestone Process

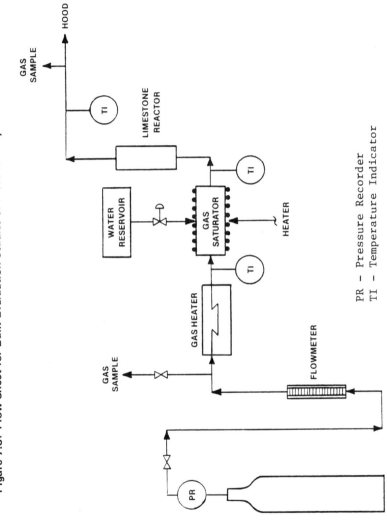

PR — Pressure Recorder
TI — Temperature Indicator

Source: EPA-600/7-80-030

In the bulk evaluation studies, simulated dry flue gas was heated to 280°F and passed through the saturator prior to entry into the bed of crushed limestone. The gas entered the bed at a temperature of 150° to 160°F and a space velocity of 500/hr. Tests were conducted using limestone beds 1-inch in diameter and 9 inches deep and $\frac{1}{2}$-inch diameter by 4 inches deep. Limestone chips were $\frac{1}{16}$-inch by $\frac{1}{2}$-inch. Results of bulk evaluation studies with 1,600 ppm inlet SO_2 at 150°F are as follows. At a saturation temperature of 150°F, a SO_2 removal efficiency of greater than 90% was maintained for over 3 hours. At lower saturation temperatures (i.e., 100° to 110°F), removal efficiency decreased rapidly with time due to the formation of a $CaSO_3/CaSO_4$ layer on the pellet after an initial period of high removal. Experiments at higher space velocity (4,000/hr) have shown that SO_2 removal efficiencies of greater than 90% can be achieved when the water vapor saturation temperature of the gas, controlled by the addition of water in the saturator, is no more than 30°F below the actual gas temperature entering the limestone bed. Sorbent utilization is reported to be "less than 90%."

Current Status: At the present time no further work is being carried out on the "modified dry" limestone process, although studies to investigate the limestone/SO_2 reaction mechanism are being conducted.

The data from Mr. Shale's test work has been verified for accuracy, but a complete interpretation of the results has not been conducted. Preliminary economic analyses, based on both the kinetic and bulk evaluation studies employing a counterflow moving-bed reactor, have shown the capital and operating costs for the MDLP to be greater than those for a conventional wet lime/limestone process. The excess cost is due in part to the large pressure drop characteristics of counterflow moving beds.

Information Sources: Shale, C.C. and Stewart, G.W., "A New Technique for Dry Removal of SO_2," DOE/Morgantown Energy Technology Center, paper presented at Second Symposium on the Transfer and Utilization of Particulate Control Technology, Denver, CO, July, 1979.

Kelly, M.E., telephone conversation with Charter Steinspring, DOE/METC, October 18, 1979.

DOE/Pittsburgh Energy Technology Center

Background: DOE's Pittsburgh Energy Technology Center (PETC) has a program under way to evaluate dry injection of various sorbents. They have a 500 lb/hr coal furnace that provides the flue gas for their testing.

Research: PETC's initial testing has focused on moderate-sulfur Pittsburgh seam coals (1.5 to 2% S). System parameters that were evaluated include sorbent type, stoichiometry, sorbent feed mechanism, and sorbent particle size. Four sorbents are being tested: sodium carbonate, sodium bicarbonate, nahcolite, and trona. Data from this test program are still under evaluation, but preliminary results have shown greater than 90 percent SO_2 removal with sorbent stoichiometries of up to 2. Pressure drop through the baghouse was reported to vary from about 10 to 14 inches of water.

Future Research: Future research plans are to continue evaluation of the system for SO_2 removal from higher-sulfur coals. PETC also has requested bids for the construction of a spray dryer which they hope to have installed by mid-1980. Testing of the spray dryer system was to begin the second half of 1980.

Information Sources: Dickerman, J.C., telephone conversation with Richard Demski, PETC, January 2, 1980.

Ecolaire

Background: Ecolaire Systems is part of the Ecolaire Corporation which was founded in 1971. The Ecolaire Corporation includes several companies which have been in the business of supplying equipment to the utility industry for many years.

Research: Ecolaire Systems has had no previous dry FGD research and development program other than Industrial Clean Air's (ICA, a subsidiary of Ecolaire Corporation) experience with dry additives for enhancement of bag filter performance. However, Ecolaire has constructed a 10,000 cfm mobile pilot plant for demonstration of a spray dryer/baghouse FGD system. When completed, the mobile plant can be trucked to any location on six semitrailers and erected in a 60-ft by 60-ft area in approximately a two-week hookup time. The unit under construction uses an Ecolaire-modified Niro Atomizer spray dryer and an ICA four-section baghouse using 12-inch diameter by 36-ft long bags. Design data for system bidding will come from the mobile demonstration unit.

The mobile unit has been set up at Nebraska Public Power District's Gerrel Gentleman Unit 1. System start-up was scheduled for late 1979. Ecolaire plans to conduct parametric and design optimization studies for up to 4 months. They will be investigating both rotary and nozzle atomization. The primary sorbent to be tested is lime and provisions have been made for investigating recycle of spent sorbent/fly ash mixtures. Other parameters that will be varied include inlet SO_2 concentration (up to 2,000 ppm SO_2), inlet gas temperature, temperature drop over the spray dryer, and sorbent concentration.

Commercial Status: Ecolaire Systems will not bid on a utility or industrial system until they have design data available from their mobile demonstration unit. Current thinking for a commercial system would call for four or five spray dryer modules, each with five rotary or nozzle atomizers. Primary control on the system would be outlet gas temperature, which sets the water rate to the scrubber. Sorbent concentration would be controlled based on inlet and outlet SO_2 concentrations, boiler load, etc. Perhaps a minicomputer would be used to calculate required sorbent concentration based on these various inputs. The control system would be designed by Ecolaire Systems, using purchased components.

Although the mobile demonstration unit uses a modified Niro spray dryer, Ecolaire has no agreement with a spray dryer manufacturer for exclusive use of their equipment. By not making direct ties with a spray dryer company, Ecolaire feels they have more flexibility in providing the best system for an individual application by choosing among any number of commercially offered spray dryers.

Information Sources: Blythe, Gary, meeting notes, meeting at Ecolaire Systems, Inc., Malvern, PA, May 22, 1979.

Kelly, M.E., telephone conversation with Carl Newman, Ecolaire, November, 1979.

Energy and Pollution Controls, Inc.

Background: EPC is a recently formed subsidiary of Flick-Reedy, Inc. They have developed a totally dry FGD reactor system for industrial applications.

Research: Development work on the EPC-designed reactor illustrated in Figure 7.9 was started in early 1978. This dry FGD reactor is intended for combination with a baghouse to result in a totally dry FGD system for industrial applications. The reactor is designed for cyclonic flow of flue gas near a hydraulically driven "slinger," which distributes dry hydrated lime in a direction countercurrent to the flue gas flow. Hydrated lime is fed to the slinger with a commercially available dry chemical feeder. Directly above the slinger is an air-operated eductor which captures and recirculates the lighter fraction of the partially spent sorbent. Below the reaction section of the reactor, a conical expansion reduces the flue gas velocity to allow dropout of heavy particulate matter before the gas flows to a bag collector.

Figure 7.9: Air Pollution Control (SO$_2$) Reactor for Dry Reagent

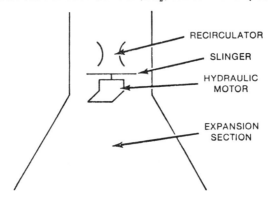

Source: EPA-600/7-80-030

Gas velocities in ductwork to and from the reactor are typically 8 to 12 ft/sec. Velocities inside the reactor vary from approximately 25 ft/sec in the cyclonic section down to approximately 5 ft/sec in the expansion section.

Development work on the reactor used flue gas from a coal-fired (1.6 million Btu/hr hot water boiler) burning a 3.3 percent sulfur Illinois coal. Flue gas sulfur content varied from 1,600 to 2,300 ppm SO$_2$. Flue gas temperatures in the reactor varied from 350° to 500°F, but temperatures to the downstream cartridge filtration device were limited to 350°F to protect the NOMEX cartridge. The speed of the slinger was not specified due to patent considerations. Pressure drop across the reactor in all tests was less than 0.5 inch of water. SO$_2$ removal efficiencies varied from 45 to 95 percent at 0.8 to 3.0 times the stoichiometric ratio.

Commercial Status: EPC will market the reactor for industrial applications. They report that capital and operating costs for SO$_2$ and particulate control using their dry system will be approximately half those for a system using a wet scrubber, on a dollars per ton of coal basis.

Preliminary design has been completed for construction of a commercial model (25,000 acfm) for Kerr Industries in South Carolina. This unit will be used in EPA-sponsored tests at Kerr to be performed by Environmental Testing, Inc. The tests were scheduled to begin March, 1980.

Information Sources: Blythe, Gary, telephone conversation with Grant Hollett, Jr., EPC, June 4, 1979.

Hollett, Grant T., Jr., "Dry Removal of SO$_2$—Applications to Industrial Coal-Fired Boilers," presented at APCA Convention, Cincinnati, OH, June 25–28, 1979.

Kelly, M.E., telephone conversations with Grant Hollett, Jr., EPC, October 5 and October 26, 1979.

EPA/Battelle-Columbus Labs

Background Information: As part of an EPA-funded program to evaluate industrial coal-fired stoker boilers, Battelle developed a limestone/high-sulfur coal pellet in an attempt to control SO_2 emissions. Initial tests were conducted in a model 20 bhp spreader-stoker boiler using a pellet with a Ca:S mol ratio of 7:1. Preliminary results indicated that 70 to 80% retention of the available sulfur in the fuel was achievable; however, the increased particulate emissions resulted in an opacity increase from a normal 4 to 7 percent up to 22 to 25 percent. In an effort to reduce the increased particulate emissions that result from firing the limestone/coal pellet, Battelle began development and testing of a 3.5:1 Ca:S pellet.

Research: The newly developed pellet (Ca:S ratio of 3.5:1) has mechanical strength, durability, and weatherability characteristics comparable to that of raw coal. Laboratory tests in the 20 bhp model spreader stoker and in a fixed-bed reactor have resulted in 60 to 80 percent retention of the available sulfur in the fuel. A high-sulfur Illinois coal was used for these tests. Laboratory tests are continuing in preparation for a 1-day test to evaluate firing the pellet in Battelle's 25,000 lb/hr steam boiler at the Columbus Lab facilities. The 1-day test will investigate combustion characteristics and sulfur retention capabilities of the pellet with a Ca:S mol ratio of 3.5:1 as well as determine the effects of the pellets on boiler operation. With regard to boiler operation, one preliminary hypothesis is that the limestone present in the pellet may result in higher ash fusion temperatures, thus helping to reduce the clinker formation tendency of the coal.

In other ongoing work, bench-scale waste disposal studies conducted by Battelle have shown that the ash produced from combustion of the coal/limestone pellet does not exhibit the same acid-leaching characteristics as fly ash from straight coal combustion. Battelle is also beginning reaction mechanism studies to determine the details of the SO_2 removal reaction in the stoker bed. Plans are to use scanning electron microscope techniques to determine what compounds are being formed and to identify their structures.

Preliminary cost estimates have been prepared and have been confirmed in further economic studies that indicate the cost of producing the coal/limestone pellet will be about $15/ton. This cost includes both capital and operating costs for grinding and pelletizing and reagent costs for the binder and limestone.

Future Research: Future work including a 30-day full-scale demonstration test on a 75,000 lb steam/hr boiler was currently under funding evaluation by the EPA. The results of the present study will be evaluated to determine their potential impact on the nation's energy supply and pollution standards.

Information Sources: Giammar, Robert D., et al, "Evaluation of Emissions and Control Technology for Industrial Stoker Boilers," in Proceedings of the Third Stationary Source Combustion Symposium, vol 1, EPA-600/7-79-050a, February, 1979.

Kelly, M.E., telephone conversation with Robert Giammar, Battelle-Columbus Labs, October 10, 1979.

Kelly, M.E., telephone conversation with J.H. Wasser, EPA, October 11, 1979.

EPA/Energy and Environmental Research Corp. (EERC)

Background: EERC conducted a preliminary feasibility study as part of a program to test EPA's concept of limestone injection into a low-NO_x burner for SO_2 removal. Initial test work at the 1×10^6 Btu/hr heat input rate was promising, and work is continuing on a smaller (70,000 Btu/hr) scale.

Research: In this process, limestone is mixed and pulverized with coal prior to combustion and fired through a B&W low-NO_x burner. Sulfur contained in the fuel reacts with the limestone present to form calcium salts which are collected with the fly ash emitted from the boiler.

The B&W low-NO_x burner system consists of two physically isolated combustion zones: a fuel-rich water-cooled primary combustion furnace and a secondary furnace where combustion products from the first furnace mix with the air necessary to complete combustion. EERC has measured the retention of sulfur resulting when limestone or trona is mixed with coal prior to combustion in the burner. The effectiveness of alkali addition to this type of burner is apparently related to the lower flame temperature resulting from the two-stage combustion as compared to conventional combustion. The lower temperature keeps the particles from approaching their melting temperatures, which can result in a glazing of the surface of the reagent particles that produce relatively unreactive particles.

EERC has noted that sulfur retention effectiveness is dependent upon good mixing of the reagent with the coal. They have found that adding reagent to the coal prior to passing the coal through the pulverizer substantially improves SO_2 removal. The pulverizer, which is designed for 75 percent minus-200 mesh, is believed to promote good mixing between the reagent and coal. EERC has also noted that sulfur retention can be greatly influenced by combustion conditions in general.

The test work so far has been very preliminary in nature.

Future Research: As a result of initially favorable performance data, EERC has proposed a full parametric study of sulfur retention in three sequential combustor sizes: 70 thousand Btu/hr, 10 million Btu/hr, and 50 million Btu/hr. There have apparently been no cost estimates yet made for this process due to its early stages of development.

Information Sources: Blythe, Gary, telephone conversation with Bill Nurick, EERC, July 25, 1979.

Jones, D.J., personal communication with Ted Phillips, Pacific Power & Light, March 1979.

Kelly, M.E., telephone conversation with Blair Martin, EPA, October 1, 1979.

EPA/Kerr Industries

Background: EPA is funding dry FGD test work at a Kerr Industries textile finishing plant in South Carolina. The major portion of the test work will focus on Energy and Pollution Control Inc.'s "dry pollution control reactor."

Research: A baghouse fire has delayed testing of this system. Test work was scheduled to begin in March, 1980, to evaluate dry FGD system which will treat 25,000 acfm of flue gas (about 4 MW). The baghouses (modified reverse air and pulse-jet) will be operated at air-to-cloth ratios of from 3 to 6, and both high and low inlet SO_2 concentrations will be tested. Lime and limestone sorbents will be used.

Information Sources: Blythe, Gary, telephone conversation with Jim Turner, EPA, June 25, 1979.

Kelly, M.E., telephone conversation with Jim Turner, EPA, November 7, 1979.

Joy/Niro Joint Venture

Background: Niro supplies some 75 percent of the world's spray dryer needs. Spray dryer applications include: instant milk, instant coffee, PVC, floor tile ceramics, kaolin, dyestuffs, copper and nickel sulfide for smelting applications, and raw cements. The kaolin plants are significant because they involve large (30-ft diameter) dryers that approach the size of those in utility FGD applications. The smelting operations are significant because some use flue gases from coal-fired generating stations for process heating, and thus operations follow boiler load as would a utility FGD spray dryer. Many of the cement plant installations also use coal-fired flue gases for process heating.

Niro began test work using a spray dryer for FGD and HCl removal applications in their Copenhagen research facility in 1974. They made several hundred test runs at 1,000 to 3,000 acfm using lime, limestone, sodium carbonate, and magnesium oxide as sorbents. As a result, Niro sold a small (2,000 to 3,000 acfm) spray dryer to Fiat of Milan, Italy, which used sodium carbonate to remove SO_2 from flue gas. In November 1977, Niro entered into a 2-year agreement with Joy Manufacturing to develop and market a spray dryer-based FGD system using baghouse or ESP particulate collection, which has since been extended another 5 years.

Research: The Joy/Niro joint venture team was invited by Basin Electric to pilot a spray dryer-based dry FGD system at their Antelope Valley Station. In mid-November 1977, design work was begun on a pilot unit to be built at Otter Tail Power's Hoot Lake station. The pilot unit included a 20,000-acfm spray dryer followed by a 3,000-acfm electrostatic precipitator and a 9,000-acfm baghouse in parallel. The baghouse had four compartments and 60 one-foot diameter bags. The pilot unit was constructed from late November to mid-February, and first operated from mid-February to April 24, 1978. During this period sodium carbonate and lime sorbents were tested, and design data were acquired for use in preparing a bid for the Basin Electric Antelope Valley Station FGD system. Parametric studies of the effects of inlet and outlet temperatures, stoichiometric ratio, recycle techniques, and particulate collector type on SO_2 removal were included in this test work. A 100-hr demonstration run of the pilot unit at Antelope Valley design conditions was also included.

Niro and Joy returned to Hoot Lake in mid-September 1978 to acquire data for preparing a bid on Basin Electric's Laramie River Station and to verify data from the previous test work. During this mid-September to mid-December 1978 test period, some 30 to 40 thousand data points on FGD performance were acquired. Also, during this period rail cars of Laramie River coal were shipped to Hoot Lake to operate the 60-MW Unit 2 boiler for several days at Laramie River conditions. The Laramie River ash was reported to be quite cementitious, but no operating problems were incurred during this period. Some tests were conducted adding water treatment sludge to the atomizer feed, and this was found to be a suitable method of disposal for the sludge. Some testing was conducted with commercially available ground limestone as a sorbent with very limited success. Other testing included SO_2 spiking up to a 4,500 ppm flue gas concentration.

All current FGD research activities are being done on a new FGD pilot unit at the Niro Copenhagen research facilities. After substantial testing at Hoot Lake, Niro feels that parametric, research-oriented testing can best be done at their facility. Site specifics, such as fly ash alkalinity effects, can be evaluated by fly ash injection. The 5,000 acfm test facility uses a propane burner as the source of flue gas. Both reverse air and pulse jet baghouses can be tested. Provisions have been made for spiking with fly ash, SO_2, SO_3 and steam to stimulate any flue gas condition.

The primary use of the Copenhagen facility is to obtain site-specific information on SO_2 removal and fly ash reactivity for responding to bids. In addition, Niro is also looking at the chemistry and reaction mechanism of the SO_2 sorption reaction. Results of this test work are for internal use only, but they may use these results to prepare papers for presentation at upcoming symposia.

Commercial Status: Following the pilot program at Fergus Falls, the Joy/Niro joint venture successfully bid on the FGD system for Basin Electric's Antelope Valley Station Unit 1, the first of two 440-MW stations to go in near Beulah, ND. The Western Precipitation Division of Joy Industrial Equipment Company was awarded the contract with Niro Atomizer Company as prime subcontractor. Construction of the Antelope Valley FGD system is reported to be on schedule with start-up planned for the spring of 1982.

The Antelope Valley installation will use five 46-ft diameter dryer modules, although any four modules could handle the total flue gas volume. Each module will be equipped with one atomizer with a direct drive motor. There will be one spare atomizer for use in any of the five modules. The sorbent used will be primarily lime slaked in a ball mill slaker, although sludge from the station's primary water treatment plant and a portion of the recycled sorbent/ash mixture will be added.

The SO_2 removal guarantee with only Unit 1 operating will be 62 percent for average sulfur fuel (0.68 percent S) and 78 percent for maximum sulfur fuel (1.22 percent S). After Unit 2 comes on line over a year later, the removal guarantee will be 81 percent for average sulfur and 89 percent for maximum sulfur. The emission limitation set by the North Dakota State Department of Health is 0.78 lb SO_2 per million Btu.

The sorbent/fly ash mixture leaving the spray dryer will be collected in a Western Precipitation baghouse. The baghouse will have 28 compartments with a total of approximately 8,000 fluorocarbon-coated fiber glass bags. The bags will be 12 inches by 35 feet and will be cleaned by reverse air. The gross air-to-cloth ratio under maximum flue gas flow conditions will be 2.19:1. The bag life guarantee for the Antelope Valley system is proprietary, but the standard Western Precipitation bag life guarantee is two years.

The lime utilization is very nearly 100 percent. This high utilization is possible because of the ability to utilize available alkalinity in the fly ash through spent sorbent/fly ash recycle. Also, the spray dryer outlet temperature will be controlled to near saturation. Bypass of up to six percent of the total flue gas flow will be used to reheat the spray dryer outlet to 185°F.

The Antelope Valley FGD system will use an innovative supervisory control scheme designed by Honeywell. The computer-controlled system will monitor boiler load, inlet SO_2 level, outlet SO_2 level, and inlet and outlet spray dryer temperatures and adjust lime concentration in the spray dryer feed for minimum lime consumption required to maintain the desired SO_2 removal. Basin Electric reported the capital cost of this system to be over $49 million in a paper presented at EPA's Fifth Flue Gas Desulfurization Symposium, March 1979 (6).

Waste disposal is not included in the Joy/Niro responsibilities. Basin will return the waste product/fly ash mixture to the mine in coal trucks for underground disposal.

To date, the Antelope Valley Station is the only FGD system sold by the Joy/Niro venture. However, they currently have two bids in evaluation, one bid in-house, and expected five to six bid requests by the end of 1979.

Although it cannot be strictly considered a commercial system, Joy/Niro has been awarded a contract to provide a 100-MW demonstration spray dryer/particulate collection system to Northern States Power for their Riverside Station. The lime-based system would treat up to 660,000 acfm of flue gas in a single 46-ft diameter spray dryer equipped with a rotary atomizer. The utility would operate the unit and participate in the demonstration R&D effort with Joy/Niro.

The flue gas to be treated is generated by burning an Eastern coal of up to three percent sulfur or a Western low-sulfur coal or a mix of the two. The spray dryer/ESP system (using an existing ESP) was to be completed during the summer of 1980, with a baghouse to be completed about six months later. This arrangement will permit comparison of ESP and baghouse particulate collection and SO_2 removal on a large-scale unit.

Joy feels the market for utility spray drying systems is favorable. They feel the market for industrial systems will develop after promulgation of industrial boiler New Source Performance Standards.

Information Sources: Blythe, Gary, meeting notes, Joy, Los Angeles, CA, June 14, 1979.

Janssen, Kent and Eriksen, Robert L., "Basin Electric's Involvement with Dry Flue Gas Desulfurization," paper presented at the Fifth EPA Symposium on Flue Gas Desulfurization, Las Vegas, NV, March 5–8, 1979.

Dickerman, J.C., telephone conversation with Jim Meyler, Joy, October 26, 1979.

Kennecott Development Company—Environmental Products Division (Formerly Carborundum)

Background: As Carborundum, the Environmental Products Division (EPD) conducted pilot-scale dry injection studies (1,000 acfm) at their University of Tennessee facility and pilot unit tests on a 15,000 acfm spray dryer/baghouse system at Basin Electric's Leland Olds Station.

Research: The Environmental Products Division is continuing pilot-scale test work (1,000 acfm) on both dry injection and spray dryer SO_2 control systems at their University of Tennessee facility. The UT unit is a small stoker-fired boiler burning a coal with an average sulfur content of 0.5% and a higher heating value of about 10,000 Btu/lb.

Dry injection tests have been run with both sodium bicarbonate and nahcolite. Temperatures of the entering flue gas ranged between 270° and 300°F, while entering SO_2 concentrations ranged between 1,000 and 4,000 ppm. (Some test work has been done at the 5,000 ppm SO_2 level.) In tests using relatively non-alkaline fly ash as the sorbent, 25% SO_2 removal was obtained. Results of the major portion of the test work are considered proprietary.

The Environmental Products Division was planning to start up a 1,000 acfm spray dryer/baghouse system in late 1979. Sorbents to be tested include lime and soda ash. Recycle of spent sorbent/fly ash mixtures and removal efficiency at high inlet SO_2 concentrations will also be investigated.

Commercial Status: The Environmental Products Division plans to emphasize both utility and industrial applications in marketing their spray dryer-based flue gas cleaning systems. They are in the process of executing an exclusive agreement with a spray dryer company with a demonstrated background in both rotary and nozzle atomization. EPD will most likely employ rotary atomization in their lime-based spray dryer systems. As Carborundum, EPD had an agreement with DeLaval for supplying the spray dryers; however, this agreement has been terminated.

The Environmental Products group recently submitted a bid to Colorado Ute Electric Association for a 450 MW flue gas cleaning system. Colorado Ute is considering only dry systems for this unit. In regard to the future market for utility and industrial dry FGD systems, the Environmental Products Division sees spray dryer-based FGD systems being, at the very least, competitive with wet systems for both low and high-sulfur coal applications.

Information Sources: Blythe, Gary, telephone conversation with Don Boyd, Carborundum, May 16, 1979.

Blythe, Gary, telephone conversation with H.M. Majdeski, Carborundum, July 6, 1979.

Majdeski, H.M., Carborundum, personal correspondence with Gary Blythe, July 17, 1979.

Kelly, M.E., telephone conversation with H. Majdeski, EPD, Kennecott Development Co., October 18, 1979.

Koch Engineering

Summary of Activities: Koch began development of a "spray dryer based dry FGD system" in early 1979. In May 1979 Koch was reportedly about one year away from having a marketable dry FGD system.

Koch Engineering has completed studies on their spray dryer system verifying published efficiency data. They have achieved SO_2 removals comparable to literature values. (Neither sorbent type nor test conditions used to achieve these removal levels were specified.)

Koch plans to begin marketing their spray dryer/baghouse system in February 1980. No details on the marketing program were available due to the somewhat unique nature of the approach they plan to take.

Koch is presently conducting in-house demonstration and design studies on the 3,000 to 5,000 acfm scale. No test results were provided. They also plan to conduct larger scale pilot tests on an 8-ft diameter spray tower. Sorbents to be investigated include sodium hydroxide, sodium bicarbonate and lime. Initially testing will be conducted with inlet SO_2 levels representative of low-sulfur coal combustion, with plans for future testing at higher SO_2 levels.

Information Sources: Blythe, Gary, telephone conversation with Ahmed Akacem, Koch Engineering, May 16, 1976.

Kelly, M.E., telephone conversation with Ahmed Akacem, Koch Engineering, October 19, 1979.

Mikropul

Background: Mikropul is a division of U.S. Filter, which includes Ducon, the wet scrubber manufacturer. Mikropul was originally in the pulverizing business and was called Pulverizing Machinery Company. Mikropul's air pollution equipment business was a result of the necessity to control pulverizing emissions. Their first equipment was a fabric blow ring collector dating back to the 40s. A version of their current continuous pulse-jet baghouse was introduced in 1957. Since then nearly 200 installations around the world have been put into service with Mikropul Pulse-jet bag collectors for recovery of the spray-dried product. In the early 1970s, Mikropul began work in the pure dry scrubbing business, designing and installing equipment using dry alumina to control fluoride emissions from aluminum smelters. In these applications, fluid bed feeders are used to introduce fresh and recycled alumina to a reactor to contact fluoride-containing flue gases. Some 70 percent of the fluoride removal occurs in the reactor; the remaining 30 percent occurs on the downstream bag collectors. Pulse-jet bags, rather than reverse air or shaker bags are employed, partly because they provide for more turbulent gas/solid contact in the vicinity of the bag.

Their largest installation at Ormet Aluminum in Hannibal, Ohio treats 2.6×10^6 acfm, using 144 pulse-jet bag modules containing some 55,000 bags and around 1,000 pulse-jet valves. Six reactors are used, and some reactant captured on the bags is recycled to improve overall sorbent utilization. Total Mikropul installed dry scrubbing equipment worldwide treats 12×10^6 acfm. Mikropul has also sold dry systems which use dry hydrated lime to remove fluorides from a gas stream in a glass manufacturing facility.

Research: A sister company to Mikropul, Filtrol, a refining catalyst manufacturer, has participated with Mikropul in identifying possible dry FGD sorbents for evaluation in pilot- or bench-scale dry injection/baghouse collection facilities. They have tested zeolites, which unfortunately selectively remove water before SO_2. Of the common dry FGD sorbents, Mikropul has only tested sodium bicarbonate; seventy percent SO_2 removal was obtained from flue gas containing 1,600 to 2,000 ppm SO_2 at $300°F$.

Mikropul began work on a spray dryer-based FGD system in January 1978. Bayliss Industries supplied the spray dryer technology. Besides lime studies, Mikropul has evaluated zinc oxide wastes as an FGD sorbent. Over 50 percent SO_2 removal with a ZnO slurry feed to a spray dryer system was obtained.

Mikropul's test facility at Summit, NJ includes a small nozzle atomization spray dryer which discharges to a small pulse-jet bag collector. A gas burner supplies flue gas. There are provisions to spike the flue gas in the 1,000 acfm unit with SO_2, moisture and fly ash from the Mercer generating station. Another U.S. Filter subsidiary, Drew Chemical, is working with Mikropul to provide lime slurry additives. Drew Chemical has provided polyelectrolyte additives which acting as surfactants, reduce lime slurry settling tendencies and improve atomization in a spray dryer.

As the result of their spray dryer work, Mikropul was awarded a contract in late 1978 to provide a spray dryer/baghouse on an industrial coal-fired boiler at Strathmore Paper Company, a subsidiary of Hammermill, Inc., at Woronoco, MA. Their agreement with Strathmore allows Mikropul to vary numerous parameters during a six-month test period in order to establish a data base for future designs. The system is lime-based and the single spray dryer will use multifluid nozzles for atomization. A pulse-jet baghouse will be used. Provisions for solids recycle will exist. Parameters to be varied include reagent type, sulfur levels, slurry density and additives, slaking techniques, atomization techniques, and baghouse parameters. In addition, other collection techniques may be investigated.

Commercial Status: As mentioned above, Mikropul has sold a lime-based spray dryer/baghouse collection FGD system to Strathmore Paper at Woronoco, MA for use on an industrial boiler. This PC boiler, which generates 90,000 lb of 675 psig steam per hour, produces 40,000 acfm at $350°$ to $400°F$ by burning 2 to 2½ percent sulfur, 9 percent ash coal. The plant must meet a Massachusetts air quality regulation of 0.55 lb SO_2 per 10^6 Btu. This fuel source can be augmented by up to 30 percent (by Btu content) No. 6 fuel oil. Mikropul's guarantee calls for 75 percent SO_2 removal on 3 percent sulfur coal with a maximum stoichiometry on the order of 2.75. Pilot studies indicate 75 percent removal can be achieved at up to 2,000 ppm SO_2 inlet concentration with a stoichiometry of around 1.2. Sorbent utilization has been on the order of 70 percent. Flue gas is withdrawn downstream of the boiler air heater at $350°$ to $400°F$, but a provision is available to bypass hot flue gas if necessary.

The Mikropul system started up in September, 1979. Mikropul personnel are operating the system on a 24-hr, 7-day-per-week basis. They have experienced some operating problems but are in the process of correcting these by instituting design changes. Performance results are not publicly available at this time. Mikropul will continue operation of the system for optimization and establishment of a design data base.

The design details of the FGD system are considered proprietary by Mikropul, but the system can be generally described as a lime sorbent spray dryer followed by a Mikro-Pulseair pulse-jet bag collector. Spent reactant and fly ash disposal will be handled by the paper company.

A third party is involved in the Strathmore Paper project: Lucien J. Luckel, an engineering constructor with an extensive background in industrial boilers, is providing the expertise for dealing with industrial boiler clients, as well as designing all interfaces and tie-ins with the boiler itself. Mikropul has no immediate plans to enter the utility boiler market, but may in the future depending on the success of the Strathmore Paper unit. If Mikropul should go into the utility business, the third party providing the interface capabilities with utility clients would be another U.S. Filter company, Resource Scientists of Tulsa, which has utility engineering and construction experience.

In a utility design, a reverse air rather than a pulse-jet bag collector may be required. The break-even point for the two types is about 200,000 acfm (applications where the pulse-jet can operate at an air-to-cloth ratio of 3.4 times that of a reverse air unit). Mikropul has installed reverse air units at a Southern Colorado Utilities power plant and on several industrial boilers.

Information Sources: Blythe, Gary, meeting notes, meeting at Mikropul, Summit, NJ, May 24, 1979.

Blythe, Gary, telephone conversation with Tom Reinauer, Mikropul Corp., October 25, 1979.

Research-Cottrell

Background: Previous FGD work by Research-Cottrell has focused on wet lime/limestone systems. The only previous purely dry FGD work done by a Research-Cottrell company has been that done by their wholly-owned subsidiary, KVB. A spray dryer/baghouse pilot system at the Texas Utilities Big Brown unit is the first major dry FGD effort by the Research-Cottrell R&D group. This spray dryer/baghouse pilot dry FGD program will evaluate several sorbents. Research-Cottrell has an exclusive agreement with Komline-Sanderson for use of their spray dryer in a dry FGD system.

Research: Research-Cottrell has completed pilot unit studies with lime at the Texas Utilities Big Brown unit (10,000 acfm). Details of the test work are considered proprietary at this time. A final report on the test work was to be completed by December 1979 and Research-Cottrell may present the results at a seminar or conference. Research-Cottrell's wholly owned subsidiary, KVB, has completed an Electric Power Research Institute (EPRI) funded study on dry injection with the results scheduled for publication in the near future.

Commercial Status: Now that pilot unit testing is completed, Research-Cottrell (RC) is ready to offer a commercial spray dryer/baghouse system. They are confident of scaling up design data from 10,000 acfm because they have scaled up wet system designs for up to 3,000 MW worth of FGD from 5,000 to 10,000 acfm pilot data. For a utility size installation, Research Cottrell anticipates using multiple atomizers per dryer vessel. The designs of the reagent feed system and the slaker, fan, dampers, and other overall system components will be based on Research-Cottrell's wet FGD system experience. Research-Cottrell has in-house nozzle atomization experience, and if a sodium system is ever required, nozzle atomization would probably be used. Rotary atomizers are believed superior in a calcium system because they are less susceptible to pluggage and erosion than nozzles. Plans are to operate the spray dryer outlet at about 50°F above adiabatic saturation to avoid flue gas buoyancy and collector bag problems.

Information Sources: Blythe, Gary, meeting at Research-Cottrell, Somerville, NJ, May 23, 1979.

Kelly, M.E., telephone conversation with Kishor Parikh, Research-Cottrell, October 18, 1979.

Rockwell International/Wheelabrator-Frye Joint Venture

Background: For the past eight years, Rockwell International has been developing their sodium carbonate-based regenerative Aqueous Carbonate Process which uses a spray dryer as a flue gas contactor. A simplification of the process excludes regeneration and instead involves operation in an open-loop manner as a "throwaway" process. Employing the open-loop portion of the Aqueous Carbonate Process (ACP), Rockwell and Wheelabrator-Frye have jointly developed a two-stage dry scrubbing process where an alkaline solution or slurry is introduced into a spray dryer contactor, and dry reaction products and fly ash are collected by a fabric filter. Rockwell successfully demonstrated the open-loop portion of the ACP in a 7-ft diameter dryer at the Southern California Edison Mohave Station in 1972. Flue gas was taken downstream of the station's electrostatic precipitators and contacted with a sodium carbonate solution in the spray dryer.

Wheelabrator-Frye piloted a dry injection/baghouse collection FGD system at the Basin Electric's Leland Olds Station using nahcolite as a sorbent. When it became evident that nahcolite supplies would not be available in commercial quantities in the near future, the Rockwell open loop ACP spray drying concept was combined with the Wheelabrator-Frye baghouse that provided spent sorbent and fly ash collection. It was found that the SO_2 removal by the open-loop portion of the Rockwell ACP was not as sensitive to sorbent type as was dry injection; yet, it still retained the desired dry collection feature. Initial testing was with sodium-based sorbents, although lime was later found to be a suitable sorbent.

In all spray dryer-based FGD development work and in commercial sales, Rockwell has used exclusively Bowen Engineering spray drying equipment. Although the Rockwell-Wheelabrator Frye joint venture developed the two-stage process, Bowen Engineering has an exclusive agreement to provide spray drying equipment.

Research: The spray dryer/baghouse pilot unit at Leland Olds was dismantled in September 1978 after 16 months of operation. During this period Rockwell/Wheelabrator studied a variety of sorbents, with some sorbent recycling, at SO_2 levels from 700 to 3,000 ppm.

There are ongoing studies at Bowen's facilities in New Jersey using a 7-ft dryer and a Mikropul pulse-jet collector. Flue gas is from an auxiliary burner with SO_2 and fly ash spiking provisions.

Rockwell has completed testing on their portable pilot plant unit at Northern States Power's Sherburne County Station and at Pacific Power and Light's Jim Bridger Station. This unit uses a 7-ft dryer with a pulse-jet bag collector and has provisions for warm gas bypass. At Jim Bridger, a pilot ESP which was originally used to design the existing precipitators is available to test a spray dryer/ESP system. Fortunately, this Flakt pilot ESP unit is about the same size as the portable test unit (approximately 5,000 cfm).

Additional pilot unit testing has begun at Commonwealth Edison's Joliet Station. The Joliet pilot unit is to be flexible enough for full parametric studies. The unit uses a 7-ft, 4,000 to 5,000 acfm spray dryer, with provisions for using either a reverse-air, off-line cleaning baghouse or a pulse jet, continuous flow baghouse as well as provisions for an electrostatic precipitator. The unit will have provisions for both warm and hot gas bypass and sorbent/fly ash recycle. Most work will be done with lime sorbent, but some soda ash studies may be made. The Joliet plant burns varying mixtures of four Western subbituminous coals, 8,000 to 9,000 Btu/lb, 0.5 to 1 percent sulfur. Testing will last 1 to 2 years.

Other test work being conducted by Rockwell includes bid support studies on their spray dryer at their California facility and periodic design studies on a 7-ft diameter dryer at Bowen Labs in New Jersey. Rockwell is also negotiating a site for testing a lime-based system for high-sulfur Eastern coal applications.

Other Rockwell research includes studies in waste disposal. These studies show that for nearby disposal (up to 1 mile), pneumatic conveying may be the most economical method of transportation of waste material. For longer distances, pelletizing or briquetting of wastes may be required to permit open trucking. This operation could be accomplished for about $1.25 per ton (operating and capital). The resulting briquette may be unleachable due to fusing. The only binder material required would be water.

Rockwell has also looked into reuse schemes. By adding a proportionate amount of water to the waste material, it sets up resulting in load bearing properties of around 500 lb/in^2. Permeability of the product is less than 1 ft/yr (10^{-7} cm/sec). Some of the waste materials may be useful as a concrete additive.

Commercial Activities: Following the research work at Leland Olds, the Rockwell/Wheelabrator-Frye joint venture was awarded a turnkey subcontract for the furnishing, fabrication, delivery, erection, and successful operation of a complete emissions control system for the Coyote plant. The Coyote Station is a 410-MW lignite-fired unit to be located near Beulah, North Dakota. The station is to be owned by a consortium of five North Dakota and Minnesota power companies. Otter Tail Power Company is to operate the plant.

The 410-MW Coyote Station FGD system is designed for 70 percent SO$_2$ removal for all fuels. Guaranteed sorbent utilization is 80 percent, a conservative value for this high utilization sodium-based system.

Design of the Coyote Station calls for an air preheater outlet temperature of 285°F. The stack gas must exit at 185°F. These conditions are not optimum for spray dryer performance; 185°F is 50° to 60°F above adiabatic saturation. The design temperature seems somewhat low as most lignite-fired boilers experience air preheater exit temperatures of 325° to 350°F. Rockwell does not expect any ash alkalinity utilization in the design since there is no ash recycle and reaction of solid phase fly ash alkalinity with SO$_2$ is minimal.

The Coyote Station will use four 46-ft dryers, each with three 150-hp atomizers (although at design conditions each will draw only 83 hp). The primary control variable of the system will be dryer outlet temperature. Outlet dew point will also be measured and used for setting approach to dew point on the outlet. Dryer outlet and stack SO$_2$ will have a narrow range of control on sorbent feed rate. In other words, if the outlet SO_2 level is well below the control point, the SO_2 input to the sorbent feed control would make a small decrease in the feed rate. If outlet SO_2 level was a primary control point, the sorbent feed rate might be shut off completely until the outlet SO_2 increased sufficiently.

Bag temperature will be limited by an internal bypass between the inlet and outlet plenums on the baghouse. These bypass dampers will be set to open if the baghouse inlet temperature exceeds a maximum set point. The baghouse will use a synthetic fabric, such as Dacron, and will operate at a net air-to-cloth ratio of 2.7 to 1. This involves a total of 28 compartments, with two off for maintenance and two off for cleaning at a given time.

Rockwell will use no control valves in slurry service. All slurry feed rates will be set by progressive cavity pumps (such as Moyno) with variable-speed drives. There will be no flue gas flow dampers or control valves. The air distribution through the FGD system is controlled by careful design of the ductwork. The dryer vessels are designed with three atomizers per vessel. The atomizers are placed a standard distance from the vessel wall to avoid wall wetting. No agglomeration problems on overlap of spray patterns from the three atomizers are foreseen. In tests using three atomizers in a 7-ft dryer with extensive spray pattern overlap, virtually the same SO_2 removal was measured as would be achieved using a single atomizer flowing the same amount of sorbent. In $\frac{1}{16}$-scale air flow testing of the Coyote ductwork design, they found they were able to distribute equal flow to four dryers within two percent and equal flow to three atomizers in a dryer to within two percent. Bowen feels the three-atomizer approach is advantageous because loss of an atomizer would not result in loss of a whole dryer module. Although the use of multiple atomizers has not been demonstrated at the 500,000 acfm level, Bowen has demonstrated flowing $\frac{1}{3}$ of the total design gas flow (170,000 acfm) to a single atomizer.

The Coyote plant will use coal trucks to return waste sorbent/fly ash to the mine. The material will be handled in the dry form, using dustless loading equipment. Design of this equipment is not within the Rockwell battery limits, however.

Construction on the Coyote plant is well under way. On site, foundations and considerable structural steel are in place. Off site, fabrication is proceeding on dryer vessels. The plant should start up by the spring of 1981. Start-up will more than likely be with commercial soda ash. Rockwell is currently looking into alternative sources of lower-quality sodium products for use as sorbent at Coyote. Sources from both the Green River, Wyoming area and the Owens Lake, California area are being pursued. One source would involve a material that is roughly 50 percent Na_2CO_3 at a cost of $13/ton. On a cost per ton of Na_2CO_3 equivalent, this is less than half the $60+/ton of commercial sodium carbonate. Of course, shipping costs for the less pure product would be greater due to the greater quantities involved. Soda ash will be stored on site at Coyote as a sodium carbonate monohydrate slurry, with water circulated through the storage tanks. Feed to the spray dryers will be saturated sodium carbonate monohydrate solution from these tanks, which is diluted to achieve the desired spray dryer temperature. Soda ash will be stored as a monohydrate slurry rather than as the anhydrous solid because the monohydrate slurry has a bulk density of 71 lb/ft^3 versus 55 lb/ft^3 for commercial dry soda ash.

The Rockwell/Wheelabrator-Frye joint venture has also sold an industrial dry FGD system to Celanese Corporation. The Celanese project involves treating flue gas from a stoker-fired industrial boiler in Cumberland, Maryland. Currently, the boiler is slated to burn 1 to 2 percent sulfur coal but may go to a 3 to 4 percent sulfur West Virginia coal. The flue gas flow rate is 65,000 acfm at 350°F. The Rockwell system will use lime sorbent with no recycle of sorbent/fly ash. SO_2 removal will be 70 to 80 percent for 1 to 2 percent sulfur coal, higher for 3 to 4 percent coal. Sorbent utilization will be on the order of 70 to 80 percent. Particulate collection will be with a pulse jet, continuous type collector. Lime slaking will be accomplished with a "pore tube" slaker. Construction of this system is on schedule with start-up planned for January 1980.

As far as other commercial sales, Rockwell has two bids under evaluation and one bid being prepared. Rockwell expected to bid on up to six more utility dry FGD systems for Western low-sulfur coal or lignite applications by the end of 1979.

Rockwell has a standard approach to a dry system design. First, determination is made whether a "standard" system will work. This involves approximately a 40°F approach to dew point, no recycle, no gas bypass, and preferably about 90 percent sorbent utilization. If this is unattainable, then warm gas bypass is tried taking a small amount of flue gas after the air preheater and routing it around the spray dryer to reheat from a closer-than-40°F approach to dew point. If this is insufficient, hot gas bypass is included taking flue gas upstream of the air preheater for reheat. There is a penalty associated with hot gas as bypass approximating 0.5 percent of the total fuel rate to the boiler if 5 percent of the gas flow is bypassed. Recycle, considered to be a supplement to any of the above designs, can be used to get an apparent sorbent utilization of greater than one. Recycle of fly ash and partially spent sorbent can result in both improved dryer performance and improved filter performance for a combined synergistic effect. Rockwell designs recycle equipment as an "add on" with a minimum of redundancy (minimum first cost). Adding recycle provisions might add about two percent to the total equipment cost.

Rockwell reports that for large Eastern applications, the cost of lime reagent is a major factor in the viability of dry systems. In utility high-sulfur applications, the cost differential between lime in a dry system and limestone in a wet system can mean sorbent savings of millions of dollars a year for the wet lime-stone system. This puts the operating economics of a dry system at a disadvantage for these applications. A scheme for using a limestone sorbent in a dry system could greatly improve the economics. Rockwell sees the application of dry FGD to Eastern applications as the focal point of future R&D efforts.

Information Sources: Blythe, Gary, meeting notes with Rockwell International, Canoga Park, CA, June 13, 1979.

Kelly, M.E., telephone conversation with Dennis Gehri, Rockwell International, October 18, 1979.

Janssen, Kent and Eriksen, Robert L, "Basin Electric's Involvement with Dry Flue Gas Desulfurization," paper presented at the EPA Symposium on Flue Gas Desulfurization, Las Vegas, NV, March 5-8, 1979

Johnson, O.B., et al, "Coyote Station—First Commercial Dry FGD System," paper presented at the 41st Annual Meeting American Power Conference, Chicago, IL, April 23-25, 1979.

Moore, K.A., et al, "Dry FGD and Particulate Control Systems," paper presented at the Fifth Annual EPA Flue Gas Desulfurization Symposium, Las Vegas, NV, March 5, 1979.

REFERENCES

(1) Kaplan, S.M. and Felsvang, Karsten, Spray Dryer Absorption of SO_2 from Industrial Boiler Flue Gas. (Presented at the 86th National AICHE Meeting, Houston, TX, April 1979.)

(2) Lui, Han and Chafee, R., Evaluation of Fabric Filter as a Chemical Contactor for Control of SO_2 in Flue Gas. (Presented at the Air Pollution Control Office Fabric Filter Symposium, Charleston, SC, March 1971.)

(3) Moore, K.A., et al, Dry FGD and Particulate Control Systems. (Presented at the Fifth EPA Symposium on Flue Gas Desulfurization, Las Vegas, NV, March 5-8, 1979.)

(4) Dustin, D.F., Report of Coyote Pilot Plant Test Program. Test Report, Rockwell International (Atomics International Division), Canoga Park, CA, November 1977.

(5) Davis, R.A., et al, Dry SO_2 Scrubbing at Antelope Valley Station. (Presented at the 41st Annual American Power Conference, Chicago, IL, April 25, 1979.)

(6) Janssen, K.E. and Eriksen, Robert L., Basin Electric's Involvement with Dry Flue Gas Desulfurization. (Presented at the Fifth EPA Symposium on Flue Gas Desulfurization, Las Vegas, NV, March 5-8, 1979.)

CITRATE PROCESS

The information in the following two sections is based on *Citrate Process Demonstration Plant Design,* 1979, BuMines IC 8806, by W.I. Nissen of Salt Lake City Metallurgy Research Center, Bureau of Mines and R.S. Madenburg of Morrison-Knudsen Co., Inc. for the Bureau of Mines.

INTRODUCTION

In 1968 the Department of the Interior's Bureau of Mines started research on FGD with particular emphasis on control of SO_2 emissions from nonferrous smelters. This research was conducted in an effort to meet the Bureau's goal of minimizing the undesirable environmental impacts of mineral processing operations. Pioneering research had indicated that flue gas desulfurization might be achieved effectively by the absorption of sulfur dioxide (SO_2) in a suitable solution, followed by reaction of the absorbed SO_2 with gaseous hydrogen sulfide (H_2S) to precipitate sulfur and regenerate the solution for recycle. After a year of screening many possible reagent combinations of inorganic and organic solutions, it was established that an aqueous solution of citric acid and sodium citrate was an effective absorbent for SO_2. Among the desirable characteristics affecting the choice of citrate were its good chemical stability, low vapor pressure, and adequate pH-buffering capacity, and the purity and physical character of the precipitated sulfur.

As first studied in the laboratory, the citrate process comprised (1) absorbing SO_2 in citrate solution, (2) reacting the absorbed SO_2 with H_2S, (3) filtering and melting the precipitated sulfur, and (4) recycling the regenerated citrate solution to the SO_2 absorption step. Preparation of the H_2S required for the precipitation step of the citrate process was studied separately. Hydrogen sulfide was generated by reacting recycle sulfur, natural gas, and steam over an alumina catalyst.

In 1970 to 1971, a pilot plant to remove SO_2 flue gas from a copper reverberatory furnace was constructed and operated jointly by the Bureau of Mines and the Magma Copper Co. This plant, located at a smelter in San Manuel, Arizona, treated 300 scfm of gas containing 1.0 to 1.5% SO_2 and consistently removed 93 to 99% of the SO_2. Results of the initial laboratory and pilot plant research were reported (1)(2).

The preliminary Bureau of Mines laboratory and pilot plant research demonstrated that the citrate process is capable of substantially complete removal of SO_2 from industrial waste gases. Most of the SO_2 is converted to sulfur with less than 1.5% converted to sulfate regardless of the SO_2 and oxygen content of the feed gas. The citrate process produces an elemental sulfur end product that can be marketed or stored with a minimum of environmental disturbance.

After the encouraging preliminary results, two other pilot plant investigations were undertaken to obtain data for engineering evaluation and cost estimates. One pilot plant was independently built and operated by Arthur G. McKee and Co., Peabody Engineered Systems, and Pfizer, Inc., at Terre Haute, Indiana (3)–(6). This operation treated a stack gas from a coal-fired steam-generating station that simulated a utility application. After several modifications to arrive at a final equipment configuration, the plant operated from March 15 to September 1, 1974, logging 2,300 operating hours. The longest sustained run was 180 hours. The plant consistently removed 95 to 97% of the SO_2 from flue gas containing 0.1 to 0.2% SO_2.

The other pilot plant was constructed by the Federal Bureau of Mines and operated jointly by the Bureau and the Bunker Hill Co. at the latter's lead smelter in Kellogg, Idaho (7)–(12). Nominal capacity of the plant was 1,000 scfm of 0.5% SO_2 gas, yielding about 600 pounds of sulfur per day. This pilot plant was operated between January 1974 and May 1976 for 5,400 hours and produced 55 net short tons of sulfur. The operation demonstrated that (1) more than 95% of the SO_2 could be removed from the lead smelter sintering furnace tail gases, (2) regeneration of the citrate solution with H_2S and precipitation of sulfur in conventional stirred vessels was readily controlled and highly efficient, (3) the precipitated sulfur could be continuously recovered as a 99.5% pure product by kerosene flotation and melting, and (4) the 77 to 79% H_2S gas used for sulfur precipitation could be readily produced from pilot plant product sulfur, natural gas, and steam.

During 1975, plans were made for the construction and operation of a large-scale plant to demonstrate the applicability of the citrate process for FGD at power plants burning high-sulfur coal. On June 1, 1976, the Bureau awarded a contract to the St. Joe Minerals Corp. for commercial-scale demonstration of the citrate process. The host site for the plant is St. Joe's George F. Weaton power plant near Monaca, Pennsylvania (13). This demonstration plant will be operated under a cooperative arrangement on a cost-sharing basis between the Bureau of Mines, the U.S. Environmental Protection Agency, and St. Joe. The design, construction, and maintenance of this plant were contracted to the Morrison-Knudsen Co., with the Ralph M. Parsons Co. providing engineering services. Radian Corp., under contract to the Bureau, will independently test and evaluate the operation.

References

(1) George, D.R., Crocker, L. and Rosenbaum, J.B., "The Recovery of Elemental Sulfur from Base Metal Smelters," *Min. Eng.,* vol 22, No. 1, January 1970, pp 75-77.
(2) Rosenbaum, J.B., George, D.R. and Crocker, L., "The Citrate Process for Removing SO_2 and Recovering Sulfur from Waste Gases," Pres. at AIME Environmental Quality Conf., Washington, D.C., June 7-9, 1971, 26 pp; available upon request from the Salt Lake City Metallurgy Research Center, Bureau of Mines, Salt Lake City, Utah.

(3) Chalmers, F.S., "Citrate Process Ideal for Claus Tailgas Cleanup," *Hydrocarbon Processing,* vol 53, No. 4, April 1974, pp 75-77.

(4) Chalmers, F.S., Korosy, L. and Saleem, A., "The Citrate Process to Convert SO$_2$ to Elemental Sulfur," Pres. at Industrial Fuel Conf., Purdue Univ., West Lafayette, Ind., Oct. 3, 1973, 7 pp; available for consultation at the Salt Lake City Metallurgy Research Center, Bureau of Mines, Salt Lake City, Utah.

(5) Korosy, L., Gewanter, H.L., Chalmers, F.S. and Vasan, S., "Sulfur Dioxide Absorption and Conversion to Sulfur by the Citrate Process," Ch. 16 in *Sulfur Removal and Recovery from Industrial Processes,* ed. by John B. Pfeiffer (Advances in Chemistry Series 139), American Chemical Society, Washington, D.C., 1975, pp 192-211.

(6) Vasan, S., "The Citrex Process for SO$_2$ Removal, *Chem. Eng. Prog.,* vol 71, No. 5, May 1975, pp 61-65.

(7) McKinney, W.A., Nissen, W.I., Crocker, L. and Martin, D.A., "Status of the Citrate Process for SO$_2$ Emission Control," Pub. in *Technology and Use of Lignite, Proceedings of a symposium held at Grand Forks, North Dakota, May 14-15, 1975,* compiled by W.R. Kube and G.H. Gronhovd. GFERC/IC-75/2, 1975, pp 148-172.

(8) McKinney, W.A., Nissen, W.I., Elkins, D.A. and Rosenbaum, J.B., *Pilot Plant Testing of the Citrate Process for SO$_2$ Emission Control,* EPA 650/2-74-126b, Proc., vol 2, December 1974, pp 1049-1067.

(9) McKinney, W.A., Nissen, W.I. and Rosenbaum, J.B., "Design and Testing of a Pilot Plant for SO$_2$ Removal from Smelter Gas," Pres. at Ann. Meeting, AIME, Dallas, Texas, Feb. 23-28, 1974, AIME Preprint A-74-85, 12 pp.

(10) Nissen, W.I., Crocker, L. and Martin, D.A., "Lead Smelter Flue Gas Desulfurization by the Citrate Process," Ch. 52 in *World Mining and Metals Technology,* ed. by Alfred Weiss (Proc. Joint MMIJ-AIME Meeting, Denver, Colorado, Sept. 1-3, 1976), Port City Press, Baltimore, Maryland, vol 2, 1976, pp 825-854.

(11) Nissen, W.I., Elkins, D.A. and McKinney, W.A., *Citrate Process for Flue Gas Desulfurization, A Status Report,* EPA-600/2-76-136b, Proc., vol 2, May 1976, pp 843-864.

(12) Rosenbaum, J.B., McKinney, W.A., Beard, H.R., Crocker, L. and Nissen, W.I., *Sulfur Dioxide Emission Control by Hydrogen Sulfide Reaction in Aqueous Solution—The Citrate System,* BuMines RI 7774, 1973, 31 pp.

(13) Madenburg, R.S., and Kurey, R.A., *Citrate Process Demonstration Plant,* A Progress Report, EPA-600/7-78-0586, Proc., vol 2, March 1978, pp 707-735.

DEMONSTRATION PLANT AT GEORGE F. WEATON POWER STATION

Power Plant Host Site

The citrate process FGD demonstration plant is being retrofitted to St. Joe's George F. Weaton power plant located on the south bank of the Ohio River at Monaca, Pennsylvania, near Pittsburgh. This power plant is a base-loaded, coal-fired electricity-generating plant operating at a unit load factor of approximately 90% on a 24-hour-per-day and 7-day-per-week basis. The power plant consists of two identical units, each capable of supplying 60 MW of electrical energy to St. Joe's nearby zinc smelter. In addition, the power plant can interchange 25 MW of electrical power with the Duquesne Light Co. The host power plant has a conventional tangentially fired, pulverized-coal-burning boiler and utilizes a direct-coupled boiler-turbine generator arrangement. The main steam flow is 450,000 lb/hr at 1000°F and 1,850 psig with integral reheat steam of 334,000 lb/hr at 1000°F and 424 psig for each boiler at maximum load. The overall power plant heat rate is approximately 10,000 Btu/kWh. Boilers operate the five stages of feedwater heating and utilize steam-turbine-drive boiler feed pumps. Each boiler has its own economizer, superheater, combustion air preheaters, draft fans, and combustion control for independent operation.

Approximately 85,000 tons of coal is stockpiled adjacent to the power plant. This coal is transported by belt conveyor to coal hoppers at the power plant. Coal from these hoppers is pulverized and fired in the power plant boilers. For the purpose of the citrate process demonstration, coal will contain from 2.5 to 4.5% sulfur and less than 15% ash.

Particulate control is achieved through a combination mechanical and two-stage electrostatic precipitator which removes over 99.6% of the fly ash. Bottom ash from the boilers and fly ash from the precipitators are conveyed to a settling pond for decantation, drying, and subsequent disposal.

Citrate Process Demonstration Plant

Construction of the citrate process FGD demonstration plant was completed at the end of April 1979. As shown in the generalized process flow sheet (Figure 8.1), the citrate process will draw SO_2-bearing flue gas from the duct between the existing electrostatic precipitator and the stack. The nominal design capacity of this plant is 156,000 scfm of 0.2% SO_2 gas, which at 90% SO_2 removal and 2% SO_2 oxidation will yield approximately 16 tons of sulfur per day. The plant will have six specific unit operations, as follows:

(1) Residual fly ash removal and gas cooling.

(2) SO_2 absorption.

(3) Solution regeneration and sulfur precipitation.

(4) Sulfur recovery.

(5) Sulfate removal.

(6) H_2S generation.

The design parameters, design features, and process equipment for these six unit operations follow.

Ash Removal and Gas Cooling: A forced-draft fan transports the flue gas through a 10-foot-diameter duct to the plant. This fan is a double-inlet mechanical-draft type which will deliver in normal operation 234,000 actual cubic feet per minute at 300°F and a discharge pressure of 9.7 inches of water. The fan is constructed of carbon steel and is driven by a 1,000-hp electric motor.

Downstream of the fan, an eductor-type venturi scrubber removes residual fly ash, sulfur trioxide (SO_3), and chlorides (Cl^-) from the flue gas. This venturi also cools the gas from 300° to 120°F by both humidification and sensible cooling. Sensible cooling requires high solution recirculation rates, and the eductor design utilizes a 4,500-gpm flow at 40 psig at the nozzle to reduce the fan requirement. Venturi scrubber solution and flue gas separate in the bottom of the vessel that contains the SO_2 absorber. Except for a bleed stream, most of the solution recycles to the venturi scrubber. Two pumps in series transport the recycle solution through a heat exchanger. These pumps are the centrifugal, horizontal slurry type and will deliver 4,500 gpm in normal operation. The pump casings and impellers are rubber-lined cast iron with Hastelloy C trim. Each pump is driven by a 250-hp electric motor. The heat exchanger cools the recycle solution from 120° to 112°F. This heat exchanger is designed to exchange 19.8×10^6 Btu/hr and is a plate type fabricated of palladium-stabilized titanium. Makeup water replaces the humidification and bleed stream losses.

Figure 8.1: Power Plant Stack Gas SO₂ Scrubbing–Citrate Process

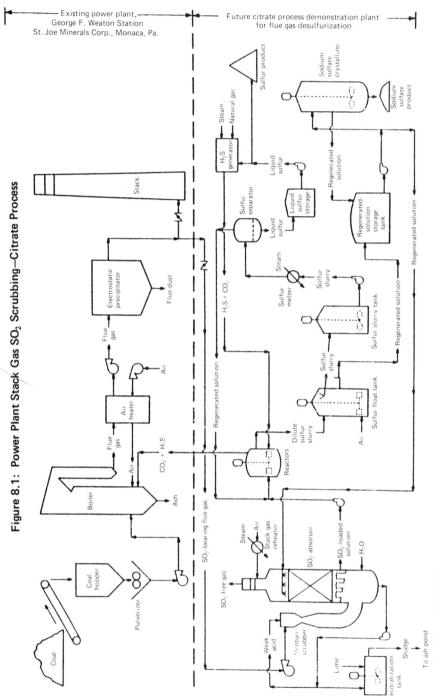

Source: BuMines IC 8806

The venturi scrubber, which is 10 feet in diameter by 45 feet high, has an acidproof brick and rubber membrane lining over a carbon steel shell. The venturi nozzle is fabricated from Inconel 625.

The bleed stream removes ash and acids and prevents their buildup. This bleed stream flows through a 2-foot-diameter by 18-foot-high scrubber effluent stripper packed with about 31 cubic feet of 1-inch polypropylene saddles. At the same time that the bleed stream flows through the stripper, a 100-scfm stream of air flows countercurrently, strips any absorbed SO_2, and returns the SO_2 to the absorption tower. The stripper is fabricated of fiberglass-reinforced polyester (FRP).

The SO_2-stripped bleed stream flows to an acid brick-lined, agitated, concrete tank where lime is added to neutralize acids. A pump transfers the neutralized bleed stream to the existing ash-settling pond.

SO_2 Absorption: The cooled, cleaned flue gas passes through a chlorinated polyvinyl chloride (CPVC) chevron entrainment separator and a chimney tray having four 6½-foot-diameter chimneys into the SO_2 absorber. The flue gas flows upward through a 26-foot-diameter by 20-foot-high packed bed of 2-inch polypropylene saddles countercurrent to citrate solution. This citrate solution is a 0.5-molar citric acid solution buffered with sodium base to a pH between 4.5 to 4.7 and contains about 0.25 molar sodium thiosulfate. In the SO_2 absorber more than 90% of the SO_2 is removed from the flue gas by the following reaction:

(1) $$SO_2 + H_2O \rightleftharpoons HSO_3^- + H^+.$$

Reaction (1) is enhanced by the buffering properties of the citrate solution illustrated in reaction (2):

(2) $$H^+ + HSO_3^- + Na^+ + citrate^- \rightleftharpoons Na^+ + HSO_3^- + citric\ acid.$$

The hydrogen ion produced by absorption of SO_2 in reaction (1) combines with a citrate ion in reaction (2) to form citric acid, a nonionized acid. The treated flue gas then passes through a CPVC chevron entrainment separator into a 10-foot-diameter by 105-foot stack mounted on top of the absorber vessel, which is 95 feet high. When required by ambient conditions, steam-heated air reheats the treated flue gas to aid dispersion.

The SO_2 absorber and stack are fabricated of carbon steel. The absorber is lined with vinyl ester reinforced with glass flakes. A Penngard lining is used in the reheat zone between the absorber and the stack. The Penngard lining consists of a corrosion-resistant asphaltic membrane covered with a lightweight heat- and corrosion-resistant insulating brick. The stack is protected with an FRP sleeve.

A pump transports the SO_2-rich citrate solution from the SO_2 absorber to the solution regeneration and sulfur precipitation section. This pump is a centrifugal, horizontal, single-stage type which will deliver 1,200 gpm in normal operation. The pump is constructed of Chlorimet 3 with Hastelloy C trim and is driven by a 40-hp electric motor. After regeneration, another pump recycles the SO_2-lean citrate solution from a citrate storage tank back to the SO_2 absorber. This lean solution pump is similar to the rich solution pump except that it is driven by a 120-hp motor. During lean solution recycle, a heat exchanger cools the solution from 131° to 123°F with river water. This heat exchanger is designed to exchange 5.4×10^6 Btu/hr and is a plate type fabricated of palladium-stabilized titanium.

Also in the SO_2 absorber, the water balance of the process is controlled by vaporization. Water additions come primarily from the solution regeneration reaction, the caustic makeup, and the gas from the H_2S generator.

To vaporize water in the absorber, saturated flue gas entering at 120°F contacts lean citrate solution entering at 123°F. The exiting flue gas, which is heated by the 3°F temperature differential, remains saturated by vaporization of water from the citrate solution.

Solution Regeneration and Sulfur Precipitation: Rich citrate solution leaving the absorber flows in series through two tandem 13,000-gallon-capacity reactors. These reactors are 12-foot-diameter by 18-foot-high carbon steel vessels, which are lined with vinyl ester reinforced with glass flakes. In the reactors, SO_2-rich solution reacts with an H_2S-CO_2 gas from the H_2S generator according to the following overall reaction to regenerate the solution and precipitate sulfur:

$$(3) \quad Na^+ + HSO_3^- + \text{citric acid} + 2H_2S \rightarrow 3S + 3H_2O + Na^+ + \text{citrate}^-.$$

The H_2S-CO_2 gas flows into the second of the two reactors through a sparger underneath an agitator which disperses the gas into the solution. The H_2S gas that does not react in the second reactor flows into the first reactor through a sparger. Again, an agitator disperses the gas. The agitators each have a 67-inch-diameter, four-blade radial impeller which turns at 68 rpm approximately 4 feet from the bottom of the reactor. The agitators are turned by 100-hp electric motors. All wetted parts of the agitators are Inconel 625.

Off-gas, which is mainly CO_2 with small quantities of H_2S, CS_2, and COS from the reactor system, flows to the boiler firebox for incineration to SO_2. Slurry flows by gravity through the reactor system, which is stepped, to a digester. The digester is physically similar to the reactors except the agitator has an axial impeller driven by a 3-hp motor. In the digester, any dissolved or entrained H_2S in the slurry reacts with a small stream of SO_2-rich solution from the absorber. From the digester, the slurry flows by gravity to the sulfur flotation tank.

Sulfur Recovery: Sulfur separates from the bulk of the solution by air flotation in an agitated 18,500-gallon-capacity tank. The flotation tank is 15 feet in diameter by 20 feet high and is made of carbon steel, which is lined with vinyl ester reinforced with glass flakes. In the tank, sulfur froth, aided by air, floats on the surface of the solution. A lobe-type blower supplies air to the tank, and a 25-hp turbo agitator, which has Hastelloy C-276 wetted parts, disperses the air throughout the solution. The sulfur froth, which is concentrated to about 10% solids, flows over a weir to an agitated 18,000-gallon-capacity sulfur slurry tank.

The sulfur slurry tank is 15 feet in diameter by 20 feet high and is made of carbon steel, which is lined with vinyl ester reinforced with glass flakes. The sulfur slurry tank agitator degasses the sulfur froth and suspends sulfur crystals in the solution. The agitator, which turns at 45 rpm, has two 43-inch-diameter, four-blade radial impellers. All wetted parts of the agitator are Inconel 625.

Regenerated lean citrate solution flows from the bottom of the flotation tank through an adjustable overflow weir box that allows adjustment in the solution level in the flotation tank. The weir box is fabricated of FRP with Hastelloy C trim. Lean citrate solution from the weir box flows by gravity to a 125,000-gallon-capacity citrate storage tank, which is made of carbon steel, which is lined with vinyl ester reinforced with glass flakes.

Vapor spaces of all tanks downstream of the reactor system are piped into an exhaust system. An exhaust fan draws vapors containing traces of H_2S and COS and discharges these vapors into the boiler firebox for incineration to SO_2. SO_2 generated from the combustion of H_2S and COS is recycled to the citrate process plant for desulfurization.

Sulfur is separated from the remaining solution by heating the sulfur slurry from 131° to about 260°F and melting the sulfur with steam in a concentric tube heat exchanger. A progressing cavity-type pump transports 64 gpm of sulfur froth from the sulfur slurry tank to the heat exchanger. This pump has Hastelloy C and rubber wetted parts and is driven by a 20-hp electric motor. The heat exchanger is designed to exchange 4.11×10^6 Btu/hr, operates at a pressure of about 35 psig, and is a jacketed-pipe type with the tube and shell fabricated of Hastelloy C-276 and carbon steel, respectively. The liquid sulfur separates from the citrate solution by gravity in an 850-gallon sulfur separator. The sulfur separator is a pressurized, steam-jacketed decanter vessel fabricated of Hastelloy C-276-clad carbon steel. Liquid sulfur, which has a purity of greater than 99.5%, flows from the bottom of the sulfur separator to a 56,000-gallon, heated, carbon steel storage tank. The citrate solution flows from the top of the sulfur separator to the first reactor in the solution regeneration and sulfur precipitation operation.

A filter removes solid impurities from the liquid sulfur, and a pump transfers one-third of the liquid sulfur to load-out for marketing or storage. Another pump transfers the other two-thirds of the liquid sulfur to the H_2S generator, where it is used to manufacture H_2S gas. The sulfur load-out pump, which will pump 50 gpm at 260°F and a 20-psi differential pressure, is driven by a 3-hp electric motor and has a cast steel case and cast iron impeller. The sulfur transfer pump will pump 15 gpm at 260°F and a 72-psi differential pressure, is driven by a 3-hp electric motor, and has a cast iron case and impeller.

Sulfate Removal: Up to 1.5% of the sulfur contained in the flue gas forms sulfuric acid (H_2SO_4) rather than sulfur in the citrate process. Sulfur trioxide, SO_3, contained in the flue gas may be absorbed in the citrate solution by the following reaction:

$$(4) \qquad SO_3 + H_2O \rightarrow H_2SO_4.$$

Absorbed sulfur dioxide, SO_2, in the form of HSO_3^- ion may be oxidized during absorption by the following reaction:

$$(5) \qquad HSO_3^- + H^+ + \tfrac{1}{2}O_2 \rightarrow H_2SO_4.$$

Thiosulfate ion, $S_2O_3^=$, which is a sulfur compound formed during solution regeneration, decomposes during sulfur melting by the following reaction:

$$(6) \qquad 6H^+ + 3S_2O_3^= \underset{\Delta}{\rightarrow} 4S + 2H_2SO_4 + H_2O.$$

Caustic (NaOH) neutralizes the H_2SO_4 and forms sodium sulfate (Na_2SO_4) in solution by the following reaction:

$$(7) \qquad H_2SO_4 + 2NaOH \rightarrow Na_2SO_4 + 2H_2O.$$

Glauber's salt ($Na_2SO_4 \cdot 10H_2O$) crystallizes from a chilled citrate solution, which provides a method to remove the buildup of Na_2SO_4 from the solution.

A vacuum-cooled crystallizer cools a bleed stream of citrate solution and crystallizes Glauber's salt. The capacity of the crystallizer is approximately 4 tons of Glauber's salt per day. Wetted parts of the crystallizer are FRP, Hastelloy, or rubber-lined carbon steel.

Hydrogen Sulfide Generation: Recovered elemental sulfur (S°), natural gas (CH_4), and steam (H_2O) as consumptive feedstocks react to form H_2S according to the following reaction:

$$(8) \qquad CH_4 + 4S^\circ + 2H_2O \rightarrow 4H_2S + CO_2.$$

An on-site H_2S generator produces about 35 tons of H_2S per day in a fixed-bed catalytic reactor. The design of the H_2S generator is based on a proprietary process developed by Home Oil Co. Ltd. of Calgary, Alberta, Canada, as modified and pilot-plant-tested by the Bureau of Mines.

The H_2S generator is designed so that reaction (8) is at least 95% efficient. Impurities besides CO_2 and H_2O include CS_2 and COS. Materials of construction for the H_2S generator process include suitable stainless steels, Alonized stainless steel, and stainless steel lined with high-temperature refractories. Additionally, the H_2S generator is designed and constructed to operate using a carbon monoxide reductant should natural gas supplies become unavailable.

Environmental Impact

The citrate process produces high-purity elemental sulfur and industrial-grade Glauber's salt. The National Environmental Policy Act of 1969 requires the identification and disposition of all the product and waste streams associated with the citrate process demonstration plant. The following list identifies those known and anticipated product or waste streams and states their disposition.

Treated Flue Gas, 156,000 scf: Flue gas will discharge to the atmosphere and will contain approximately 0.02 vol % SO_2.

Product Sulfur, 16 Tons per Day: High-purity elemental sulfur will be used as a feedstock in St. Joe's sulfuric acid plants.

By-Product Glauber's Salt, 5 Tons per Day (80% Solids): Industrial-grade Glauber's salt will be marketed or disposed of in an environmentally acceptable manner (chemical landfill).

Venturi Scrubber Bleed Stream, 72 Tons per Day (5.5% Solids): The bleed stream will contain about 1½ tons of fly ash per day and will be neutralized with lime. The neutralized sludge, which will contain about 1 ton of $CaCl_2$ and 1½ tons of $CaSO_4$ per day, will join the power plant's fly ash disposal stream, which contains about 120 tons of fly ash per day. Fly ash and other solids settle in ponds.

Cooling and Condensate Water Streams: Water streams that are not recycled will be discharged in an environmentally acceptable manner.

H_2S Gas During Startup and Transient Conditions: H_2S gas not utilized by the process will be incinerated to SO_2. The maximum SO_2 emission to the atmosphere during startup will be 3,300 pounds of SO_2 per hour for 30 minutes for a total emission of 1,650 pounds of SO_2. The maximum SO_2 emission to the atmosphere during upset conditions (failure of the process chemistry, failure of the plant equipment, or interruption of process flows) will be 6,600 pounds of SO_2 per hour for 10 minutes, or a total emission of 1,100 pounds of SO_2. During the demonstration year of operation, two startup periods are expected. Upset conditions are estimated to take place no more than once a week and are expected to be of short duration. The emissions resulting from startup or transient conditions will be eliminated or minimized as experience is gained in operating and effectively using the installed instrumentation for process operation and control.

Combustion Gases from the Sulfur Superheater: These gases result from firing No. 2 fuel oil containing 0.3% sulfur and will be discharged to the atmosphere after waste heat recovery.

Vent Gas from the Sulfur Precipitation Reactors: This gas, containing mostly CO_2 with traces of H_2S, COS, CS_2, and CH_4, will flow to the power plant boiler for incineration to SO_2 for recycle to the citrate process demonstration plant.

Vent Gases from the Flotation Tank, Sulfur Slurry Tank, and Citrate Storage Tank: These vent gases may contain traces of COS, CS_2, and H_2S and will flow to the power plant boiler for incineration to SO_2 for recycle to the citrate process demonstration plant.

Sulfur Filter Cake, 300 Pounds per Week: This filter cake, containing mainly sulfur and diatomaceous earth with minor amounts of citrate salts and fly ash, will be disposed of in an environmentally acceptable manner.

Citrate Solution, Leaks or Spills: Leaks and spills of citrate solution will be collected in a common sump and recycled to the process.

Operational Testing and Evaluation

The citrate process demonstration plant will be tested and evaluated for 1 year following acceptance. The objectives of these tests are:

(1) To characterize the citrate system with respect to the operating parameters.

(2) To determine the citrate system's optimum operating conditions.

(3) To determine long-term system reliability.

(4) To determine the environmental impact of the citrate system.

(5) To determine the technical and economic feasibility of the citrate system.

(6) To document the results of the test program.

The Bureau has retained the Radian Corp. to serve as an independent, unbiased test and evaluation contractor. Radian will develop and implement a test plant that includes:

(1) Baseline testing, which will be divided into two distinct segments. One segment will observe actual boiler operation over about 4 weeks, while monitoring all available boiler operating parameters and effluent discharges. The second segment of the baseline testing will investigate the historical limits of boiler operation. This segment will be executed during the same testing period. The coordination of these two segments will insure that the boiler is operating at historically normal conditions during the test period.

(2) Acceptance testing, which will be performed to certify that the citrate process demonstration plant will meet performance guarantees. The acceptance test will require the system to operate for a period of not less than 10 consecutive days while burning coal containing not less than 2.5% sulfur and meeting both the New Source Performance Standards and a 90%-SO_2-removal standard.

(3) Optimization of the system, which will be investigated through the 1-year demonstration period. Data collected during the demonstration testing will be used to determine what adjustments are required to optimize plant operations (a) to minimize absorber liquid-gas ratio, system pressure drop, reagent and feedstock consumption, power and utility consumption, and process costs and (b) to maximize SO_x removal, particle removal, system reliability, and system availability.

(4) Documentation of the citrate system's performance, which will include reliability, availability, and utilization information. For proper overall power station operation, FGD system availability should be maintained at high levels. Information concerning boiler and citrate system operation will be gathered daily and reported monthly. System performance will be reported in terms of availability, reliability, and utilization. Availability is defined as the percentage of a given time period that the FGD system is available for operation. Reliability is the percentage of instances that the FGD system operated during a time period and reflects the load demand imposed on the FGD system.

Special studies to investigate sulfate formation, system corrosion, materials evaluation, stack gas reheat, and other items of interest will also be conducted during this 1-year period. In addition, process economics will be developed based upon the demonstration project information.

PILOT PLANT OPERATION AT BUNKER HILL COMPANY

The information in this section is based on *Citrate-Process Pilot-Plant operation at the Bunker Hill Company,* 1979, BuMines RI 8374, by L. Crocker, D.A. Martin and W.I. Nissen of Salt Lake City Metallurgy Research Center, Bureau of Mines.

This section describes a Bureau of Mines second generation pilot plant located at the Bunker Hill Company's lead smelter in Kellogg, Idaho (1)-(3). At that smelter, lead concentrates are sintered in a Lurgi updraft sintering machine. Strong and weak SO_2 gases are emitted from the sintering operation. The strong gas containing from 4 to 5 vol % SO_2 is sent to a contact sulfuric acid plant for SO_2 recovery. The weak gas containing from 0.5 to 1 vol % SO_2 is filtered in a baghouse and vented through a stack to the atmosphere.

In an effort to improve air quality, the Bunker Hill Company expressed interest in testing the citrate process on the sintering machine gas. At the same time, the Bureau of Mines needed an industrial cooperator to further develop the citrate process. On October 2, 1972, the Bunker Hill Company and the Bureau of Mines executed a cooperative agreement to pilot-test the citrate process at the Bunker Hill lead smelter.

General Process Description

The following process steps are shown in Figure 8.2 which is a simplified flow sheet of the citrate process after modifications that evolved from both laboratory and pilot-plant testing.

(1) The SO_2-bearing gas is cooled to between 113° and 149°F and cleaned of sulfuric acid (H_2SO_4) mist and solid particles.

(2) The SO_2 is absorbed from the cooled and cleaned gas by a solution of sodium citrate, citric acid, and sodium thiosulfate.

(3) Absorbed SO_2 is reacted with H_2S at about 149°F and atmospheric pressure to precipitate elemental sulfur and regenerate the solution for recycle.

(4) Sulfate is removed from a slipstream of regenerated recycle citrate solution by cooling the solution and crystallizing Glauber's salt ($Na_2SO_4 \cdot 10H_2O$), which is removed by filtration and then washed.

(5) Sulfur is separated from the solution by oil or air flotation and melting.

(6) The H_2S for step (3), if not otherwise available, is made by reacting two-thirds of the recovered sulfur with natural gas (CH_4) and steam.

Process Chemistry: Three steps of the process include the following chemical operations: (1) absorption of SO_2 in a buffered citric acid solution, (2) reaction of sulfur, CH_4, and steam to produce H_2S, and (3) reaction of the dissolved SO_2 with H_2S to produce sulfur.

The absorption of SO_2 in water may be expressed as follows:

(1) $$SO_2 + H_2O \rightleftharpoons H^+ + HSO_3^-.$$

The salt of a weak acid (buffer) such as sodium citrate, glycolate, or maleate dissolved in the system reacts accordingly

(2) $$H^+ + HSO_3^- + Na^+ + glycolate^- \rightleftharpoons Na^+ + HSO_3^- + glycolic\ acid.$$

The hydrogen ion produced by absorption of SO_2 in reaction (1) combines with a glycolate ion in reaction (2), to form glycolic acid, a nonionized acid. This increases the capacity of the solution for SO_2 absorption. The effect of SO_2 loading on solution pH for three buffer concentrations is shown in Figure 8.3.

H.W.F. Wackenroder investigated the reaction between H_2S and SO_2 in H_2O in 1847 (4). The chemistry of these reactions is complex. Polythionic acids and thiosulfate ion, as well as sulfur may be formed. The stoichiometry is usually expressed as follows:

(3) $$SO_2 + 2H_2S \rightarrow 3S + 2H_2O.$$

This reaction does not proceed to elemental sulfur above pH 6. The reaction is favored by a low pH. However, below pH 3 the sulfur tends to be colloidal.

Half-molar citric acid solution buffered with a sodium base to a pH between 4.5 to 4.7 is satisfactory for the reactions (1)-(3). During SO_2 absorption, the solution pH decreases as illustrated in Figure 8.3.

Figure 8.2: Generalized Citrate Process Flow Sheet

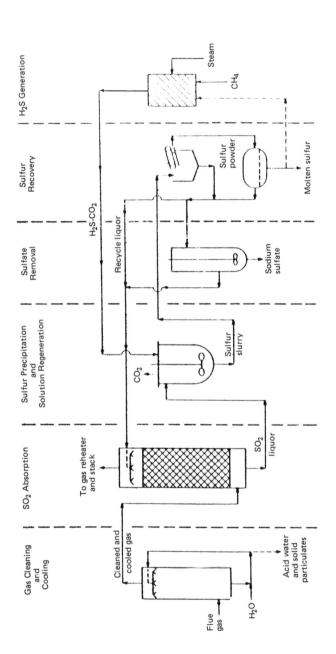

Source: BuMines RI 8374

Figure 8.3: Effect of SO_2 Loading on Solution pH

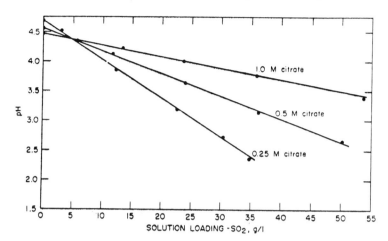

Source: BuMines RI 8374

When the SO_2 has been reduced to sulfur, the solution pH returns to the starting value prior to SO_2 absorption. Thus, pH is one indicator of the degree of regeneration. However, sulfuric acid is formed by oxidation of SO_2 to SO_3 in the solution and by absorption of SO_3 from the waste gas. Thus, base must be added to maintain the pH of the buffer because H_2S will not reduce sulfuric acid. Eventually sodium sulfate must be removed to prevent crystallization in the process. Sodium thiosulfate inhibits the oxidation of SO_2 to SO_3. However, the pH of the buffer should be maintained around 4.5 because thiosulfate will decompose with acid and contribute to the sulfate problem.

The regeneration temperature was chosen to be 149°F for the following reasons: (1) The regenerated liquor would carry sufficient heat to the kerosine flotation section for satisfactory sulfur recovery, (2) it did not require excessive heating or cooling between regeneration and absorption, and (3) in laboratory studies the reaction was about 30% faster at 149°F than at 113°F.

The hydrogen sulfide for reaction (3) was produced by two modes at the Bunker Hill plant.

Natural gas (CH_4) was reacted with sulfur vapors at 1202°F (5) over a proprietary catalyst

(4) $CH_4 + 4S \rightarrow CS_2 + 2H_2S.$

The carbon disulfide (CS_2) in the product was hydrolyzed with steam in a second stage at 608°F,

(5) $2H_2S + CS_2 + 2H_2O \rightarrow 4H_2S + CO_2.$

The final product was about 80-dry-vol % H_2S and 20-dry-vol % CO_2.

A small amount of undesirable carbonyl sulfide (COS) may be present,

(6) $CS_2 + H_2O \rightarrow COS + H_2S.$

In the second mode, reactions (4) and (5) were effected simultaneously in one reactor as shown [according to reference (6)].

(7) $CH_4 + 4S + 2H_2O \rightarrow 4H_2S + CO_2$.

The reaction is exothermic. Ten kilocalories per gram-mol of H_2S is produced at 40 psi and 1202°F.

Absorption of SO_2 was monitored and controlled by an "in-line" commercial SO_2 analyzer on the feed and off-gas streams. Also, SO_2 in the scrubber solution was determined by standard iodometric titration.

Gas chromatography was used to determine the H_2S content of the feed and off-gas streams in the regeneration reactors. Iodometric titration for SO_2 and thiosulfate, plus pH measurements were used to follow regeneration of the SO_2-loaded solution.

H_2S generation was controlled by gas chromatographic analyses for H_2S, CO_2, COS, and CH_4.

Design and Construction: The Bunker Hill pilot-plant design was based on the 300-scfm pilot plant at Magma Copper Co. and laboratory investigations. Contracts to design, fabricate, and install the plant were awarded to Morrison-Knudsen Co., Inc., of Boise, Idaho. A major subcontractor, Home Oil Co., Ltd., of Canada designed the H_2S generator section. This plant was designed to treat 1,000 to 1,500 scfm (60°F and 1 atm) of 0.5 vol % SO_2 from either one of two possible sources, either acid plant feed gas diluted with air approximately eightfold or weak gas from the lead smelter sintering machine that was reported to be about 0.9 vol % SO_2.

Generally, only materials known to resist the corrosive effects of the SO_2-loaded gas and liquid streams were used in the plant. Fiber glass-reinforced polyester (FRP), polyvinyl chloride (PVC), or bitumen-lined carbon steel were used to fabricate process piping and vessels where temperature permitted. Type 316 stainless steel was used where high temperature or structural strength were factors. Carbon steel was used where the gas temperatures were above the sulfuric acid (H_2SO_4) dew point. The H_2S generator process piping and vessels were all fabricated from suitable stainless steels because of high temperature and corrosive effects of feedstocks and products.

The Bunker Hill pilot plant was constructed in the following three phases: Phase 1—SO_2 absorption, solution regeneration, and sulfur recovery section, Phase 2—H_2S generator section, and Phase 3—Flue gas cooling and cleaning section.

Phase 1 of the project was constructed between March and December 1973, and acceptance testing began on January 9, 1974. The plant operated on H_2S from a 13-ton capacity tank trailer during construction and operational checkout of the H_2S generator section. Construction of Phase 2 was started in December 1973, and after some design modifications of the H_2S generator was ready for acceptance testing in April 1975. Phase 3 of the pilot plant was constructed during 1974. It was brought on-stream to supply a cool and clean lead sintering machine dilute gas to the absorber in January 1975. By the end of April 1975, the completed plant was in operation and the tank trailer of H_2S was no longer needed.

Pilot Plant Description and Operation

The block diagram in Figure 8.4 shows the process when all three sections of the plant were operating. Variations from Figure 8.4, which would be required in a full-scale plant, and recommendations based on a pilot plant operation are discussed as the different process steps are considered.

Figure 8.4: Bunker Hill Pilot Plant

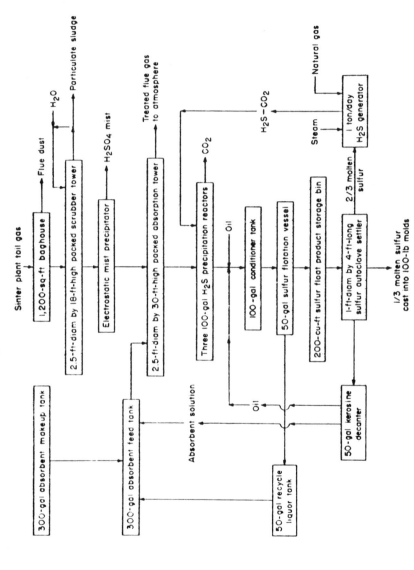

Source: BuMines RI 8374

The Bunker Hill pilot plant began operation in January 1974, logged over 5,400 hours, and produced about 55 net short tons of sulfur until completion of testing in May 1976. During that time, the H_2S generator was operated to supply H_2S gas to the citrate process for 3,090 hours. The gas cooling and cleaning section was operated as needed to test the respective components and to provide cool, clean, humidified gas to the SO_2 absorption section.

Gas Cooling and Cleaning: The SO_2 absorption, solution regeneration, and sulfur recovery section of the pilot plant operated for about 1 year starting in January 1974, on a precleaned sulfuric acid plant feed gas diluted about 8:1 with air to obtain the required 0.5 vol % SO_2 feed gas. After installation and acceptance of the gas cooling and cleaning section, the source of the pilot plant feed gas was switched to the weak gas from the Bunker Hill Company's lead smelter sintering machine. An induced draft fan was used to move the gas through the gas cooling and cleaning section and into the SO_2 absorber. The weak gas flow was controlled by a valve located at the main weak gas duct and was drawn through about 150 ft of 10-inch insulated carbon steel pipeline to the first unit in the pilot plant, a 3.5-ft-diameter by 24-ft-high dust knockout tank that was originally intended to cool and condition the weak gas with a water spray for subsequent sulfuric acid mist removal. However, the weak gas temperature averaged only 266°F and heat losses from the weak gas pipeline and dust knockout tank were such that the weak gas had to be heated to prevent acid mist condensation and corrosion of carbon steel pipelines and process equipment. A natural gas-fired preheater was installed in a branch pipeline before the dust knockout tank to heat the weak gas. The weak gas also proved to be higher in SO_2 than was reported and more variable in concentration; thus it was necessary to dilute weak gas with air and combustion gas to achieve on average 0.5 vol % SO_2 feed gas for the pilot plant. Some of the SO_2 fluctuation was reduced by adjusting the valve at the weak gas duct takeoff point. As the valve was closed, more dilution air was drawn in at the gas preheater.

The next unit in the gas cooling and cleaning section was a carbon steel baghouse containing 1,200 ft² of Orlon bags that were periodically cleaned by a pulsed air jet. The gas-fired preheater was automatically controlled to keep the baghouse inlet temperature at about 248°F. The baghouse dust hopper was periodically discharged to remove accumulated dust and determine dust loading in the weak gas. The baghouse dust removal efficiency averaged 99.6% resulting in an exit gas dust loading of about 0.26 mg/ft³ when the inlet gas contained 65 mg/ft³.

The baghouse was backed by a 2.5-ft diameter by 18-ft prescrubber tower constructed of FRP and packed with a 6-ft bed of 1-inch polypropylene Intalox saddles. This prescrubber removed about 95% of the remaining dust in the gas, resulting in exit gas containing about 0.013 mg/ft³. Also, the prescrubber removed about 91% of the sulfur trioxide from a prescrubber inlet gas containing 0.23 mg SO_3/ft³, resulting in an outlet gas containing about 0.02 mg SO_3/ft³. The countercurrent scrubber solution flow was 25 gal/min. A small purge flow of about 0.2 gal/min was maintained to keep the solution pH above 1.5 and avoid solids buildup. A level-controlled solenoid valve replaced the purge with plant water. This purge joined other neutralized wastes from the smelter acid plant. In a full-scale plant this acid waste would have to be neutralized with lime or limestone and ponded to prevent contamination of water supplies.

The prescrubber tower was used for all operations except for a 1-month test. During this test, the sulfate accumulation increased slightly from about 1.2 to 1.4 wt %. However, this was not considered significant. There was no other noticeable effect on the rest of the pilot plant operation.

Although operation without the prescrubber caused no problems at the Bunker Hill pilot plant, it was a special situation because of the low gas temperature and high efficiency of the bag filter. In a commercial installation it is expected that a baghouse or a dry electrostatic precipitator followed by a wet prescrubber would be the minimum requirement for gas cooling and cleaning. The presence of SO_3 mist in the gas may also require a wet electrostatic mist precipitator following a wet prescrubber. A commercial plant would also require a cooler for the circulating dilute sulfuric acid to the prescrubber. Although such a cooler was installed in the pilot plant, it was never used.

The induced draft fan was installed downstream of the prescrubber. The fan impeller was fabricated of type 316 stainless steel and the fan casing was fabricated of cast iron and coated with Bisonite. Although the location of the induced draft fan at the pilot plant was satisfactory, some corrosion of the Bisonite-coated casing did occur. A better fan location would have been downstream of the baghouse where most particulates had been removed but the gas temperature was still hot enough to prevent acid condensation and corrosion. By moving this fan the need for corrosion-resistant alloy construction or casing line would be eliminated.

Finally, to control the temperature of the SO_2 absorber tower for testing and process evaluation, it was necessary to humidify and heat the gas after leaving the prescrubber. This was done by adding steam directly and then passing the gas through a shell and tube steam heat exchanger to achieve enough superheat to prevent condensation in the gas flow meter. This ability to humidify and heat the gas permitted evaporation and temperature control in the SO_2 absorber.

SO$_2$ Absorption: The main unit in the SO_2 absorption section of the pilot plant was a 2.5-ft-diameter by 30-ft-high absorber tower. This tower was constructed of FRP and was packed with 18 ft of 1-inch polypropylene Intalox saddles. Entrainment of absorber solution in the gas was controlled by a type 316 stainless steel mesh demister in the top of the tower. The absorber solution was pumped to a distributor above the packing by a sealless magnetic drive pump at rates ranging from 5 to 15 gal/min. The solution was clarified by a dual backwash, porous type 316 stainless steel filter having a surface area of about 6 ft^2. The solution was cooled with water, when necessary, by a type 316 stainless steel shell and tube heat exchanger. The solution flow was measured by a float-type flow meter and automatically controlled to the desired flow, usually 10 gal/min.

Among the conditions varied in the pilot plant tests were type and concentration of the absorbent solution, liquid flow rate and resultant SO_2 loading, feed gas SO_2 concentration and flow rate, and feed gas and solution temperatures. Operation of the absorber was generally satisfactory and trouble-free. Some problems did occur in the auxiliary system, however. The most troublesome of these were blinding of the clarification filters and plugging of the transfer pumps. This was due to upsets in the sulfur recovery section, which at times allowed more sulfur to carry over with the solution than could be handled by the filters. Blinding of the filters also occurred when too much kerosine was used in the flotation step and allowed to get into the filters. These problems were minimized when the solution regeneration and sulfur recovery sections were operating properly. The normal amount of sulfur carryover (less than 1% of the sulfur float product) was handled by occasional backwashing of the filters. The backwash sulfur slurry was returned to the conditioner tank in the sulfur recovery section of the pilot plant.

SO₂ Absorption Results: The results of operating tests are summarized in Tables 8.1 to 8.4. Table 8.1 shows the effect of citrate solution concentration, absorption temperature, and solution and gas flow rates on the SO_2 removal efficiency. The SO_2 removal efficiencies were based on feed and off-gas analyses. The tests were made using pure H_2S to regenerate the solution in a single 100-gal reactor operating at an agitator speed of 220 rpm and a temperature of about 149°F. Citrate solution concentrations of 0.25 M and 0.5 M were tested. The absorption temperature was varied from 95° to 147°F. The solution and feed gas flow rates were varied from 5 to 10 gal/min and from 1,000 to 1,500 scfm, respectively. The concentration of SO_2 ranged from 0.30 to 0.54 vol %, the regenerated solution pH ranged from 4.2 to 4.7, and the solution SO_2 loadings ranged from 6.3 to 21.3 g/ℓ. These loadings represent from 50% up to 95% of the equilibrium loading values. Only in tests where the absorption temperature was 147°F or the solution flow rate was 5 gal/min did the absorption efficiency decrease to any extent. The design SO_2 loading for the pilot plant was 10.8 g SO_2/ℓ when treating 1,000 scfm of 0.5 vol % SO_2 gas with 10 gal/min of 0.5 M citrate solution at a temperature of 113°F. This design loading represents about 60% of the equilibrium loading. The test that came closest in loading and conditions to this design resulted in a 99.4% SO_2 removal efficiency.

Table 8.1

. . . . . Feed Gas.		. Citrate Solution .			
Concentration % SO₂	Flow scfm	Flow gal/min	Loading g/ℓ SO₂	Absorption Temperature, °F	SO₂ Removal Efficiency, %
0.32	1,000	10*	6.3	97	98.4
0.30	1,500	10*	9.0	99	98.1
0.42	1,000	10	8.9	102	99.2
0.47	1,000	10	9.8	113	99.4
0.47	1,000	10	10.1	131	98.9
0.37	1,000	10	7.5	147	95.6
0.44	1,000	8	12.0	97	97.5
0.47	1,000	7	14.5	99	97.8
0.54	1,000	5	21.3	95	92.5
0.46	1,200	10	11.5	104	99.1

*0.25 M citrate solution.

Source: BuMines RI 8374

Although the pilot plant was designed to test SO_2 feed gases in the 0.5 vol % SO_2 range, a few tests were run with SO_2 feed gas concentration of 0.10 vol % or less. Such a gas might be typical of the flue gas from a coal-burning power plant firing medium-sulfur-content coal (1 to 2% sulfur).

Table 8.2 shows the effect of low SO_2 feed gas compared with results using higher SO_2 feed gas. All tests in Table 8.2 were run using 78 dry vol % H_2S gas from the H_2S generator to regenerate the solution in two 100-gal reactors operating at an agitator speed of 220 rpm and a temperature of about 149°F. The citrate solution concentration was between 0.4 and 0.5 M. The most significant results of these tests were an SO_2 loading of 4.1 g SO_2/ℓ, which represents 83% of the equilibrium loading value, and an SO_2 removal efficiency of 95%. The results in Table 8.2 show that the process is applicable to flue gases from coal-fired power plants.

Table 8.2

| .... Feed Gas .... | | .....Citrate Solution ..... | | | Absorption | SO_2 Removal |
Concentration % SO_2	Flow scfm	Flow gal/min	Loading g/ℓ SO_2	Regener- ated pH	Temperature °F	Efficiency %
0.09	1,000	7.5	2.4	4.8	99	99.9
0.12	1,300	7.5	4.1	4.9	135	95.0
0.46	1,000	10.0	9.6	4.3	97	98.6

Source: BuMines RI 8374

Table 8.3 shows the effect of H_2S excess on the SO_2 removal efficiency and pH of the regenerated solution. The term "H_2S excess, %" in Table 8.3 means that the stoichiometric H_2S feed rate plus the stated excess was fed to the solution regeneration section. Although this excess was lost to the incinerator, it raised the partial pressure of H_2S in the reactors.

Table 8.3

| Feed Gas Concentration % SO_2 | .... Citrate Solution ... | | Regeneration Reactor H_2S Excess, % | SO_2 Removal Efficiency % |
	Loading g/ℓ SO_2	Regenerated pH		
0.38	9.0	4.3	15.1	98.4
0.31	7.3	4.1	6.1	97.4
0.44	10.3	4.1	2.5	94.0

Source: BuMines RI 8374

Tests were made using 78 dry vol % H_2S gas from the H_2S generator to regenerate the solution in either two or three 100-gal reactors operating at an agitator speed of 220 rpm and a temperature of about 149°F. Ten gallons per minute solution flow was used and the solution concentration ranged from 0.45 to 0.60 M citrate. The absorption temperature was between 93° and 97°F. The flow rate of the feed gas, which was either lead smelter sintering machine weak gas or commercial SO_2 containing from 0.31 to 0.44 vol % SO_2, was about 1,100 scfm.

Table 8.4 shows the effect of solution concentration, absorption temperature, solution and gas flow rates, and excess H_2S on the SO_2 removal efficiency when using an absorbent made up with glycolic acid instead of citric acid. Tests were made using 78 dry vol % H_2S to regenerate the solution in two 100-gal reactors operating at an agitator speed of 220 rpm. The regeneration reactor temperature was about 149°F. The flow rate of the feed gas, which was commercial SO_2 except for one test on lead smelter sintering machine weak gas, was varied from 1,000 to 1,200 scfm and the feed gas concentration ranged from 0.25 to 0.39 vol % SO_2. The solution flow rates were varied from 7 to 12 gal/min and the regenerated solution pH ranged from 4.1 to 4.6. The glycolate molarity, except for one test in which a 0.23 M glycolate was used, was between 0.8 and 0.9 M. The results shown in Table 8.4 confirm two important process parameters:

(1)   Glycolate solution may be substituted for citrate solution.

(2)   From 3 to 4% excess H_2S is required for efficient operation.

Table 8.4

| ..... Feed Gas ..... | | Glycolate ... Solution ... | | Absorption | Regeneration | SO_2 Removal |
Concentration % SO_2	Flow scfm	Flow gal/min	Loading g/ℓ SO_2	Temperature °F	Reactor H_2S Excess, %	Efficiency %
0.25	1,050	10	5.4	100	3.3	97.1
0.39	1,000	10	8.5	113	4.0	96.2
0.38	1,000	10	7.8	129	3.8	94.1
0.32	1,050	7	10.5	113	3.6	96.6
0.37	1,050	12	7.1	113	2.7	97.4
0.30	1,200	10	7.7	115	2.8	93.7
0.28	1,060	10	6.4	100	12.3	98.9
0.39	1,040*	10**	7.9	100	7.5	89.4

*Part lead smelter sintering machine waste gas.
**0.23 M glycolic acid.

Source: BuMines RI 8374

Solution Regeneration: The solution regeneration section consisted of three 100-gal, baffled, type 316 stainless steel reactors. Each was stirred with a 17.5-inch-diameter, 6-blade radial turbine-type impeller. These variable-speed agitators were capable of 230 rpm and a tip speed of 1,050 ft/min.

The first reactor was fed by gravity through a variable level control and siphon breaker from the absorber tower bottom. The three reactors were cascaded for gravity flow of sulfur slurry. The H_2S could be fed into any of the three reactors countercurrent to the flow of slurry and was introduced under the reactor impeller through an open 3-inch pipe. Reactor 1 was steam-jacketed for temperature control. The other reactors were insulated to conserve heat.

When using one reactor and pure H_2S, H_2S was supplied on a demand basis. That is, the H_2S supply system was automatically controlled to maintain 0.5 psi of H_2S pressure over the solution. This mode of operation did not waste H_2S. However, at startup, it was necessary to vent the reactor to displace the air from the system to an H_2S incinerator.

When using H_2S generator product (78% H_2S), the H_2S requirement was determined by following the degree of solution regeneration and analyzing the reactor off-gas for H_2S. During operation, it was necessary to vent the reactor to displace the CO_2 and unreacted H_2S from the system to the H_2S incinerator.

Table 8.5 presents some typical regeneration data with glycolate and citrate scrubbing solutions. Agitation was the same in all tests, 1,050 ft/min impeller tip speed. The SO_2 loadings are concentrations in the reactor feed. They are not steady-state SO_2 concentrations in the reactors. The final reactor effluent solutions averaged less than 0.5 g of $SO_2/ℓ$. The pH of the loaded solutions in these tests averaged 4.4 for the glycolate and citrate solutions. These solutions also contained 0.25 M $Na_2S_2O_3$.

Table 8.5

Solution Type	SO$_2$ Loading g/ℓ	Reactors in Use	Feed Rate gal/min	Average Retention Time, min	. . . . . . H$_2$S . . . . . . Feed Gas Concentration dry vol %	Excess %	Reduction Rate SO$_2$ lb/hr	SO$_2$ lb/hr-gal
Citrate	10.2	1	8.3	12	100	—	41	0.41
Citrate	15.6	1	7.1	14	100	—	55	0.55
Citrate	21.3	1	5.0	20	100	—	53	0.53
Citrate	11.0	2	10.0	20	78	9	55	0.27
Glycolate	7.6	2	10.0	20	78	2	38	0.19
Glycolate	8.5	2	10.0	20	78	3	42	0.21
Glycolate	10.5	3	10.3	29	78	4	54	0.18
Glycolate	12.7	3	7.0	43	78	3	45	0.15

Source: BuMines RI 8374

The data indicate that 100 gal of reactor volume are sufficient to treat 50 lb of SO$_2$/hr with pure H$_2$S. With 78% H$_2$S, 200 gal of reactor volume were required. Glycolate scrubber solutions regenerated as readily as the citrate solutions. Therefore, glycolate was assumed to be a satisfactory scrubber solution for the process.

One of the most serious problems in this section of the plant was the formation of sulfur plugs in the gas vent lines and slurry lines between the reactors. Although this problem remained unsolved, these plugs were usually removed and flow restored by injecting steam at various points along the lines.

A second and less serious problem, from the operational standpoint, involved the presence of dissolved H$_2$S and some colloidal sulfur in the regenerated solution. The H$_2$S would evolve from solution in the sulfur flotation cell and contaminate the air. The colloidal sulfur could not be removed by flotation.

The problem was solved by injecting about 5% of the SO$_2$-loaded solution from the absorption tower into the conditioner tank. This eliminated the excess H$_2$S and dissolved the colloidal sulfur.

Sulfur Separation and Solution Recovery: After the reactors, the regenerated solution containing about 1.5% sulfur overflowed to a 50-gal stirred, type 316 stainless steel conditioner feed tank. From this tank the sulfur slurry, at a temperature of about 127°F was pumped to a 100-gal FRP conditioner tank. The conditioner tank was stirred by a variable-speed agitator equipped with two 16-inch-diameter axial impellers. The lower impeller mixed the hydrocarbon flotation reagent and the upper impeller, located just below the surface, broke up the accumulations of floating sulfur. The mixture of floating sulfur and regenerated solution overflowed through a 3-inch-ID sloping line to the sulfur flotation vessel. An agitator tip speed of 700 ft/min was sufficient for good sulfur flotation.

By design, 6 to 12 g of kerosine per minute were added to the conditioner by a chemical feed pump. This quantity of kerosine amounted to about 1 to 2% of the gross sulfur production. The average sulfur content of the float product was 48% solids under these conditions. Later the kerosine addition point was changed from the bottom of the conditioner tank to a point just above the overflow line. At the same time the kerosine feed pump was replaced with a smaller pump. These changes improved kerosine dispersion. The kerosine consumption using this new addition point was only about one-fifth of the previous requirement or 0.2% of the gross sulfur production. The percent solids of the float product was the same when using the new addition point and less kerosine.

The sulfur flotation vessel was a 50-gal capacity, rectangular, type 316 stainless steel tank with a bottom outlet and an inclined chute with a drag-type belt conveyor. The sulfur, in the form of 40 to 50% solids powder, floated on top of the vessel and was continuously skimmed off by the blades of the conveyor. The end of the conveyor discharged into a 200-ft³ capacity, type 316 stainless steel storage bin. The clear solution flowed from the bottom of the sulfur flotation vessel through a level controller into a 65-gal-capacity cone bottom FRP settler tank. Here, a small amount of sulfur that failed to float (usually less than 1%) was settled and recycled to the conditioner tank by a slurry pump. A vertical baffle in the settler tank separated a small amount of floating sulfur that escaped from the sulfur flotation vessel. This floating sulfur was periodically removed manually and added to the sulfur storage bin. The carryover of the floating and sinking sulfur indicated deficiencies in the design of the sulfur flotation circuit. The clarified solution, after leaving the settler tank, overflowed by gravity into a 300-gal capacity FRP absorber feed tank, thus completing the cycle for the mainstream of absorbent solution.

Flotation of sulfur with air alone (no addition of kerosine to the conditioner) was tried because it resulted in improved product sulfur purity. Sulfur produced with air flotation contained 0.04% carbon, but that produced with kerosine flotation contained 0.15% carbon. Flotation with air also reduced reagent costs and eliminated the need for a kerosine recovery system. The flotation of sulfur with air involved the addition of 5 to 10 liters of air per minute through a porous stainless steel diffuser tube located in the bottom of the sulfur flotation vessel. The sulfur pickup conveyor worked just as well as the oil-floated sulfur and the clarity of the underflow solution was also very good. However, the percent solids of the separated sulfur was from 8 to 20%. Assuming an average percent solids of 14, the weight of solution that would have to be heated in the process of melting sulfur is increased by about six times. The cost of the additional heat, plus the decomposition of citrate and formation of sulfate that occurs in the autoclave must be weighed against the advantages of the air-flotation method.

Sulfur Melting and Solution Recovery: The sulfur melting was done during day shift only. The sulfur float product accumulated in the 200-ft³ storage bin for the other 16 hours. This mode of operation allowed the use of a commercial-size progressing, cavity-type slurry pump for moving the sulfur into the melter. This pump was installed directly on the bottom of the storage bin and was fed through an 8-inch gate valve into the bridge-breaker of the pump. Feed to the pump was aided by a bin vibrator and by a timed blast of air that discharged into the sulfur every 30 seconds near the bottom of the bin to prevent bridging. Downstream of the pump, the sulfur was melted with steam at a rate of 300 to 400 lb/hr in a 14-ft² area concentric tube heat exchanger. After exiting from the heat exchanger the melted sulfur was separated from the remaining absorbent solution in a 57-gal-capacity, type 316 stainless steel autoclave settler, operating at a pressure of about 35 psig, and a temperature of 284°F. The pressure in the heat exchanger and autoclave settler was controlled by automatic release of solution from the top of the autoclave settler, through a water-cooled concentric tube heat exchanger and a knockout pct into a 50-gallon-capacity cone-bottom FRP kerosine separator tank.

The cooled solution was withdrawn from the bottom of this tank through a level controller and flowed by gravity to the absorber feed tank. Kerosine that floated to the surface was allowed to accumulate and then was periodically floated out through an overflow line to a tank for reuse. In practice, there was little recovery of kerosine when its usage was held below 1% of the gross sulfur produced because of vaporization losses. The kerosine dissolved in the sulfur was about 20% of the total kerosine based on carbon analysis of the sulfur product. No attempt was made to recovery kerosine lost out of the tank vents because of the small amount involved, about 15 lb/day. However, some could be recovered by collecting and chilling the vent gases or bypassing them through a cold absorption tower.

Sulfur withdrawal from the autoclave settler was accomplished by a steam-jacketed, automatic level control valve on the bottom of the settler. Two-thirds of the sulfur produced was transferred by pressure of the settler through steam-jacketed lines to the H_2 generator sulfur storage tank where it was maintained as a molten liquid for use. The remaining sulfur was cast into 100-lb blocks and stored. Since the melting was only done for 8 hours of the 24 hours, all sulfur was drained from the autoclave settler each day, leaving the autoclave settler free of sulfur and preventing the buildup of the impurities that could have caused plugging of the drain valves. This mode of operation led to a small unaccounted loss of solution each time the autoclave settler was drained.

The sulfur melting operation required the close attention of an operator. The sulfur float product occasionally bridged in the bin and failed to pump. Also, the pressure letdown valve, lines, and solution cooler heat exchanger would plug due to kerosine-dissolved sulfur that crystallized upon cooling. This material analyzed about 91% sulfur, up to 1% lead and zinc, and smaller percentages of iron, copper, cadmium, silicon, and aluminum. The minor components are a carryover of flue dust into the system. Most of the material eventually ended up in the kerosine separator tank where it collected in the cone bottom, requiring cleanout of about 40 lb/month. An alternate route for pressure letdown from the autoclave settler was to the bottom of the conditioner tank. This route was used when the line to the kerosine separator tank became plugged and had to be cleaned. It also could be used to bypass the pressure letdown cooler and control valve when these needed cleaning. The plugging problem was minimized by operating with less kerosine for flotation and was largely eliminated when no kerosine was used.

The product from the autoclave was high-purity, bright yellow sulfur, and it was used successfully for H_2S generation. However, the carbon and other impurities combined with sulfur and formed a black flakelike material that caused sticking of the sulfur pump check valves. This condition caused loss of sulfur flow and frequent backflushing was required to keep the sulfur pumps operating. Probably a filter before these sulfur pumps would have solved this problem. Table 8.6 shows the quality of sulfur made with kerosine and with air flotation. As shown, the solution used, citrate or glycolate, had little effect on sulfur quality.

Table 8.6: Sulfur Product Analysis

System	Type of Flotation	Ash, %	H_2O, %	Carbon, %	Elemental Sulfur, %
Citrate	Kerosine	0.05	0.25	0.17	99.6
Citrate	Kerosine	0.07	0.25	0.15	99.5
Citrate	Air	0.03	0.28	0.04	99.5
Citrate	Air	0.08	0.15	0.04	99.7
Citrate	Air	0.006	0.03	0.024	99.9
Glycolate	Kerosine	0.17	0.08	0.19	99.6
Glycolate	Air	0.04	0.00	<0.01	99.9

Source: BuMines RI 8374

Solution Makeup: The absorbent solution was prepared in a 300-gal stirred FRP absorbent makeup tank. The ingredients and order of mixing for a 250-gal batch of 0.5 M citrate solution were as follows: water, 150 gal; citric acid, 200 lb; soda ash, 100 lb; sodium thiosulfate, 128 lb; and water to final volume. The soda ash was added slowly to prevent boilover due to CO_2 release and the thiosulfate was added last to prevent decomposition. About 1,000 gal were needed to fill the pilot plant system. The drain lines of the two tanks, makeup and feed, were also interconnected to provide extra system surge capacity.

Chemical Usage: *Inventories* — The important chemical makeup requirements for the SO_2 absorption, solution regeneration, and sulfur recovery section of the pilot plant included citric or glycolic acid, sodium carbonate, and kerosine. The pilot plant was designed so that solution losses from spills, leaks, etc., could be colllected in a common sump. The plant concrete floor was sloped to trenches which, in turn, drained into the sump. The sump pump was set to pump to a 300-gal stirred FRP spill tank where all spilled solution was measured and sampled for analysis before dumping to the chemical sewer. Any future plant should provide for the cleanup and recycle of known major spills of absorbent solution. Solution inventories and samples were taken at the end of operating periods (usually about every 2 weeks) and citric or glycolic acid losses were calculated.

The average of many citrate loss tests during runs when upsets or solution losses were minimized showed that about 20 lb of citric acid were lost per net ton of sulfur produced. The average glycolic acid consumption was less than 10 lb per net ton of sulfur. A sodium carbonate addition of about 90 lb per net ton of sulfur was required to neutralize sulfuric acid and to make up fresh citrate solution. Toward the end of the pilot plant operation the kerosine requirements were reduced to about 2 gal per net ton of sulfur produced.

H_2S Generation: The H_2S generator indicated as a single block in Figure 8.4 is depicted in some detail in the flow diagram in Figure 8.5. Solid lines in this diagram indicate flows for a two-stage reaction considered to be a proven process. The dashed lines indicate provision to bypass reactor 2 for later experimentation with a single-stage reaction. The design capacity of the plant was 0.4 to 1.25 tons of H_2S per day.

Molten sulfur produced in the pilot plant was transferred to a sulfur feed tank from which it was pumped through a filter to remove carbon and ash impurities and to a gas-fired superheater. The sulfur was vaporized and super-heated to about 1350°F. Natural gas containing about 92% methane at 1200°F from a second superheater was combined with the hot sulfur vapor ahead of reactor 1. The sulfur-methane reaction produces H_2S and CS_2 over a catalyst in this reactor according to reaction 4.

The H_2S-CS_2 product gas was first air-cooled to about 800°F and then further cooled in a steam-cooled heat exchanger to about 300°F. Excess sulfur was condensed in the reactor product cooler and was removed from the gas mixture in the first sulfur knockout tank.

Steam at 800°F from a third superheater was combined with the cooled H_2S-CS_2 gas prior to entering reactor 2. In this reactor, the CS_2 was hydrolyzed to H_2S and CO_2 in a catalyst bed at 700°F according to reaction 5.

A side reaction occurs during CS_2 hydrolysis to form COS according to reaction 6. In the laboratory it was determined that reaction 6 was depressed by feeding excess steam and favored by high reaction temperatures.

Figure 8.5: Bunker Hill Citrate Pilot Plant H₂S Generation Section

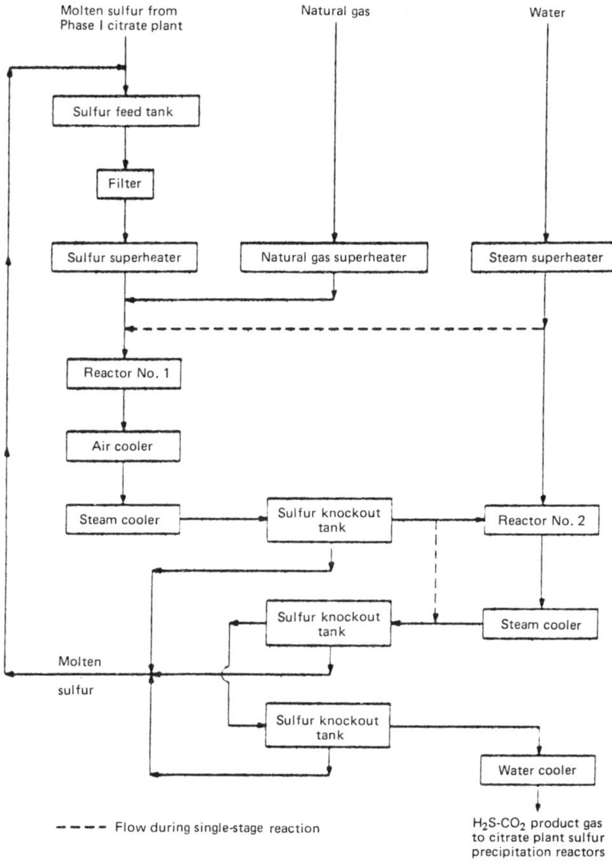

Source: BuMines RI 8374

A steam-cooled heat exchanger cooled the reaction 2 product gas to 300°F to condense free sulfur. The condensed sulfur was removed in two additional knockout tanks. Sulfur collected in all three knockout tanks was periodically drained to the sulfur feed tank. Product gas from the second reactor, containing H_2S, CO_2, and some water vapor, was cooled to about 149°F in a water-cooled heat exchanger and then fed to the solution regeneration section of the citrate pilot plant.

The major problem experienced during the pretesting of the H_2S generator was the inability to heat reactor 1 to the desired temperature. Thus, the sulfur vaporizer piping was damaged when attempting to provide the necessary heat for the reaction. This was solved by shortening the process piping between the superheaters and reactor 1, adding insulation to process piping and reactor 1, and feeding excess sulfur to carry more heat to reactor 1.

After pretesting, the operation of the H_2S generator was integrated with the operation of the SO_2 absorption, solution regeneration, and sulfur recovery section. Two operating problems developed during continuous operation. The first, and probably the most chronic, was plugging of the cold H_2S product lines and valves with sulfur. This was partially solved by modifying the final sulfur knockout tank to cool the product gas with a water jacket. This mode of operation required periodic steam heating to clear the knockout tank of sulfur. The second problem, plugging of the sulfur superheater coil with metal sulfides, was due to high-temperature corrosion.

Reactor 2 was bypassed as indicated in Figure 8.5 after achieving reliable operation with two stages. This allowed reaction 7 to take place in reactor 1. The operation proceeded without any additional problems; the only noticeable difference was that the product H_2S contained more COS and CH_4.

Next, both the steam and natural gas superheaters were eliminated. All reactant feed materials were superheated in the sulfur superheater. Again the plant worked satisfactorily; the only noticeable difference was that the temperature of the superheater could be reduced about 100°F without affecting the H_2S product quality. This difference is probably due to an increase in the overall reaction time. The advantages of single-stage operation would be the elimination of capital equipment and possibly reduced corrosion in high-temperature equipment.

Between April 1975 and May 1976, the H_2S generator operated for about 3,090 hours and provided the H_2S required for SO_2 reduction in the citrate process. During the latter 2,369 hours of this time, the one-stage operation was used.

Table 8.7 summarizes results of the H_2S generator operation under reasonably steady-state continuous operation. During this time sulfur was fed to the H_2S generator while varying the natural gas and steam flows to change the H_2S production rate. Throughout most of the H_2S generator operation, the sulfur flow was about 160 lb/hr. This flow provided between 150 and 300% excess sulfur, which was recovered and recycled. The large amount of excess sulfur was used to prevent overheating of the sulfur vaporizer coil and to transfer heat to the first reactor. During the two-stage operation, a 25% excess of steam was used to prevent COS formation. During the one-stage operation, where the CS_2 hydrolysis occurs at a higher temperature, a 50% excess of steam was necessary to achieve low COS in the product gas. The first reactor inlet and outlet skin temperatures both averaged about 1050°F. While operating with two stages, the second reactor inlet and outlet temperatures averaged 430° and 655°F, respectively. In addition to the principal gas analyses listed in Table 8.7, the H_2S product gas contained about 1% N_2 and traces of CS_2, CO, and O_2.

Table 8.7

Reactor Stages	Number of Super- heaters	Natural Gas Flow, lb/hr	H_2S Pro- duction, tons/day	H_2S Product Analysis, ... dry vol %			
				H_2S	CO_2	COS	CH_4
2	3	12.2	1.15	79	19	0.7	0.4
2	3	8.1	0.75	79	19	0.5	0.2
2	3	6.0	0.55	79	19	0.3	0.2
1	3	10	0.95	78	18	1.5	2.0
1	2	9	0.85	77	17	2.4	2.1
1	1	7	0.65	78	18	1.6	1.0

Source: BuMines RI 8374

The test results show that the two-stage operation of the pilot plant makes the most efficient use of natural gas and produces the least amount of undesirable COS. The decrease in CH_4 utilization during one-stage operation, as indicated by the higher CH_4 in the product gas, may be caused by the increased space velocity through the first reactor. The COS in the H_2S product gas was vented from the sulfur precipitation reactors and incinerated to SO_2 and CO_2. By all methods of operation, the H_2S generator produced a gas that analyzed between 77 and 79% H_2S and was suitable for the SO_2 reduction.

The quantities of sulfur, CH_4, and steam required for the H_2S generator for each short ton of H_2S produced were sulfur, 1,900 lb; CH_4, 5,600 scf; and steam, 1,000 lb. These quantities include 1% COS, 0.5% CH_4, and 50% excess steam in the H_2S product.

Sulfur dioxide reduction was tried with a gas of lower H_2S content. This gas could be produced with a CO-H_2 mixture from a coal gasifier. This would eliminate the need for natural gas reductant in the citrate process. The H_2S reactor 1 was operated at 900°F which reduced the reaction rate. This produced a gas that contained on a dry vol % basis, 56 H_2S, 15 CO_2, 24 CH_4, 1 COS, 0.5 CS_2, 2 N_2, and trace CO and O_2, that was suitable for SO_2 reduction.

Under a cooperative agreement, Morrison-Knudsen Co., Inc. briefly tested other gaseous reductant feeds to replace natural gas for H_2S generation. Table 8.8 summarizes results of the H_2S generator operation using propane, methanol, and carbon monoxide as reductant feeds. The tests used the one-stage operation in which the sulfur superheater vaporized and heated the sulfur, reductant, and steam. The test conditions were similar to those used in producing H_2S from sulfur, natural gas, and steam.

Table 8.8

Reductant	Reactor Skin Temperature, °F		H_2S Production lb/hr	H_2S Product Analysis . . . . dry vol %					
	In	Out		H_2S	CO_2	COS	CS_2 .	CO	N_2
Propane	1,080	1,030	54.2	76.1	20.2	1.7	1.4	Trace	0.6
Methanol	1,100	1,100	54.2	74.2	21.6	2.6	0	Trace	1.6
Carbon monoxide	1,000	1,050	27.7	50.1	46.8	2.2	0	Trace	0.9

Source: BuMines RI 8374

The test results showed that all processes resulted in the production of an H_2S gas that could be used in the citrate process. The lower temperatures needed for H_2S production from sulfur, carbon monoxide, and steam may reduce the energy requirement and the corrosion in the sulfur superheater coil and high-temperature process piping.

In general, the H_2S generator worked reasonably well after initial shake-down problems were eliminated. With additional refinements it should be able to produce suitable H_2S at a rate to match the normal requirements of the citrate process. However, corrosion and plugging of lines may be a continuing maintenance problem, even though probably reduced in a larger-scale operation.

Safety

Safety was an overriding consideration in design, construction, and operation of the pilot plant. Trace quantities of hydrogen sulfide may not be harmful and are easily detected by the characteristic foul odor. However, high concentrations destroy the ability to detect the odor and concentrations of only 0.1% H_2S are lethal. For this reason, the pilot plant emergency ventilation system was designed and installed for a complete air change in the building every 2 minutes. The emergency ventilation fan took suction from a ducted hood in the immediate vicinity of the reactors where H_2S leaks were most likely to occur.

The pilot plant was equipped with an H_2S detector alarm system that consisted of remote H_2S detection sensors, one in the solution regeneration reactor area, two in the H_2S storage tank areas, three in the H_2S generator area, and one in the control room adjacent to the central H_2S alarm system control panel. An audible bell system was set to alarm at an H_2S concentration of 10 ppm followed by a wailing horn that sounded if a concentration of 25 ppm was reached. The emergency ventilation fans were set to start and an automatic valve was set to close if the 25-ppm level was reached in the process area. During actual pilot plant operation, the H_2S concentration rarely reached a level that set off the alarms. However, smaller concentrations in the range of 0 to 4 ppm were frequently encountered. These concentrations were rarely high enough to show on the H_2S sensors; nevertheless, they were objectionable.

A program was instituted for regularly checking H_2S concentration in the plant by visual inspection of H_2S indicating cards that are sensitive to a fraction of 1 ppm H_2S. These cards were useful for detection and repair of very small leaks before the alarm system ever became activated. A commercial syringe-type detector, which involves drawing an air sample through a sensitized material in a glass tube and observing the length of stain produced, was also used for pinpointing H_2S and SO_2 leaks. This monitoring was not as sensitive as the indicator cards and was more time-consuming. A simple substitute for the cards consisted of small strips of filter paper saturated with lead acetate solution that were also sensitive to a fraction of 1 ppm H_2S.

Several types of air breathing masks were available for working where a serious leak could occur. For small jobs such as disconnecting lines containing residual H_2S, a 3-minute backpack unit having a full face mask was used. A 7-minute, self-contained, over-the-shoulder-type escape breathing mask was carried at all times for routine inspection of the H_2S storage and vaporization area or the H_2S generator. It was not worn unless there was a known or suspected leak of H_2S. This type of mask could also be used for a longer period of time by attaching the mask to the hose on a larger compressed breathing air container. A refill station was provided so that when empty, the air containers could be recharged with breathing air.

The control room and laboratory section of the pilot plant, which was a modified 40-ft trailer, had a separate ventilating system. It consisted of a fan, a heater, and a carbon-filled air purification filter on the fan suction. The carbon filter removed SO_2 and smelter dusts from the air.

Provisions must be made for disposing of H_2S during upset periods in plant operations. Also, a small continuous stream of carbon dioxide containing 5 to 15 vol % H_2S was discharged from the reactor vent when using 78 vol % H_2S for regenerating the absorbent solution. This gas was vented through a fan acting as a flame arrester to a natural gas-fired incinerator. The H_2S was burned to

SO_2 before being discharged to the stack. In a larger plant striving for zero air polution, it would be necessary to feed the off-gas from the incinerator back into the gas cleaning system. The incinerator was supplied with two burners as a safety factor. Only one was required during routine operation; however, both were used to bring the incinerator up to operating temperature. The entire H_2S generator product could be burned in the incinerator. This permitted operation of the H_2S generator when the SO_2 absorption section was down.

Corrosion Testing

Except for corrosion already mentioned and some problems with FRP joint leaks, all materials of construction used for the gas cooling and cleaning and the SO_2 absorption, solution regeneration, and sulfur recovery section were satisfactory. To evaluate other stainless steels and alloys, corrosion test specimens were placed in various locations in the Bunker Hill pilot plant prior to the start of operation in January 1974, and were examined after the plant was shut down in May 1976.

For equipment handling citrate or glycolate solution, it appears that type 304 stainless steel would be the most economical choice of metals, even in the sulfur melting equipment. Materials to avoid in the sulfur melting equipment appear to be the Hastelloy G, Hastelloy C-276, and type 316 stainless steels. Although none of the metals were severely attacked, the most corrosive service was the humidifier-scrubber where both the type 304 and type 316 stainless steels showed pitting. The best resistance to corrosion in this area was shown by the Hastelloy G, Hastelloy C-276, Inconel 617, Inconel 625, Inconel 617 sensitized, and type 317 stainless steels. As stated before, FRP and PVC plastics were satisfactory where temperature and mechanical strength requirements permitted. It should be mentioned that the presence of fluorides, chlorides, or other ions not found in the pilot plant solutions might result in different requirements for other installations.

References

(1) McKinney, W.A., Nissen, W.I. and Rosenbaum, J.B., "Design and Testing of a Pilot Plant for SO_2 Removal from Smelter Gas," Pres. at AIME Ann. Meeting, Dallas, Texas, Feb. 23-28, 1974, AIME Preprint A-74-85, 12 pp.

(2) Nissen, W.I., Elkins, D.A., and McKinney, W.A., *Citrate Process for Flue Gas Desulfurization, A Status Report,* EPA-600/2-76-1366, Proc., Vol II, pp 843-864 (May 1976).

(3) Nissen, W.I., Crocker, L. and Martin, D.A., "Lead Smelter Flue Gas Desulfurization by the Citrate Process," Ch in *World Mining and Metals Technology,* ed. by A. Weiss, Port City Press, Baltimore, Maryland, Vol 2, pp 825-854 (1976).

(4) Wackenroder, H.W.F., Liebig's *Ann.,* Vol 60, p 189 (1846).

(5) Thacker, C.M., "What's Ahead for Carbon Disulfide," *Hydrocarbon Process.,* Vol 49, No. 4, pp 124-128 (April 1970).

(6) Bacon, R.F., and Boe, E.S., "Hydrogen Sulfide Production from Sulfur and Hydrocarcarbons," *Ind. Eng. Chem.,* Vol 37, No. 5, pp 469-476 (May 1945).

PEABODY CITRATE PROCESS COMBINED WITH FOSTER WHEELER RESOX PROCESS

The information in this section is based on Comparative Economics of Advanced Regenerable Flue Gas Desulfurization Processes, March 1980, EPRI CS-1381, prepared by M.R. Beychok for Electric Power Research Institute.

The Citrate Process, as proposed by Peabody Engineered Systems, uses an aqueous solution of sodium citrate to absorb and remove SO_2 from flue gases. The sodium citrate absorbent is then stripped with steam to remove and recover concentrated SO_2 (about 95%) as an end product.

Since the end product of the Citrate Process proposed by Peabody would be SO_2, EPRI required information on the additional plant facilities that would be needed to convert the SO_2 into elemental sulfur. This section combines Peabody's Citrate Process with a RESOX Process unit. The RESOX Process, developed by the Foster Wheeler Energy Corporation, uses coal as a reductant for the conversion of gaseous SO_2 into elemental sulfur.

Process Description

A schematic flow diagram of Peabody's Citrate Process combined with Foster Wheeler's RESOX Process is presented in Figure 8.6. Flue gas from an electrostatic precipitator (ESP) in a coal-burning power plant is first washed with water in a venturi scrubber to cool the flue gas and to remove chlorides, sulfur trioxide and residual fly ash. The flue gas then enters the SO_2 absorber.

Absorption Section: The flue gas flows upward through the SO_2 absorber where it is contacted by the downward flowing sodium citrate solution which absorbs and removes SO_2 from the flue gas. The absorption and dissociation of SO_2 in water can be represented by this reaction:

$$SO_2 + H_2O \rightleftharpoons HSO_3^- + H^+$$

As SO_2 is absorbed and dissolved in water, it dissociates into HSO_3^- and H^+ ions. The extent of dissociation is governed by an equilibrium which is specific to the above chemical reaction. As more SO_2 dissolves, the concentration of hydrogen ions (H^+) in the solution increases. Eventually, the hydrogen ion concentration increases to the equilibrium point where no further absorption of SO_2 can occur.

However, if the hydrogen ion concentration is reduced by reaction with some other chemical species present in the solution, then additional SO_2 can be absorbed. This can be accomplished by introducing some "buffering agent," such as the sodium citrate used in Peabody's Citrate Process, which reacts with hydrogen ions and thereby reduces their concentration in the solution. In other words, the buffering action of sodium citrate enhances the solubility of SO_2 in water which permits a lower ratio of absorbent solution flow to flue gas flow.

Stripping Section: The SO_2-rich absorbent solution from the absorber is pumped through a heat exchanger and then into a steam-heated stripping tower where the SO_2 is distilled out of the solution. The SO_2-lean solution from the stripper is then recirculated back to the top of the absorber, to be reused for SO_2 absorption, after being cooled by exchanging heat with the SO_2-rich absorbent and with cooling water.

The mixture of SO_2 and water vapor distilled overhead from the stripper is condensed at a pressure of 60 psia. The condensate forms two immiscible liquid phases: a wet SO_2 phase and a water phase. The water phase is returned to the stripper and the wet SO_2 liquid is sent to intermediate storage. At this point:

- The wet SO_2 could be dried, to reduce its water content, and be sent to an acid manufacturing plant for the production of sulfuric acid.

- Alternatively, the wet SO_2 could be revaporized and further processed for conversion to elemental sulfur.

Figure 8.6: Peabody Citrate/Foster Wheeler Resox Process

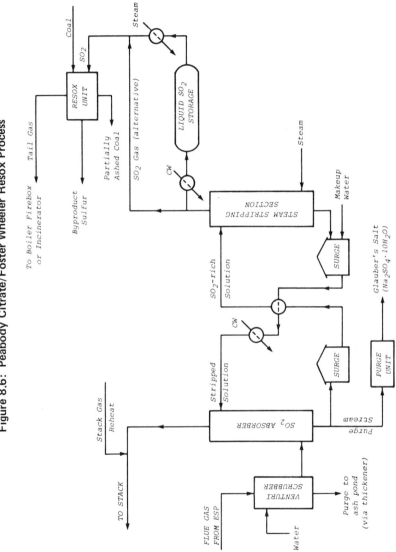

Source: EPRI CS-1381

RESOX Unit

The RESOX process is a proprietary development of the Foster Wheeler Energy Corporation. The major components of the RESOX unit are a reactor and a condenser. Crushed athracite coal and gaseous SO_2 are contacted in a vertical, countercurrent reactor at atmospheric pressure and temperatures within the range of 1100° to 1500°F. The carbon in the coal acts as a reducing agent to convert the SO_2 to gaseous elemental sulfur as represented by this reaction:

$$C(s) + SO_2(g) \rightarrow S(g) + CO_2(g)$$

The reactor temperature is maintained and controlled by the injection of air and steam into the reactor.

The sulfur in the reactor product gas is condensed in a shell-and-tube exchanger and the resulting molten sulfur is sent to heated storage tanks. The heat of condensation is used to generate steam from boiler feedwater in the shell side of the exchanger.

The residual tail gas from the sulfur condenser consists mostly of CO_2 and water vapor, but contains small amounts of SO_2 and other gaseous sulfur compounds which are not suitable for venting to the atmosphere. Therefore, all of the tail gas sulfur compounds are incinerated to SO_2 by either of these options:

- The tail gas can be incinerated in the fireboxes of the power plant's boilers. The incinerated gas then becomes part of the flue gas flow from the boilers to the SO_2 absorption section.

- A separate thermal incinerator can be used to burn the tail gas, with the incinerated gas being routed through the SO_2 absorption section.

A mixture of ash and unreacted coal is discharged from the bottom of the RESOX reactor and sent to storage. This material has been evaluated by Foster Wheeler and they feel it can be used as boiler fuel. Foster Wheeler is currently developing other possible uses for the unreacted coal, including use as replacement char for use in Bergbau-Forschung adsorbers.

Purge Streams: A slipstream of the SO_2-rich absorbent solution is sent to a crystallizer where sulfates are removed and purged as Glauber's salt (which is $Na_2SO_4 \cdot 10H_2O$). The mother liquor from the crystallizer is returned to the absorbent solution circulation loop.

A slipstream of the wash water from the inlet flue gas venturi scrubber is discharged to a thickener which concentrates the suspended solids (primarily fly ash). The underflow from the thickener is purged to the power plant's ash pond. The overflow is recirculated for reuse in the venturi scrubber.

COPPER OXIDE PROCESS

The information in this chapter is based on the NATO-CCMS study, *Flue Gas Desulfurization Pilot Study–Phase I–Survey of Major Installations–Copper Oxide Flue Gas Desulfurization Process, Appendix 95-L,* January 1979, NTIS PB-295 013, prepared by F. Princiotta of U.S. Environmental Protection Agency and R.W. Gerstle and E. Schindler of PEDCo Environmental.

BACKGROUND

The Shell Corporation and the Bureau of Mines began development of the copper oxide FGD process in the early 1960s. The objective was to develop a process using a dry selective adsorbent to avoid complications characteristic of wet systems. A further objective was to keep energy requirements low. Early work concentrated on using CuO sorbent granules precipitated from solution or a CuO powder. When this approach proved unsuccessful, efforts were directed toward fixing copper in a strong porous support. Alumina and macroporous silica gave the best results. Shell investigated various types of reactors, but found that a parallel-passage reactor with isolated regeneration cycling between two reactors had the most advantage.

In 1967 Shell began demonstrating the copper oxide system on a slipstream (670 to 1,000 Nm^3/hr) from a high-sulfur fuel-oil process heater at their Pernis refinery near Rotterdam in the Netherlands. 90% SO_2 removal was achieved over a period of 20,000 operating hours. Sorbent life was estimated at 1.5 years.

A commercial unit was installed in 1973 at the Showa Yokkaichi Sekiyu (SYS) oil-fired 40-MW boiler (125,000 Nm^3/hr and 2,500 ppm SO_2). 90% SO_2 removal was achieved. The SO_2 was processed in a Claus plant with normal refinery acid gas and also in an absorber stripper. By injecting ammonia, NO_x was removed simultaneously at an efficiency up to 70%.

To test this process on a coal-fired power plant, the U.S. Environmental Protection Agency (EPA) and Tampa Electric Co (TECO) have jointly funded a demonstration plant on a slipstream of TECO's Big Bend Station.

The following are advantages and disadvantages of the copper oxide system compared with other FGD systems.

Advantages: Since the system uses a dry process, wet materials are not handled, and water input requirements are minimal. Acceptance and regeneration occur at approximately the same temperature, obviating any heating or cooling of the absorption beds. The flue gas passes over the surface of the acceptor material rather than through it. Plugging of the acceptor bed does not occur. By alternating the acceptor and regeneration units, continuous processing can be maintained. Operating costs are reduced, since flue gas reheating is unnecessary and water requirements are low. The process has the potential for conversion to a combined SO_2 and NO_x removal system by ammonia injection into the acceptor bed. The CuO will act as a catalyst for NO_x absorption. The reactor can be operated either upstream or downstream of an electrostatic precipitator.

Disadvantages/Problems: Equipment and installation costs are high. A hydrogen source is needed for regeneration, and the large hydrogen requirements may result in high raw material costs. Damper operation becomes critical when operational modes are switched at high temperature. The stripper uses larger quantities of low pressure steam, thus requiring high energy input. The high temperature (400°C) required for operation may make installation on existing boilers difficult. The system can be installed between the boiler and air preheater in new plants.

GENERAL PROCESS DESCRIPTION

The copper oxide FGD process employs a dry acceptor, CuO on alumina, in a fixed packed-bed reactor, to extract sulfur oxides from flue gas containing oxygen and particulate matter. Desulfurization efficiency decreases as the acceptor becomes loaded with sulfur. The material is regenerated in situ and at the same temperature at which it was accepted to release the sulfur as sulfur dioxide in the regeneration off-gas. Two or more identical reactors are applied in swing operation to provide continuous processing of flue gas. The SO_2 in the off-gas from regeneration, free from oxygen and particulate matter, can be processed to yield liquid SO_2 or elemental sulfur. Figure 9.1 presents a flowsheet of a typical Shell/Universal Oil Products system as it might be applied to a 500 MW coal-fired boiler.

Process Chemistry

Cupric oxide was chosen among various metal oxides because CuO readily reacts with SO_2 in the presence of O_2 at 400°C to yield $CuSO_4$. The $CuSO_4$ can easily be reduced at the same temperature to yield SO_2; $CuSO_4$ is also stable to thermal dissociation at the acceptance/regeneration temperature.

The process is easily divided into three reaction steps: oxidation, acceptance, and regeneration.

Figure 9.1: Flow Diagram for the Copper Oxide Process

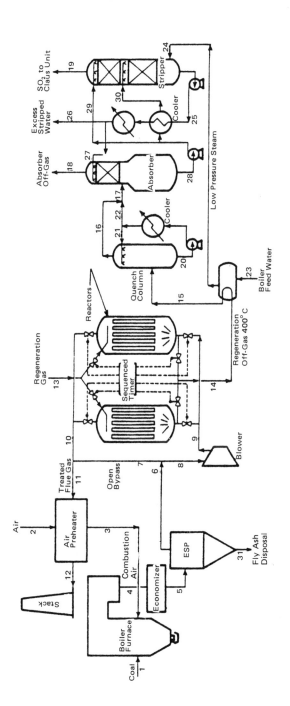

Source: NTIS PB-295 013

Oxidation: The copper in the acceptor is present primarily as elemental copper, but small quantities occur as various copper sulfides. Upon contact with oxygen in the flue gas, the copper and any Cu_2S and CuS react rapidly to form CuO, $CuSO_4$, and SO_2:

$$Cu + \tfrac{1}{2}O_2 \rightarrow CuO$$
$$Cu_2S + \tfrac{5}{2}O_2 \rightarrow CuO + CuSO_4$$
$$CuS + \tfrac{3}{2}O_2 \rightarrow CuO + SO_2$$

The formation of Cu_2S is undesirable because one-half the Cu_2S is converted to $CuSO_4$ during subsequent oxidation and therefore becomes unavailable for reaction with SO_2.

Acceptance: SO_2, O_2, and CuO react to form $CuSO_4$:

$$CuO + \tfrac{1}{2}O_2 + SO_2 \rightarrow CuSO_4$$

The SO_2 is reacted until the conversion of the CuO to $CuSO_4$ has proceeded to the extent that the unconverted CuO reaches a low level and the SO_2 that escapes capture reaches the maximum acceptable limit. The loaded acceptor is then regenerated. This acceptance reaction is shown graphically in Figure 9.2 by expressing the SO_2 content of the treated gas as a function of elapsed time after flue gas is contacted with the regenerated acceptor. The high SO_2 level at the beginning of the cycle represents the low SO_2 reaction rate that occurs during the first few minutes while the copper is being converted to CuO. If the acceptance is continued past 120 minutes, the concavity of the curve reverses and the SO_2 content of the treated gas approaches that of the feed asymptotically. Thus, at practical levels of desulfurization, a portion of the copper exists as CuO when the acceptor is regenerating.

Figure 9.2: SO_2 Concentrations in Treated Flue Gas

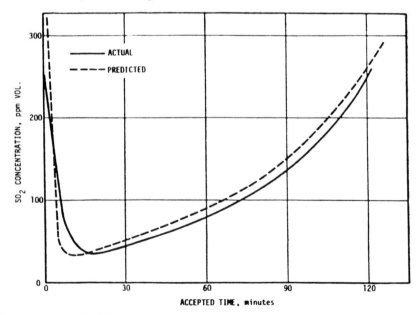

Source: NTIS PB-295 013

Regeneration: Copper sulfate releases the accepted sulfur in the form of SO_2 when regenerated with reducing agents such as H_2, CO, and light hydrocarbons or when raised to a temperature of about 700°C. Temperature is not used for this purpose because it requires excess energy and places a thermal stress on the sorbent. Hydrogen seems to be the best reducing agent. The postulated reactions are:

$$CuSO_4 + 2H_2 \rightarrow Cu + SO_2 + 2H_2O$$
$$CuSO_4 + 4H_2 \rightarrow CuS + 4H_2O$$
$$2CuSO_4 + 6H_2 \rightarrow Cu_2S + 6H_2O + SO_2$$
$$Cu_2S + H_2 \rightarrow 2Cu + H_2S$$
$$Al_2(SO_4)_3 + 12H_2 \rightarrow Al_2S_3 + 12H_2O \text{ (from alumina)}$$

The use of hydrogen promotes sulfide formation, as do lower regeneration temperatures. Higher temperatures promote $Al_2(SO_4)_3$ reduction. Reaction with SO_3 causes $Al_2(SO_4)_3$ to form in the alumina support material. Close control of temperature to approximately 400°C is required to reduce the H_2 consumption and to better reactivate the copper. Another advantage of using H_2 is that regeneration can occur at the same temperature as the oxidation and acceptance step.

The use of methane reduces sulfide formation, but it increases either the time or the temperature of regeneration. The reaction postulated is shown below:

$$CuSO_4 + \tfrac{1}{2}CH_4 \rightarrow Cu + SO_2 + \tfrac{1}{2}CO_2 + H_2O$$

Higher hydrocarbons reduce the $CuSO_4$ acceptably, but tend to leave cokelike deposits on the sorbent. These are readily burned off during the oxidation step; however, they tend to cause high temperatures, which increase temperature control problems and thermal stress.

Depending on the exact reactions involved, 2 or 4 mols of H_2 are required to reduce the $CuSO_4$ to Cu and CuS, respectively. Assuming 25% excess H_2 is required, the estimated average H_2 requirement is approximately $[(3 \times 2)/64] \times 1.25 = 0.117$ kg H_2/kg SO_2 removed. Actual experience has shown H_2 requirements of 0.125 kg H_2/kg SO_2 removed.

Absorber Stripper: After regeneration, the SO_2 and water vapor pass through a waste heat boiler, a quench tower, and finally to the absorber stripper section where the following simplified reactions occur:

Absorber
$$SO_2 + H_2O \rightarrow H_2SO_3 \text{ (aqueous)}$$

Stripper
$$H_2SO_3 \xrightarrow{\Delta} SO_2 + H_2O$$

This section also serves to level out the fluctuations of SO_2 production from the reactor.

End Product: The end product choice is liquid SO_2 or elemental sulfur, which is produced by the Claus process. Liquid SO_2 is made by drying, compressing, and then condensing the gas from the stripper. The resulting liquid SO_2 is collected and stored in pressurized vessels.

Elemental sulfur is recovered in two steps. In the reduction section, a portion of the SO_2 reacts with methane and produces a mixture of elemental sulfur, hydrogen sulfide, carbon dioxide, water vapor, and unreacted SO_2.

$$2CH_4 + 3SO_2 \rightarrow 2CO_2 + 2H_2O + S + 2H_2S$$

In the Claus section of the process, which follows, the hydrogen sulfide and remaining SO_2 from the reduction section react to produce additional elemental sulfur and water vapor:

$$2H_2S + SO_2 \rightarrow 2H_2O + 3S$$

The tail gas is incinerated and recycled to the reactor.

Description of Equipment Components

A copper oxide FGD unit may be applied to a boiler in either of two ways. It can be integrated between the economizer and air preheater to obtain the temperature required for the acceptance reaction, or it can be installed as an add-on downstream from the air preheater. The latter application requires additional fuel to increase the gas temperature to 400°C and a heat exchanger to recover some portion of the added thermal energy.

Process Particulate Removal: Particulate loading of flue gas from oil-fired boilers normally ranges from 50 to 300 mg/Nm³, which is low enough not to interfere with reactor operation. If space allows, or if the system is newly designed, the flue gases are removed after they are treated in the economizer and then returned to the boiler preheater.

A coal-fired boiler presents a different problem. Because particulate loading of the flue gas is normally 50 to 100 times the 300 mg/Nm³ acceptable limit of the reactor, an electrostatic precipitator (ESP) must be installed before the reactor. (New evidence from the EPA-TECO demonstration project suggests that the reactor will not plug if it is placed before the ESP, and the abrasive nature of the fly ash may actually scour the reactor of any deposits.) In a retrofit situation an ESP is normally already in place after the preheater. The flue gas must be heated to 400°C before it enters the reactor. The cleaned gas should then be passed through a heat recovery system before venting to the atmosphere. In a newly designed system, a hot ESP and the reactor can be integrated between the economizer and the preheater.

Reactor: The reactor in the copper oxide system consists of two parallel flow units that alternately adsorb and regenerate i.e., while one is adsorbing the other is regenerating. The purpose of this design is to allow continuous operation and overcome sorbent attrition. Tests indicate that when a pelletized sorbent is in a moving bed or fluidized bed, agitation will destroy sorbent. In a parallel passage packing adsorber, the flue gas flows in channels parallel to beds of sorbent. The sorbent is held in thin sheets to present a large surface area, a design that alleviates the dust accumulation problem that would occur in a normal packed bed design, where the sorbent bed is perpendicular to the gas flow.

The reactor at the Yokkaichi refinery is modular and 68 unit cells are used to control the 36 MW output of the boilers. The unit cell design permits a scale-up to larger plants simply by adding more cells in parallel. Adding cells in series increases SO_2 removal efficiency. Replacement of unit cells, maintenance, and inspection are all relatively easy.

The reactor body is made of a chrome-molybdenum steel. The sorbent is CuO supported by activated alumina held in thin sheets by a fine stainless steel gauze.

The flue gas is switched from the "full" reactor (loaded with SO_2) to the freshly regenerated reactor by an automatic sequence timer. The sequence timer determines when the reactor is full, switches reactors, and then sets the regeneration sequence in motion. Because the regeneration reaction uses H_2 and the acceptance reaction contains O_2 at high temperatures, an explosive mixture could be formed. Three steps are taken to prevent such an occurrence: (1) the reactor is purged with steam at $400°C$; (2) special precautions are taken to use gas valves and dampers that provide a 100% seal (e.g., a continuous purge of the valve seal with high-pressure steam); and (3) explosion limit calculations are made on the gases present to assure that an explosive mixture will not occur under normal operating conditions.

The cleaned flue gases pass out of the reactor to the air preheater. The air preheater is able to extract more heat from the flue gases than normal because most of the SO_3 is removed, thereby reducing the dew point of the flue gas. This prevents visual emission limits from being exceeded and enables flue gas to be released at a much lower temperature.

The off-gas from regeneration, which contains SO_2, water vapor, some H_2, and inerts, is forced through a waste heat boiler to remove usable heat and then to a quench tower to cool it even further in preparation for the absorber stripper section.

Absorber Stripper: The absorber stripper is used at the Yokkaichi refinery to reduce fluctuations of SO_2 gas and separate SO_2 from water vapor so that the Claus conversion is not suppressed. The flow fluctuations of the stripper overhead are reduced by incorporating surge capacity in the bottom of the absorber. Liquid pumped from the stripper contains SO_2 concentrations of 20 to 50 ppm, 75% as sulfates. This water could be used as boiler feedwater after it is processed or, because of its small quantity, be released to the atmosphere in the flue gas.

Because the absorber stripper method of processing the SO_2-rich off-gas is very inefficient, other systems of processing are being developed.

Operating Parameters

The operating temperature of the reactor is $400°C$ and the optimum copper content of the acceptor is from 4 to 6%. A 500 MW unit treating gas containing 2,000 ppm SO_2 would require acceptor and regenerator volumes ranging from 50 to 100 m^3 each. About 50% of the off-gas from the regenerator is SO_2. Pressure drop across the reactor is about 2,100 pascals.

The absorber stripper section has an efficiency of 99.9%. The total sulfur content of the stripped water is 20 ppm, and the wastewater contains very low impurities. Overall pressure drop for the system ranges from 4,500 to 5,600 pascals.

STATUS, ASSESSMENT, AND PLANS

At present only one oil-fired boiler in the world is operating with a full-scale copper oxide FGD system. The system has been studied in three pilot plants, two of which were for the purpose of ascertaining the effects of coal combustion gases on the system.

The full-scale operation (at the Yokkaichi refinery in Japan) has shown an SO_2 removal rate of 90%, but high availability has not been achieved. The pilot plant at the Galileistraat Power Station in Rotterdam in the Netherlands demonstrated the effectiveness of the parallel passage principle in coping with high particulate loadings from a coal-fired utility boiler and suggested that the system could function successfully through oxidation-reduction cycles. The results of the jointly funded EPA-TECO pilot study at TECO's Big Bend power station will have a big impact on the future of copper oxide systems in the United States.

The copper oxide FGD process is applicable to new boilers located in industrialized areas where a market exists for the sulfur based by-products, land is at a premium, and an inexpensive hydrogen source is available. Control of SO_2 above 95% appears to be a problem inherent in the cyclical nature of the system. The system is highly developed, but as yet has not demonstrated continued full-scale operation on coal-fired boilers. Effective NO_x control has been obtained with the addition of ammonia gas. It is an attractive system for limited application, but has a high capital cost.

The future of the copper oxide process is difficult to predict. No new installations are presently planned, but the system could become a good choice if its potential for simultaneous NO_x removal were developed and regulations made such removal necessary.

RECENT TECHNOLOGICAL DEVELOPMENTS AND SOLUTIONS TO PROBLEM AREAS

New Developments: *NO_x Removal* — The Shell copper oxide process at the Yokkaichi refinery is reported to have an NO_x removal efficiency as high as 70%. Ammonia added to flue gas reacts with NO at the reactor temperature to form nitrogen and water. The ammonia concentration in the flue gas released to the atmosphere is 2 ppm.

The study recently completed at TECO's Big Bend Station shows that the commercial acceptor manufactured by Ketjen has been demonstrated to be chemically and physically stable. The parallel passage reactor is an acceptable design for flue gas fly ash loadings as high as 25 g/Nm^3. Pressure drop and SO_2 removal were unaffected after 2,210 acceptance-regeneration cycles over 3 months. In case they were found to be necessary, in situ cleaning techniques were developed. It was determined that it is not necessary to have fly ash removal upstream of the CuO reactor when used on a coal fired boiler.

Solution to Problem Areas: The Yokkaichi refinery has experienced three problems: (1) dust deposition around the inlet of the reactor; (2) plugging of the waste heat boiler; and (3) corrosion of the quench towers. The first problem was solved by installing steam soot blowers where the deposits collect. Plugging of the waste heat boiler and corrosion of the quench tower are expected to be solved by changing the material and the structure of these pieces of equipment.

CARBON ADSORPTION PROCESS

The information in this chapter is based on the NATO-CCMS study, *Flue Gas Desulfurization Pilot Study–Phase I–Survey of Major Installations–Carbon-Adsorption Flue Gas Desulfurization Process, Appendix 95-K,* January 1979, NTIS PB-295 012, prepared by N. Haug of Umweltbundesamt and G. Oelert and G. Weisser of Battelle-Institut e.V.

BACKGROUND

The use of activated carbon or other adsorbents to remove sulfur oxides from flue gases is a promising way to bypass problems connected with wet scrubbing techniques, such as scaling, pH control, reheating and waste disposal.

Although the sorption temperature should be as low as possible in order to achieve maximum sulfur oxide pickup, the adsorption processes allow operating at higher temperatures than the wet scrubbing techniques, thus retaining sufficient stack gas buoyancy. The loaded adsorbent—predominantly activated coke—may be regenerated by thermal stripping at relatively low temperatures which results in an SO_2-rich gas suitable for the production of sulfuric acid or elemental sulfur.

Sorption of SO_2 by activated carbon can be both chemical and physical. The degree of each sorption mode is determined by the presence of oxygen and water vapor in the gas. The active adsorbent surface catalyzes oxidation of SO_2 to SO_3, which forms H_2SO_4 in the adsorbent pores with the water vapor, thus increasing the degree of chemisorption at the expense of physicosorption. With O_2 and H_2O in the flue gas, a sulfur loading of the activated carbon ranging between 5 and 15% is reported to be feasible.

The first of the major carbon processes, the Reinluft Process, was developed in Germany, beginning in about 1957. The first version of this process was characterized by sorption and regeneration being carried out in a single tower with the carbon circulating through the relevant sections of the tower. Between 1959 and 1968 four test units with capacities up to 55,000 m^3/hr were in operation in Germany.

After self-ignition problems of the carbon arose in the larger test units, Chemiebau (Davy Power Gas) continued process development and proposed separate sorption and regeneration vessels with a screening device in the loaded carbon stream and regeneration heat input by hot gas.

Following the Reinluft activities, the carbon adsorption technology was advanced by the process development of Bergbau Forschung, Germany. The Bergbau Forschung Process differs from the Reinluft Process mainly in that it uses hot sand instead of gas in the regeneration system, thus easing local over-heating problems. Moreover a specifically developed activated coke is used as adsorbent.

After a small-scale plant (2,000 m^3/hr), a semiscale plant was installed in 1974; it serves for desulfurizing 150,000 m^3/hr of flue gas from the Gemein-schaftskraftwerk-Ost at Lünen, Germany. The Lünen unit also allows investiga-tions into the efficiency of the Bergbau Forschung Process in retaining fly ash and NO_x from the flue gas.

In the USA, Foster Wheeler constructed a 63,000 m^3/hr unit at Gulf Power's Scholz station under license of Bergbau Forschung in 1975. The Foster Wheeler design differs from the original Bergbau Forschung design mainly in that it com-prises a two-stage adsorber.

Reducing SO_2 to gaseous sulfur by means of crushed coal characterizes the RESOX Process subsequently developed by Foster Wheeler.

Another development activity was undertaken in Japan by Sumimoto Ship-building and Machinery Co. and the Kansai Electric Power Co. This process employs cross-flow adsorption as the Bergbau Forschung Process and regeneration by hot inert gas as the Reinluft Process. A pilot plant (10,000 m^3/hr) was tested in 1969. Since 1972 a 260,000 m^3/hr demonstration unit has been operating at Kansai's Sakai Port power station.

The main drawback in the development of carbon adsorption processes with dry regeneration was the behavior of the carbon with respect to attrition, poison-ing of the active surface by fly ash and unreactive compounds, and self-ignition of the carbon in the sorption section.

In the Reinluft Process the respective carbon cost was reduced by using relatively cheap coke from peat and lignite, whereas for the Bergbau Forschung Process a coke was made from hard coal, which is relatively attrition-resistant and has a high ignition temperature.

The advantages and disadvantages of the carbon adsorption process are as follows:

Advantages:

(1)   No scaling or plugging in the adsorber.

(2)   Most of the construction material is carbon steel.

(3)   Clean fan operation.

(4) NO_x removal (partial).

(5)   Fly ash removal.

(6)   No need for stack gas reheating.

(7)   No waste disposal pond required.

Disadvantages:

(1) Deterioration of circulating sorbent through physical and chemical influences.

(2) Large equipment (sorption and desorption vessels) compared with lime scrubbing.

GENERAL PROCESS DESCRIPTION

The principles of all dry adsorption systems are essentially the same. The use of activated carbon or other sorbents allows removing of adsorbed compounds by simple thermal stripping. A problem is the oxidation of SO_2 catalyzed by the highly activated surface of the adsorbent. SO_2 reacts with moisture in the gas forming H_2SO_4 in the pores of the adsorbent. Thus one of the main problems is getting the acid out of the pores.

The sorption process of SO_2 by activated carbon can be both chemical and physical. The main chemical reactions and operating parameters are collected in Table 10.1. A generalized flow diagram is shown in Figure 10.1.

Table 10.1: Basic Chemistry and Process Data

Chemical Reactions	Adsorption Section:
	$SO_2 + H_2O + O_2 \rightarrow$ adsorption
	$SO_2 + H_2O + \frac{1}{2}O_2 \xrightarrow[\text{charcoal}]{120-140^\circ C} H_2SO_4$
	Regeneration Section:
	$2H_2SO_4 + C \xrightarrow[\text{sand}]{260-700^\circ C} CO_2 + 2H_2O + 2SO_2$
Control Theory	Commercial application of hot sand causes a heating rate of $500^\circ C/min$.
Reaction Kinetics	The maximum reaction rate occurs in the temperature range of 520° to $680^\circ C$. The reaction in the regeneration section is practically complete at $450^\circ C$.
Agent Requirement	Loading: 5 to 15 wt % SO_2
	Regeneration: $SiO_2:C = 4:1$
Quality of Flue Gases (in→out)/ Removal Efficiencies	SO_2: 1000. . . 2300 → 200. . . 600 ppm. particulate: 150 → 25 mg/m^3
	NO_x: R 15. . . 60%
	Cl^-, F^-: R 50%
Adsorption Agent	carbon > 90%, special activated coke, prepared by air oxidation of a fine ground hard coal.
Product	Regenerator off-gas: 20-30% SO_2, 50-60% H_2O, 20-30% $(N_2 + CO_2)$
Effluents, Waste Streams	No waste
Gas Reheat	No reheat

(continued)

Table 10:1: (continued)

Subsequent Processing   Bergbau Forschung: Claus Process
             Foster-Wheeler: RESOX Process: Crushed coal is the reducing agent at 650-815°C.

$$SO_2 + C \xrightarrow[\text{1 bar}]{650-815°C} S + CO_2$$

$$5SO_2 + H_2O + 7C \rightarrow 5CO_2 + S_{liq} + H_2S + COS + CS_2$$

$$(SO_2\text{-rich gas} \rightarrow SO_2 \text{ liq or } H_2SO_4)$$

Note: R is removal efficiency.

Source: NTIS PB-295 012

Figure 10:1 Generalized Flow Diagram/Carbon Adsorption

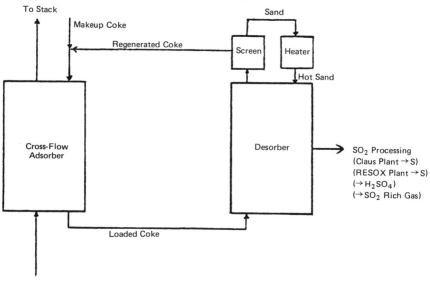

Source: NTIS PB-295 012

Coming from the boiler the flue gas enters the adsorber where—in the modern processes—it is brought into contact with activated coke. The desulfurized flue gas passes to the stack directly, while the loaded carbon is transferred to the desorber vessel. By means of heat the SO_2-rich gas is set free and flows to a Claus plant. The regenerated carbon is filled up with makeup coke and recycled to the adsorber.

There are different ways of bringing heat into the desorber: one is to introduce a hot gas into the desorber, another to mix the loaded adsorbent with hot sand (the latter is done in the Bergbau Forschung Process). As auxiliary equipment there is a screening device which separates friction fines and—in the case of the Bergbau Forschung Process—the heat carrier sand from the recycle coke.

STATUS AND PLANS

The design of the desulfurization reactors as applied in the demonstration plants has proved its value through extensive tests performed since 1974. The circular silo desulfurization reactor for 150,000 m^3/hr tested at Lünen may be increased in size for a capacity of 650,000 m^3/hr and the two-stage reactor (Foster-Wheeler) in a modular fashion may be increased to any size.

Heat transfer by means of solid matter for regeneration of activated char loaded with sulfuric acid was successful. The extension of the sand circuit for flue gas desulfurization units, matched to a power station capacity of 350 MW does not constitute a technical problem.

Since the adsorption and regeneration techniques were given first attention during the initial phases of experimental operation, experiments with the subsequent processing of the SO_2-rich gas have been emphasized more recently.

The design and operating mode of the Lünen Claus unit has been adjusted to the particular composition of the SO_2-rich gas.

Dry separators for chlorine, fluorine and dust have solved corrosion and catalyst poisoning problems. Also with respect to catalyst saving the Claus plant does not fully follow the peak-shaving operations of the power plant. It continues operating under partial load during weekend shutdowns of the power station using the char charge of the desulfurization unit as a buffer of loaded adsorption agent.

The basic feasibility of directly reducing SO_2 to S using coal as reductant (as done in the RESOX Process) was demonstrated. Current plans extend to investigating the general processing of SO_2-rich gas and also the possibility of NO separation under conditions prevailing in flue gas desulfurization.

RECENT DEVELOPMENT AND SOLUTIONS TO PROBLEM AREAS

Through extensive development and test programs in recent years with carbon adsorption units in the U.S., Japan and Germany, good progress was achieved on solving the classic problems of this technology like excessive char consumption, char and gas distribution, plant instrumentation and solids handling systems.

Bergbau Forschung developed a suitable activated char, based on hard coal. This activated char is characterized by a high abrasion resistance, a particle size (diameter 9 mm) that guarantees low pressure drop in the adsorber, a very low reactivity with oxygen and a high SO_2 adsorptive capacity.

The travelling bed adsorber with traverse gas flow as applied in the Bergbau Forschung Process along with the A.M. low-pressure drop char eased the solids/gas distribution problems.

Modification or change of gas shut-off valves and solids-handling components, both of which had been subject to severe wear by mechanical, thermal and corrosive influences and plugging problems improved the reliability of the systems.

BERGBAU FORSCHUNG PROCESS

The Bergbau Forschung (BF) process makes use of the well-known fact that any carbonaceous adsorbent will remove sulfur dioxide from a gas containing that pollutant. Bergbau Forschung has developed a special activated coke for this purpose, which is outstanding in its overall performance, i.e., it is characterized by its high SO_2 adsorption rate, low wear factor and high ignition temperature. This coke can easily be regenerated and is therefore recycled in the BF process, as can be seen from the basic flow diagram (Figure 10.2).

In the adsorption system **1** sulfur dioxide is removed from polluted gas streams by adsorption on activated coke. The SO_2-charged adsorbent is then regenerated in the desorption system **2** thus producing a gas stream of high SO_2 concentration. For the final step, the processing of the SO_2-rich gas **3**, one has the option either to produce sulfur of high purity, liquid sulfur dioxide or sulfuric acid.

The polluted gas stream is brought in contact with the activated coke in the adsorption reactor **1**, which is a cross-stream moving bed reactor. The clean gas is then directly blown to the stack **4**, i.e., no reheating is required. Since the coke not only removes the SO_2 from the gas but in addition flue dust which has passed the electrical precipitator, this dust together with a small amount of coke fines resulting from wear in spite of the great hardness of the coke, has to be separated from the adsorbent by a screening machine **5**. By bucket elevators **6** the coke, laden now with sulfur dioxide, is then conveyed to the desorption plant **2**. The heart of this subsystem is the desorption reactor **2**, likewise a moving bed reactor. Here the regeneration takes place by heating the adsorbent to an average temperature of 600°C.

Several ways of heating up the coke can be thought of. At present it is being done by mixing the coke in the reactor directly with sand as a solid heat carrier. This sand is heated and at the same time conveyed by combustion gases of a separate combustion chamber **7** in the so-called sandlift **8**. The waste heat of these combustion gases is partly recovered in a heat exchanger for air and in the boiler of the power plant.

The regenerated coke discharged from the desorption reactor has to be cooled down **9** to 100°C once it is separated from the sand by a screening machine **10**. This is being done in two stages with water and air on cooling conveying tracks **9**. The waste heat of this process step is not recovered due to capital investment considerations. Finally the coke is recycled to the adsorption reactor by bucket elevator **11**.

During the regeneration the SO_2-charged coke undergoes a modified reversal of the adsorption reaction yielding an effluent gas with a sulfur dioxide concentration of 20 to 30% by volume. This gas, after passing a dust separator, is blown to the SO_2 processing system **3**. This can be a modified Claus plant for the production of sulfur, a sulfur dioxide liquefaction plant producing SO_2, if necessary of quality required by the food industry, or a sulfuric acid plant. Each of these plants comprise standard technology of the chemical industry.

REFERENCES

(1) Noack, R., Oberländer, K., "Large Scale Flue Gas Desulfurization Plants Using the Babcock-BF Process," Deutsche Babcock AG Report No. 66, October 1976.

(2) Aufbau und Funktion der Grossversuchsanlage in Lünen (Bergbau Forschung-Steag), Attachment 1 to NATO-CCMS questionnaire.

Figure 10.2: Flow Diagram of the Bergbau Forschung Process (1) (2)

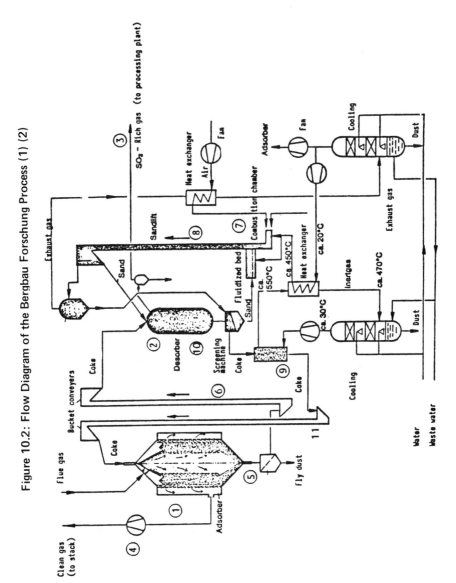

SEAWATER SCRUBBING PROCESS

The information in this chapter is based on the NATO-CCMS study, *Flue Gas Desulfurization Pilot Study–Phase I–Survey of Major Installations–Sea Water Scrubbing Flue Gas Desulfurization Process, Appendix 95-D,* January 1979, NTIS PB-295 005, prepared by R.I. Hagen and H. Kolderup of Foundation of Scientific and Industrial Research, Norway.

BACKGROUND

This chapter will deal with the Fläkt-Hydro flue gas desulfurization process, which has been developed jointly by Norsk Hydro a.s. and A/S Norsk Viftefabrikk, the Norwegian subsidiary of AB Svenska Fläktfabriken (Fläkt).

The principle of the Fläkt-Hydro FGD-process of employing seawater as a scrubbing agent is not new. The Electrolytic Zinc Company in Tasmania has used tidal water as the absorbent for the removal of sulfur dioxide from smelter gas since around 1949 (1). Scrubbers for removing fluorides at the A/S Årdal and Sunndal Aluminum plant in Norway also remove 80 to 90% of the sulfur dioxide content utilizing seawater. The scrubbers are installed in the pot-exhaust system (2).

Laboratory investigations on the absorption of sulfur dioxide in seawater have been made by L.A. Bromley (3) (4).

Research and development of the Fläkt-Hydro process were started in 1971 with laboratory investigations. The basic development of the process was completed during the summer of 1973 after having carried out tests on a pilot plant with a gas capacity of approximately 10,000 Nm³/hr.

The first commercial Fläkt-Hydro unit was installed in August 1975 on an existing 10 MWe boiler. At present there are two new commercial units in operation each serving 20 MWe boilers. A fourth unit of the same size has been commissioned. All the commercial units mentioned are located at Norsk Hydro, Porgrunn Fabrikker.

GENERAL PROCESS DESCRIPTION

Chemistry

The Fläkt-Hydro process is based on the utilization of the alkalinity of seawater to absorb and neutralize SO_2. Seawater contains bases corresponding to weak acids, such as bicarbonate, carbonate, borate, phosphate, arsenite and sulfide and has a considerably higher buffer capacity than freshwater. This is illustrated in Figure 11.1 which shows the effect on pH by the addition of sulfuric acid to seawater and freshwater, respectively.

Figure 11.1: Buffer Capacity of Pure Water and Seawater (5)

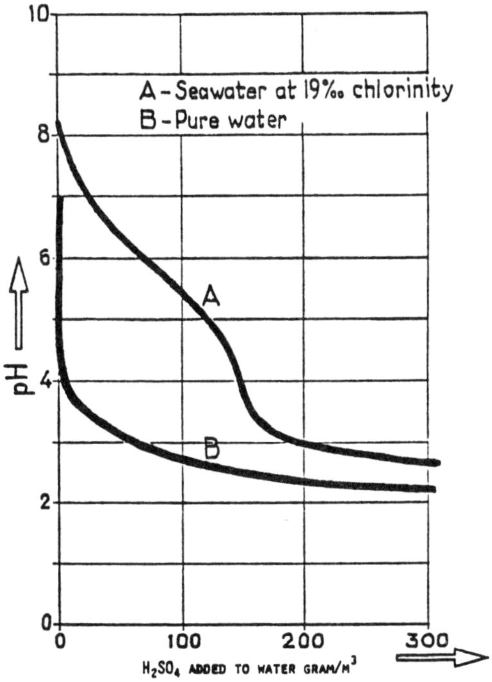

Source: NTIS PB-295 005

The fundamentals in the process are shown by the following chemical equilibria and reactions [References (2) and (5)] :

(1) $$SO_2 \ (g) \rightleftharpoons SO_2 \ (aq)$$

(2) $$SO_2 \ (aq) + 2H_2O \rightleftharpoons HSO_3^- + H_3O^+$$

(3) $$HSO_3^- + H_2O \rightleftharpoons SO_3^{2-} + H_3O^+$$

(4) $$CO_2 + H_2O \rightleftharpoons H_2CO_3$$

(5) $$H_2CO_3 + H_2O \rightleftharpoons HCO_3^- + H_3O^+$$

(6) $$HCO_3^- + H_2O \rightleftharpoons CO_3^{2-} + H_3O^+$$

(7) $H_x A + H_2 O \rightleftharpoons H_{x-1} A^- + H_3 O^+$

(8) $SO_3^{2-} + \frac{1}{2}O_2 \rightarrow SO_4^{2-}$

In reaction (7), A is arsenate, borate, phosphate or sulfide.

The buffer capacity results mainly from the equilibrium between carbonic acid and its corresponding bases ($CO_2/H_2CO_3/HCO_3^-/CO_3^{2-}$). The total alkalinity for uncontaminated oceanic seawater is normally 2.3 to 2.5 meq/ℓ, and the pH is about 8.3.

In this process, sulfur dioxide is absorbed in seawater and converted mainly to bisulfite (HSO_3^-) according to the equilibria (1) to (3). The oxidation reaction (8) will finally convert the bisulfite into sulfate.

Commercial Installations at Porsgrunn

Plant Description: A general flow diagram of the present installations is shown in Figure 11.2. The installations comprise a flue gas fan, cyclones for collection of particulates (only on the first installation), a multi-venturi quencher, a countercurrent packed tower, demister and a reheat unit for stack gas.

The multi-venturi quencher and the reheat unit are made of corrosion resistant steel, and the scrubber tower is constructed in GRP (glass-fiber-reinforced polyester).

The three installed FGD systems take care of 35,000, 60,000 and 60,000 Nm³/hr, respectively. The pressure drop for the quencher is 2.5 mbar, for the packed tower 7.5 mbar and for the demister 0.5 mbar. These pressure drops refer to maximum load. For the total FGD system the pressure drop is approximately 24 mbar. The cyclones in the first installation result in a higher total pressure drop (35 mbar).

The purpose of the separate collection of particulates has been to reduce the amount of potentially objectionable materials, which otherwise might be collected in the scrubber and discharged to the fjord. The cyclones were installed before it was determined that particulates containing heavy metals were of an extremely small size.

The reheat unit is a simple unit for mixing in a small, adjustable bypass stream of hot flue gas taken from the inlet to the economizer. This is done to achieve sufficient plume rise and dispersion of the emissions. The reheater is designed to give a maximum temperature increase in the flue gas of 25°C. Experience has shown, however, that the plume rise and dispersion is better than expected. The reheater is therefore used only during unfavorable weather conditions. Stack heights are 38 m above ground level.

The seawater supply for the three FGD systems is 270, 540 and 540 m³/hr, respectively. Approximately one-tenth of the supply goes to the venturi-quenchers.

The effluent liquid from the scrubbers is added to an alkaline sewer system containing calcium hypochlorite. In this system the absorbed sulfur dioxide is neutralized and oxidized to sulfate.

The boilers and the FGD systems are equipped for automatic operation and the FGD system is controlled by the boiler load. On the first unit the boiler load can be changed from 20 to 100% within 1 minute, while the two other units need 3 minutes for a corresponding change in the load. The FGD system and the boilers can operate down to 20% of maximum load. The FGD system, however, theoretically can operate at a lower load.

Figure 11.2: Fläkt-Hydro System–Steam Boiler Station, Porsgrunn Fabrikker

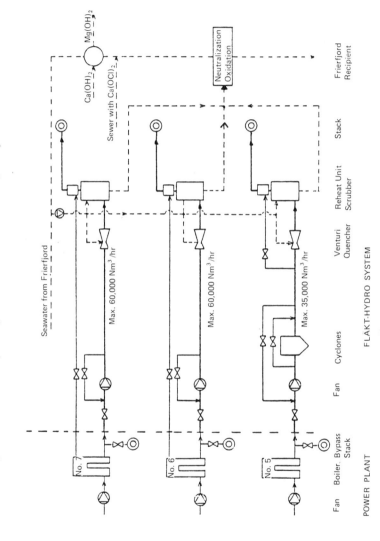

Inlet guide vane controls are installed on each of the forced-draught fans up-stream of the boilers for control of combustion air.

The pressure downstream of the boiler is set to –1 mbar (±1.5). This pressure is controlled by an inlet guide vane control and a recirculation shunt at the flue gas fan.

The damper in the shunt is controlled by the boiler load (amount of combustion air) between 20 to 60% of maximum boiler load.

Measurements by SINTEF: *Summary* – SINTEF and the respective companies have carried out measurements of several components both in the flue gas at the inlet and the outlet of the scrubber and in the scrubber liquid. A summary of the results obtained by SINTEF for the scrubber on boiler no. 7 (20 MWe) at Norse Hydro a.s., Porsgrunn Fabrikker is given in Tables 11.1 through 11.4 (6). Similar results have been obtained by the companies through measurements on boiler nos. 5 and 6.

Table 11.1: Data for Oil Used

. . . . Analysis of Components in Oil . . . .

Sulfur	2.1–2.5% S (by wt)
Ash*	0.06–0.1%
Vanadium*	132–160 mg V/kg oil
Nickel*	39–63 mg Ni/kg oil
Water*	2.0–2.45%

Note: Operating data–oil consumption at maximum load is 4,500 to 4,700 kg/hr.

*4 to 5% soot from ammonia production is mixed into the heavy oil used.

Table 11.2: Results from Measurements in the Flue Gas

	Unit*	Scrubber Inlet	Scrubber Outlet
Flue gas stream	Nm^3/hr	57,000	50,000
Flue gas temperature	°C	210	5–7
Sulfur dioxide concentration**	mg S/Nm^3	1,960–2,080	12.7–14.4
Particulate concentration	mg/Nm^3	~100	70–80
Nickel in particulate, concentration in flue gas	mg/Nm^3	2.4–2.9	1.6–2.1
Vanadium in particulate, concentration in flue gas	mg/Nm^3	6.3–7.4	2.9–3.5
Polynuclear aromatic hydrocarbons concentration (Sum of 3 fractions)	μg/Nm^3	7.3–38.3***	—
Nitrogen oxides concentration	ppm NO_x	220–250	210–230
	ppm NO_2	≤10	

Note: Boilers operated at 100% load and no reheating.

*Nm^3 = unit volume of gas referred to standard conditions (0°C and 1.01325 bars).
**Both sulfur trioxide/sulfuric acid and sulfates in particulates are measured (6).
***The high value is obtained under an operating condition with too low air excess in the boiler (sooting). The results are shown in detail in a measuring report (6).

Source: NTIS PB-295 005

Table 11.3: Results from Measurements in Scrubbing Liquid

	Unit	Scrubber Inlet	Scrubber Outlet
Seawater flow	m³/hr	530-540	—
Sulfate concentration	mg S/ℓ	—	213-218
Sulfate concentration	mg S/ℓ	871-884	—
pH	pH	—	2.23-2.32
Nickel concentration	μg/ℓ	8-9	80-310
Vanadium concentration	μg/ℓ	0-11	250-540
Polynuclear aromatic hydrocarbon concentration	μg/ℓ	<1.0*	≤4.6*

*Data are uncertain as concentrations are close or below detection limit for the measuring equipment used. Comparing the individual measurements on the gas and liquid side, the mass balance does not comply.

Source: NTIS PB-295 005

Table 11.4: Absorption and Collection Efficiencies

	Percent Based on . . . Measurement in. . . .	
	Flue Gas	Effluent Water
Sulfur dioxide absorption efficiency 100% load	99.4	—
Sulfur dioxide removal with maximum reheating 25°C	93	—
Particulate removal in scrubber	30-60	—
Nickel removal in scrubber		
100% boiler load*	31-41	32-34
50% boiler load**	40-53	33-45
Vanadium removal in scrubber		
100% boiler load*	52-66	40-42
50% boiler load**	47-55	47-54

*At 100% boiler load it was found that approximately 43 to 52% of the nickel and 32 to 40% of the vanadium content in the oil was retained by the boiler between soot blowing.
**At 50% boiler load 13 to 23% of the nickel and 20 to 35% of the vanadium content in the oil was retained by the boiler.

Source: NTIS PB-295 005

The procedure of the measurements compares closely to the EPA standards. The main difference is the temperature of the filter holder of the sampling train where SINTEF used 90°C instead of 121°C. Corrections for sulfuric acid mist on the filter are made.

Measurements were performed at two different operating conditions for the boiler, namely 100% and 50% of the maximum boiler load. Mainly, the results from 100% boiler load are given in this summary. For extensive information and results from the 50% boiler load conditions, reference is made to the separate report on measurements (6). Complete results of the particle size distribution and distribution of heavy metals in the different fractions are also given (6).

The concentrations of heavy metals and polynuclear aromatic hydrocarbons (PAH) in the scrubber liquid differ significantly between measurements. The removal efficiencies are therefore based on measurements in the flue gas. For the sulfur mass balance the different results match fairly well with a maximum deviation of 4% from the sulfur content in the oil.

As to the particles size distribution (investigated with an in-stack cascade impactor) it may be concluded that 60 to 80% by wt of the particles belong to the fraction with particle size less than 5 μm. Furthermore, approximately 70% of the nickel and 60 to 70% of the vanadium are found in the particles with smaller size than around 1 μm. Even particles with smaller size than 0.25 to 0.4 μm still contain around 50 to 60% of the heavy metals.

STATUS AND PLANS

Summary of Status of the Surveyed FGD Systems

The Fläkt-Hydro process has been in commercial operation since 1975. Boiler no. 5, an old boiler, was equipped with the FGD system in 1975 and the system was in operation from August 1975 to late fall 1976. Boiler no. 5 was then shut down for overhauls and exists today as a stand-by installation.

Two new boilers no. 6 and no. 7 have been installed and equipped with the FGD system. These units are in operation. A new unit has been commissioned for boilers and process gas at a magnesium plant. All these installations are in the same industrial area in Porsgrunn. Presently, installations at oil refineries and oil or coal-fired power stations are seriously considered.

Operational Experience

The installations on boilers nos. 6 and 7 have proved troublefree after the startup period. The first installation on boiler no. 5, however, did cause some problems. The main problem was the regulation of the gas flow and control of the pressure in the duct system. This problem was caused by rapid and considerable changes in load since the boilers take care of the variations in steam demand for the Norsk Hydro plant. After a preliminary period it was found necessary to install a shunt or recirculation loop over the flue gas fan. The quantity of recirculated gas is determined by a damper which in turn is controlled by the flow rate of combustion air.

As to corrosion problems, the absorption towers are constructed of GRP (glass-fiber-reinforced polyester) and all tower internals are also made of plastic materials. The quencher, however, is constructed of corrosion resistant metal because of contact with hot gases (160° to 380°C). In the opening phase there were corrosion problems in the cooling zone in the quencher for the installation on boiler no. 5. The corrosion problem for this unit was solved by changing the design. This modification caused an availability factor of 97.0% during March 1976 for unit no. 5. There have been no corrosion problems for the two new units.

For all the units there have been mechanical problems with the inlet guide vane control on the flue gas fans which gave a small decrease in the availability factors during several months. During soot-blowing the FGD system occasionally has cut out, due to interlock system, in order to protect the scrubber materials against too high temperatures. The reason for the temperature rise was the use of steam for soot-blowing in the boiler.

The bypass valve for the reheater has had a tendency to stick in boiler system no. 5. There is a small leakage stream ensuring the ducts temperature is above the dew point of the gas.

Furthermore, some negligible overhauls in the seawater supply and discharge system such as the seawater pump and collection tank for the three units proved necessary.

All taken into consideration, it may be concluded that there is no major problem in the operation of the Fläkt-Hydro process. Reference is made to the average availability factors of 98.5% and 99% for units nos. 6 and 7, respectively.

Fog Formation

The fog emission from the Fläkt-Hydro process appears to be small. During measurements by SINTEF in March and June 1977 the light plume became invisible at a distance of 15 to 30 meters from the stack. No reheat was used, and stack gas temperature was 7°C. Total water content (vapor + mist) was average 10 g/Nm^3 and maximum 12 g/Nm^3. Saturated air at 7°C contains 8 g/Nm^3 water vapor.

Applicability of the Process

The seawater scrubbing process is applicable to fossil fuel fired power plants using seawater as a coolant and to fossil fuel fired industrial boilers located at a reasonable distance from the sea.

As far as excess seawater temperature is concerned there should generally be no restrictions for boilers of 20 MWe size. The applicability of all types of thermal power plants using seawater as a coolant, however, is restricted by the allowable excess temperature in the sea recipient.

A 1,200 MWe thermal power plant using seawater as a coolant would require a recipient capacity equivalent to the outer part of the Oslofjord at Slagentangen or even better the Skagerak sea at Tromoya outside Arendal. The excess temperature caused by the seawater scrubbing process itself would be small compared to the effect of effluent cooling water from the condensers.

Plans for Installation in Power Plants

The seawater scrubbing process installed at Norsk Hydro a.s. is of a special type with respect to the size, separate particulate collection and treatment and discharge of the used seawater.

The Fläkt-Hydro-FGD-process will generally consist of the following main elements for the installation in a power plant: a fan, separate particulate collector, quencher, scrubber, gas reheater, dilution and aeration unit. See Figure 11.3.

The problem of scaling up the process will mainly be related to the distribution of liquid and gas flow in the scrubbing tower.

The demand for particulate collection (precollector) will probably be the same as required with regard to atmospheric pollution in order to minimize effluent of objectionable materials which otherwise might be released.

In order to reduce the chemical oxygen demand (COD) in the used scrubbing liquid, it is necessary to have an oxidation unit. The seawater from the scrubber constitutes 14 to 20% of the cooling water stream from the condensers. The used scrubbing liquid will be mixed in the return cooling water to give a higher pH and lower concentrations. The dilute scrubbing water will then be aerated in an oxidation unit before discharge.

Figure 11.3: Main Elements in the Fläkt-Hydro-Process in a Power Plant

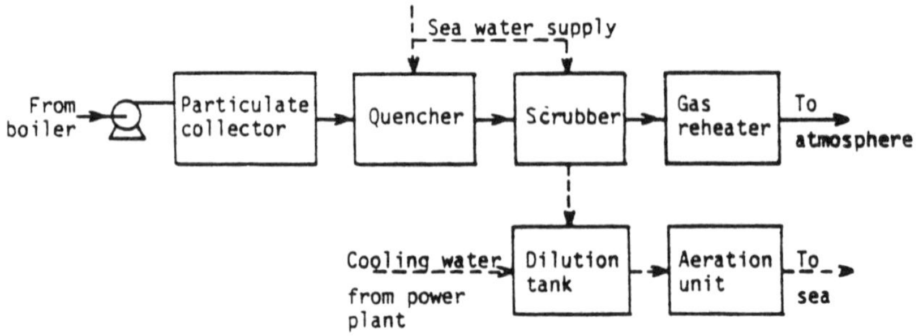

Source: NTIS PB-295 005

The Water Resources and Electrical Board State Power System (NVE) is evaluating an oil- or coal-fired power plant installation in the Oslofjord region. Planned capacity is 600 MWe. A/S Norsk Viftefabrikk has projected and given a budget offer for the Fläkt-Hydro-FGD system. The NVE has estimated total investment and operating costs based on this offer. It is assumed that the same electrostatic precipitator and the same stack height is required even if oil with low sulfur content is used.

The main objections to the process deal with the discharge of sulfites, nitrogen oxides and heavy metals to the sea. Furthermore the degree of reheat requirements is being discussed (7). Several recent investigations conclude that the effluent water will cause only marginal local effects (8)-(10).

RECENT DEVELOPMENTS AND SOLUTIONS TO PROBLEM AREAS

Liquid Disposal and Dispersion

The seawater process depends fully on discharge of large amounts of used seawater and a thorough knowledge of the extent of the water pollution caused by this process is imperative. The effluent water from the scrubber contains sulfite and trace elements from the fuel, such as nickel and vanadium. The seawater temperature also increases when passing through the scrubber.

An assessment of the environmental impact is being made by the VHL (River and Harbour Laboratory affiliated with SINTEF) (10). A summary of the most important basic conditions and the results obtained is given below.

Basic Conditions: It is difficult to base a general environmental assessment of the seawater scrubbing process on the three scrubbers at Porsgrunn. This is because the scrubbers use a special alkaline sewer for the effluent water. Also extra soot is being mixed with the oil used in the boilers.

Instead of the existing scrubbers, two assumed installations, one of 20 MWe and one of 1,200 MWe size, each working with heavy oil and seawater effluent, have been chosen in the further study. A typical size of an industrial boiler would be 20 MWe, while an oil based power plant of 1,200 MWe is relevant as a supplement to existing hydroelectric power in Norway. The recipients chosen are a small fjord (Frierfjorden), a large fjord (Oslofjorden) and a sea area (Skagerak). The locations are shown in Figure 11.4.

Figure 11.4: Actual Locations and Recipients in the
Environmental Assessment Study (10)

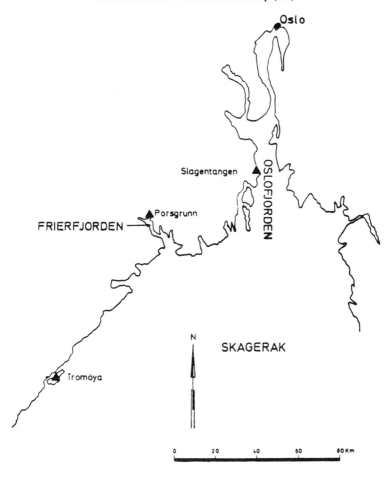

Note: ▲ Locations referred herein.

Source: NTIS PB-295 005

The basic conditions used in the calculations of concentrations of pollutants in the effluent water and in the recipient, are shown in Table 11.5.

For the 20 MWe industrial boiler, data on nickel and vanadium retained in the boiler and the efficiency of the scrubber for the collection of these metals, are based on SINTEF data from the 20 MWe boiler no. 7 at maximum boiler load (6). The information on frequency and duration of soot-blowing are taken from material published by AB Svenska Fläktfabriken (2). The calculated concentrations of nickel and vanadium during soot-blowing vary strongly with the frequency and duration of soot-blowing. Concentrations will decrease rapidly if the frequency and duration of soot-blowing are increased. The concentrations used in the calculations should be regarded as a worst case. 20 m below surface has been taken as a reasonable depth for intake and outlet of seawater. This arrangement usually prevents recirculation for small units. pH of the effluent water is estimated using data computed by Abdulsattar (1977) (11). His data of pH versus sulfite concentrations added to seawater is based on a closed oxygen free sulfur dioxide-carbon dioxide-seawater system. The carbon dioxide produced by reaction of sulfur dioxide with water is assumed to remain in the liquid phase. Abdulsattar's model seems reasonably valid when the effluent water is not being diluted and aerated.

Table 11.5: Basic Data in the Dispersion Study

Size	20 MWe Industrial Boiler	1,200 MWe Power Plant
Oil consumption, t/hr	4.5	270
Total amount of effluent water, m^3/hr	540	180,000 (includes cooling water from the power plant)
Retained Ni and V in the boiler	50% of Ni and 40% of the V in the oil are retained by the boiler during the 24 hours, which is the time sequence of soot blowing. During the 10 minute soot blowing period, all the accumulated Ni and V is released to the scrubber.	Continuous soot blowing is used and all the Ni and V in the oil is released continuously.
Collection efficiency of Ni and V in the scrubber	50% efficiency for the collection of Ni and V in the scrubber, is assumed during ordinary operation and soot blowing.	Only 27% of both Ni and V from the boiler is assumed collected by the scrubber, because an electrostatic precipitator is installed between the boiler and the scrubber.
COD of the effluent water	COD of the effluent water is equal to half of the sulfite concentration mg S/ℓ	COD of the effluent water is decreased by 75% with aeration of the mixture of scrubber and condenser cooling water.
Level of water intake	20 m below sea level	30 m below sea level
Level of water outlet	20 m below sea level	At sea level
Discharge velocities, m/sec	2 and 4	2
Number of outlets	1	1

Note: Impurity levels in oil used in the boiler and power plant: 2.5% S, 20 ppm Ni and 60 ppm V. Sulfur dioxide removal efficiency of the scrubber is assumed to be 100%. No oxidiation of sulfite is assumed, except during aeration. Operation at maximum load. No extra soot is mixed with the oil.

Source: NTIS PB-295 005

For the 1,200 MWe power plant continuous soot-blowing is assumed. Furthermore, a precollector for the particulates will be required. For the 1,200 MWe power plant aeration of the effluent water will be required. The choice of oxidation efficiency should be based on an evaluation of the recipient capacity. An oxidation efficiency of 75% (defined as 75% oxidation of the sulfite) is assumed as a reasonable value based on aeration time. A higher efficiency would require a considerable increase in the aeration time. The intake for seawater is placed 30 m below sea level and the outlet at the surface in order to minimize the recirculation of liquid. It has been suggested that this type of arrangement will be most suitable for a 1,200 MWe power plant, at least for the Norwegian coastal waters.

pH of the effluent water after dilution and aeration is estimated from Bromley's (1972) experimental data obtained after simple spray into air of seawater containing sulfuric acid (4). His graph of pH of seawater as a function of alkalinity change caused by addition of sulfuric acid (i.e. oxidized SO_2) is shown in Figure 11.5.

Figure 11.5: pH of Seawater as Function of Alkalinity Change

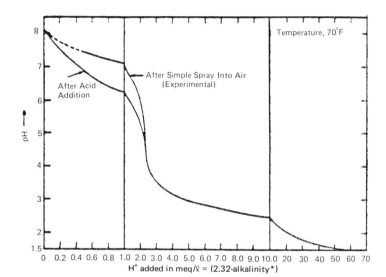

*Alkalinity must be in meq/ℓ.

Source: NTIS PB-295 005

The COD expresses the sulfite ion concentration due to the following reaction:

$$SO_3^{2-} + \tfrac{1}{2}O_2 \rightarrow SO_4^{2-}$$

The COD (mg O_2/ℓ) is accordingly equal to half of the sulfite ion concentration expressed in mg S/ℓ.

Table 11.6 shows the calculated concentrations of pollutants added to the total effluent water based on the conditions as given in Table 11.5.

Calculations of the COD are based on the sulfite ion concentrations only. The oxygen content in seawater, which is used in the scrubbing process and for cooling water to the condensers, has not been taken into account. The COD value of the effluent water might therefore be too high and should be regarded as a worst case.

Table 11.6: Calculated Concentrations of Pollutants

	. . . .Power Plant Size, MWe. . . .	
	20	1,200
Total effluent water, m³/hr	540	180,000*
Concentration of nickel, μg Ni/ℓ	42 (6,000)**	8
Concentration of vanadium, μg V/ℓ	150 (14,400)**	24
Concentration of sulfur, mg S/ℓ	210	37
COD, mg O₂/ℓ	105	4.6
pH	2.6	~5***
Temperature increase, °C	10	10†

*Effluent water from scrubber mixed with effluent cooling water from the power plant.

**Numbers in parenthesis are calculated for 10 min of soot blowing each 24 hr. Recent measurements indicate considerably less retention of Ni and V in the boilers.

***Based on complete oxidation of sulfite. The calculated pH is possibly too low because more CO_2 is stripped off during aeration and because the oxidation of sulfite is not complete.

†For the 1,200 MWe power plant the temperature rise of the effluent water is mainly caused by the cooling water from the condensers.

Source: NTIS PB-295 005

The estimated pH value is not completely verified. For the scrubber no. 5, pH varies with the load between 2.2 and 5.2. A pH value between 2.2 and 2.3 was registered at maximum load. For the 1,200 MWe plant, a more detailed evaluation of the effect of oxidation is necessary. With no oxidation and no stripping of carbon dioxide, the pH according to Abdulsattar's model would be equal to 5.9. After oxidation a resulting effluent pH of 4.6 might be estimated on the basis of complete oxidation to sulfate using the lower graph with the text "after acid addition" in Figure 11.5. Because carbon dioxide will be stripped off during aeration, the upper curve with the text "after simple spray into air," is more realistic, however. pH then increases to approximately 5. At higher pH, however, the two graphs differ considerably. The effect of increasing pH due to removal of carbonic acid is quite desirable in moving the pH to a safe level.

The FGD system supplier refers to test results and theoretical calculations indicating an increase of pH during aeration provided pH prior to aeration exceeds 5.5 to 6. Stripping of carbon dioxide has been considered in these calculations.

The Recipients: An evaluation of the dilution in the so-called near-field-zone and the far-field-zone has been carried out at three different locations in southern Norway.

The near-field-zone is defined as the zone where the main dilution of the effluent water with the recipient is caused by the momentum of the discharge itself. The far-field-zone is defined as the zone where the main dilution is caused by the general mixing process, which occurs in the recipient itself and is independent of the momentum of the discharge.

The locations of the effluents are Porsgrunn at Frierfjorden, Slagentangen at Oslofjorden and Tromöya outside Arendal at Skagerak as shown in Figure 11.4.

Locations were chosen representing three different types and sizes of recipients. Sufficient data for rough estimates at these locations existed already.

In Table 11.7 are given some recipient characteristics for the three places. The terms upper and lower layer being used in the table are defined as follows. The upper layer is located between the surface and the density jump. The lower layer is located between the sea bottom and the density jump. The density jump is located to the depth where a significant change in the vertical density gradient occurs. The density jump separates the upper brackish water masses from the lower and heavier saline water masses.

Table 11.7 Recipient Characteristics for Three Different Recipients in Southern Norway

Effluent Location	Porsgrunn	Slagentangen	Tromöya Outside Arendal
Recipient	Frierfjorden	Oslofjorden	Skagerak
Location of the density jump, m below water surface	1-7	5-15	0-30
Salinity of the upper layer, %	0.5-10	17-32	23-32
Water exchange in the upper layer, m³/sec	90-580	1,000-1,800	Larger than 2,000
Water exchange in the lower layer, m³/sec	60-110	1,000-1,800	Larger than 2,000

Source: NTIS PB-295 005

Results: The results for the 20 MWe industrial boiler are summarized in Table 11.8.

Table 11.8: Results Related to the 20 MWe Industrial Boiler

Constituents	Excess Concentration at the End of the Near-Field-Zone	Average Daily Excess Concentration in the Far-Field-Zone Frierfjorden	Slagentangen	Concentration in a Nonpolluted Recipient
Ni, µg/ℓ	1-2 (200-300)*	0.1-0.2	<0.1	~2
V, µg/ℓ	5-8 (400-800)*	0.3-0.6	<0.1	~2
Excess temperature, °C	0.5	<0.1	<0.1	—
COD, mg O₂/ℓ**	3-5	0.1-0.2	<0.1	0
pH	6.6-6.7	—	—	7.5-8.3

*The numbers in the parenthesis refer to concentrations calculated for the periods of soot-blowing of the boilers.
**COD defined as the equivalent sulfite content, not considering the O₂ content.

Source: NTIS PB-295 005

The results from the table indicate that the Ni and V concentrations obtained by the near-field dilution are of the same order of magnitude as the concentrations found in natural seawater (8) (9). However, during the periods of soot blowing, considerably higher concentrations are reached. These will, on the other hand, only persist for very short periods of time.

The lowest oxygen concentration which biologists claim is critical for marine life is 3 mg O_2/ℓ. The actual oxygen concentration in the recipient is not directly shown by Table 11.8. This value depends on the oxidation rate of the sulfite ion. No oxidation is assumed in this calculation, however, to give a worst case example. The lowest possible net oxygen concentration might be estimated by subtracting the chemical oxygen demand from the concentration of oxygen in a nonpolluted recipient.

The results for the 1,200 MWe power plant are summarized in Table 11.9.

Table 11.9: Results Related to the 1,200 MWe Power Plant

Constituents	Excess Concentration After Dilution in the Near-Field-Zone	Average Excess Concentration in the Far-Field-Zone Slagentangen	Tromöya	Concentration in a Nonpolluted Recipient
Ni, $\mu g/\ell$	1–2	0.2–0.4	<0.2	~2
V, $\mu g/\ell$	5	0.7–1.2	<0.5	~2
Excess temperature, °C	0.8–2.9	0.5	<0.5	—
COD, mg O_2/ℓ*	1**	0.4–0.8	<0.4	0
pH	6.8***	—	—	7.5–8.3

Note: Electrostatic precipitator is assumed to be installed prior to the scrubber. Soot-blowing is assumed to be carried out continuously.

*COD defined as the equivalent sulfite content only, not considering the O_2 content (see Table 11.5).

**These numbers are based on the assumption that the discharge has been aerated before entering the recipient. Without aeration, the numbers would be considerably higher (with a factor of max. 4).

***The number is the same for both aeration and no aeration of the discharge, according to Bromley (4).

Source: NTIS PB-295 005

When entering the far-field-zone, both the Ni and V concentrations from the discharge are reduced to the order of magnitude that is found in natural seawater. Continuous soot blowing is here assumed (see Basic Conditions). With other soot blowing procedures this situation might change.

The excess temperatures may be as high as 3°C when entering the far-field-zone. This value is considerably in excess of the 1°C temperature rise which biologists claim may be critical for marine ecosystems. This problem is, however, common for all types of thermal power plants using seawater as a coolant.

Oxygen Content in Effluent Water: The aerated effluent water will contain free oxygen. Tests made by the supplier indicate an oxygen content of 5 mg O_2/ℓ in compliance with 75% oxidation as used in the recipient evaluation. By dilution in the near- and far-field-zone, COD and O_2 will approach the background concentrations.

Reheating of Stack Gas

The water vapor content in the stack gas is particularly low, because the gas temperature at the scrubber outlet is close to the inlet seawater temperature. The need for reheating is therefore mainly concerned with the NO_x emission. The Norwegian Institute for Air Research (NILU) has studied the effect of the stack heights and the degree of reheating to satisfy a maximum ground level concentration of 320 μg NO_2/m^3 (7). The report concludes that two stacks of 150 m for a 2 x 600 MWe power plant will not require reheat if the EPA regulations on NO_x emission is met for the boiler operation.

Current Research

The current research on the Fläkt-Hydro process is mainly concerned with the treatment of the effluent scrubber liquid. A/S Norsk Viftefabrikk has carried out laboratory investigations on the dilution and aeration of the used scrubber liquid. More detailed studies are planned to be carried out in a pilot plant.

REFERENCES

(1) Kelly, F.H., *Proceedings Australasien.* IMM, Inc., No. 152–153, p.17–39, 1949.

(2) Böckmann, O.K., Gramme, P.E., Terjesen, S.G., Thurmann-Nielsen, E., and Tokerud, A., "Process for removal of sulfur dioxide from flue gases by absorption in sea water." *Int. Scand. Congr. Chem. Eng. (Proc.)* 1974 and Brochure from AB Svenska Fläkt-fabriken. (Undated)

(3) Bromley, L.A. and Read, S.M., "Removal of sulfur dioxide from stack gases by sea water." Research Project S-15, The University of California, Sept. 1, 1970.

(4) Bromley, L.A., "Use of sea water to scrub sulfur dioxide from stack gases." *Int. J. Sulfur Chem.*, Part B7, No. 1, p. 72, (1972).

(5) Böckmann, O.K. and Tokerud, A., "The Fläkt-Hydro process for SO_2 removal. Environmental seminar.

(6) Hagen, R.I., Report of the measurements done by SINTEF on the Fläkt-Hydro-FGD-process (Norwegian text), STF21 F78015. SINTEF. 1978-01-24.

(7) Sivertsen, B. and Gotaas, Y., "Skorsteinshoyder og royktemperatur etter rensing av avgassene fra et varmekraftverk." (In Norwegian) *Oppdragsrapport* nr. 27/76. NILU.

(8) Föyn, E., "Problems of fossil fuel combustion and waste." *Env. Poll. Management.* March/April, p. 48 (1977).

(9) Nilsen, G., "En vurdering av utslipp fra et sjovannsvaskeanlegg for roykgasser fra olje-kraftverk." No. -0-135/75. Norsk Institute for Vannforskning, 22, Jan. 1976. (In Norwegian).

(10) Rye, H. and Thendrup, A., "Grov vurdering av utslipp fra Fläkt-Hydro-prosessen for rensing av avgasser ved forbrenning av olje." Preliminary report dated August 15, 1977, VHL/SINTEF. (In Norwegian)

(11) Abdulsattar, A.H., Sridhar, S. and Bromley, L.A., "Thermodynamics of the sulfur dioxide sea water systems." *A.I.CH.E. Journal,* 23, p. 62 (1977).

TECHNICAL FEASIBILITY
AND ECONOMIC EVALUATION
OF NINE PROCESSES

The material in this chapter is based on *NATO-CCMS Flue Gas Desulfurization Pilot Study—Phase II—Applicability Study,* August 1980, EPA-600/7-80-142, prepared by R.L. Torstrick, S.V. Tomlinson, J.R. Byrd and J.D. Veitch of Tennessee Valley Authority, Division of Energy Demonstrations and Technology, Office of Power, and R.W. Gerstle of PEDCo Environmental, Inc.

INTRODUCTION

The Phase I reports on 12 flue gas desulfurization processes for the North Atlantic Treaty Organization Committee on the Challenges of Modern Society (NATO-CCMS) FGD Study Group were reviewed by the NATO-CCMS delegates in April 1978 and nine of the processes were selected for comparative economic evaluations as Phase II of the study. The purpose of the Phase II study is to provide procedures and technical and economic data for the selection of FGD processes for specific applications. The study consists of technical feasibility and economic evaluations developed by the U.S. Tennessee Valley Authority (TVA) and a decision-chart selection procedure developed by PEDCo Environmental, Inc.

The basis of the economic evaluations is an FGD system for a new midwestern U.S. 500-MW power plant. The FGD system is designed for removal of 90% of the SO_2 in the flue gas. Six fuels, consisting of bituminous coals, lignite, and oil, are evaluated. Projection of costs for other conditions such as different power plant sizes and fuels with different sulfur levels are also provided.

The design and economic bases of this study differ from the Phase I study because of updated information and standardizations to provide comparable results between processes. It should be recognized that the data are based on U.S. conditions and that simplifying assumptions are made in the design model. These factors must be borne in mind when using the cost comparisons. Site-specific conditions may substantially alter the cost relationships of the processes.

The decision-chart system consists of an elimination procedure which rates the applicability of the processes relating the FGD process characteristics to specific site conditions. The system is an initial selection procedure which allows FGD selection efforts to be focused on the most promising processes.

PREMISES FOR THE ECONOMIC EVALUATIONS

The design and economic premises used were developed by TVA and others for comparative FGD cost studies using representative U.S. power plant conditions. The base case is a frontal-fired, balanced-draft 500-MW power plant constructed in the period 1977 to 1980 in the U.S. Midwest for startup in 1980. The plant is assumed to have a 30-year lifetime of 117,500 operating hours and to operate 6,000 hours in the first year. Fly ash removal of 99.2% by electrostatic precipitators (ESP) upstream of the FGD system is not included in FGD costs. The base case fuel is a 3.5% sulfur, 16% ash, 5,830 kcal/kg high heat rate bituminous coal. For this fuel and all other solid fuels it is assumed that 95% of the sulfur and 80% of the ash are emitted in the flue gas. For oil all of the sulfur in the fuel is assumed emitted in the flue gas.

The FGD systems are assumed to be installed downstream from the ESP units. For the wet systems the flue gas is supplied from a common plenum to four parallel trains of SO_x removal equipment including booster FD fans and reheat provisions. Presaturators to cool the gas from 149° to 53°C and mist eliminators to reduce liquid entrainment to 0.1% are provided. Indirect steam reheat (or direct-fired or oil-fired units used in the FGD process) is provided to reheat the gas entering the stack plenum to 79°C. The dry adsorption process is similar except two trains are used and reheat is not required. Costs for chloride removal facilities are included for the regenerable processes; however, costs for chloride disposal facilities are excluded. The FGD system consists of the common plenum and all equipment downstream to the stack plenum, including all raw material and effluent processing equipment and land requirements. Removal efficiencies are assumed to be 90% of the SO_2, 50% of the SO_3, 95% of the chloride, and 75% of the remaining fly ash in the flue gas.

The economic premises are based on U.S. regulated utility economics and financing. The costs estimated consist of capital costs for construction of the FGD system and annual revenue requirements for the first-year operation. All costs are based on midwestern U.S. costs using mid-1979 as the basis of capital costs and mid-1980 as the basis for annual revenue requirements. Capital costs consist of all direct and indirect costs for equipment, land, materials, labor, fees, services, and other construction costs required to install the FGD system. Annual revenue requirements, based on a first-year, 6,000-hour operation, consist of all raw material, labor, utility, and other conversion costs and indirect costs such as capital charges, taxes, and overheads. By-product sales for processes producing elemental sulfur, H_2SO_4, gypsum and sodium sulfate are included as credits. For the gypsum-producing processes, however, costs for disposal of gypsum may be obtained where necessary by substituting a disposal charge for stacking or landfill of the gypsum in place of the credit. These costs are projected to range between $4 and $9 per ton of dry solids, depending on site-specific conditions.

Case Variations: Case variations, in which design assumptions are varied to determine effect on costs, are also evaluated. Except for the seawater process, 0.8, 1.4, and 2.0% sulfur bituminous coal, 0.5% sulfur lignite, and 2.5% sulfur oil fuels are included. The seawater process is not evaluated for the 2.0 and 3.5% sulfur coals because those coals would require additional seawater in addition to condenser water. Other case variations consist of 200-, 700-, and 1,000-MW power plant sizes for the limestone sludge process, reheat to 53°C for the seawater process, and a sulfur production case for the Wellman-Lord process.

PROCESSES EVALUATED AND ACCURACY OF RESULTS

The nine process evaluations are based on data from a number of sources. Because of differences in the stage of development and amount of information available, the accuracy ranges of the economic results differ. Stage of development is difficult to quantify and is not considered in these evaluations. The accuracy ranges can, however, be related to the amount of information available. Normally, actual investment costs may be expected to depart from those shown in the economic evaluation by a factor of 0.7 to 1.5. Because of the extensive information available for some of the processes evaluated, however, smaller ranges can be projected. The accuracy ranges of each process, based on the amount of information available, are shown in the tabulated results. The nine processes evaluated are discussed below.

Limestone Sludge Process

Process Description: The limestone slurry process is designed to use a mobile-bed absorber, with presaturator and mist eliminator. The mist eliminator is equipped for upstream and downstream wash with fresh makeup water. The flue gas from the common plenum is cooled in the presaturator, scrubbed with limestone slurry in the absorber, passed through the mist eliminator, and is reheated before being vented to the stack plenum.

Limestone slurry is recirculated through the absorber and an external surge tank. The slurry is maintained at 15% solids by withdrawal of a purge stream and addition of fresh slurry. The reaction of SO_2 with $CaCO_3$ in the limestone is assumed to produce 80% calcium sulfite hemihydrate ($CaSO_3 \cdot \frac{1}{2}H_2O$) and 20% gypsum ($CaSO_4 \cdot 2H_2O$). The purge stream containing these salts, unreacted limestone, and minor impurities is pumped one mile to an earthen-diked, clay-lined pond where it settles to a sludge of about 40% solids. Supernate water is returned from the pond for reuse in the process.

The feed preparation area consists of two trains of crushers and wet ball mills serving all four absorber trains. As-received 40 mm maximum-size crushed limestone is further reduced in crushers and processed in wet ball mills to a 60% solids slurry with a particle size of about 70% less than 0.003 mm that is metered to the scrubber slurry loop.

Specific Process Premises:

- The flue gas is assumed to be cooled from 149° to 53°C (300° to 127°F) in the presaturator at an L/G ratio of 0.5 ℓ/m^3 (4 gal/10^3 aft^3).

- The absorber is a mobile-bed-type turbulent contact absorber with a flue gas superficial velocity of 3.8 m/sec (12.5 ft/sec) and a pressure drop, including the mist eliminator, of 2.14 kPa (8.6 inches H_2O). An L/G ratio of 6.7 ℓ/m^3 (50 gal/10^3 aft^3).

- Stoichiometry is 1.4 mols of $CaCO_3$ to 1.0 mol of SO_2 removed and 1.0 mol of $CaCO_3$ to 2.0 mols of HCl removed.

Energy Requirements: For base-case conditions, reheat of the cleaned gas from 53° to 79°C requires 42.2 x 10^3 kg/hr (93,050 lb/hr) of 243°C (470°F) steam at 3.55 x 10^3 kPa absolute pressure (500 psig) equivalent to approximately 17.62 x 10^6 kcal/hr.

The electrical power demand for the base-case limestone sludge process is about 7,995 kW or 1.6% of the rated output of a 500-MW power plant. For 6,000 hours of operation the annual electrical energy consumption is 48.0×10^6 kWh.

The total equivalent energy consumption for the base case is approximately 37.7×10^6 kcal/hr or 3.3% of the input energy required for the 500-MW power unit.

By-Product Management: ESP units remove 99.2% of the fly ash from the flue gas and, therefore, only a small amount of fly ash is found in the FGD process sludge. (Fly ash emission from oil-fired units does not exceed the EPA particulate emission standard and fly ash collection facilities are not included in oil-fired power plant design.) Projected mass flow rates of wastes for the base case are shown below:

Component	Kilograms per Hour	Pounds per Hour
$CaSO_3 \cdot \frac{1}{2}H_2O$	16,550	36,480
$CaSO_4 \cdot 2H_2O$	5,670	12,500
$CaCO_3$	6,448	14,215
$CaCl_2$	433	955
Mg	39	85
Fly ash	149	329
Inerts	1,236	2,726
Total	30,525	67,290

Based on a 30-year life for both the power unit and the FGD units, the sludge disposal pond for the base case requires approximately 123 ha (305 acres). It is designed for an optimum depth of approximately 6.1 m (20 ft).

Lime Sludge Process

Process Description: Lime slurry scrubbing differs from limestone scrubbing in this study only in the raw material used as SO_2 absorbent and in the method of preparing the scrubbing slurry.

Pebble lime is slaked in two parallel slakers at a slurry concentration of 60% solids and combined with scrubber effluent slurry and recycle pond water to control the concentration of the recirculating slurry at approximately 15% solids. The flue gas is cooled in a presaturation chamber and passed through a mobile-bed absorber. The lime slurry circulates through the absorber where it reacts with the SO_2 in the cooled flue gas. Mist eliminators equipped for upstream and downstream wash with fresh makeup water control entrainment carryover in the gas stream. A bleedstream from the recirculation tank is pumped to an earthen-diked, clay-lined pond one mile away where it settles to form a sludge containing approximately 40% solids. The sludge is assumed to be 80% $CaSO_3 \cdot \frac{1}{2}H_2O$ and 20% gypsum. Pond supernate is recycled to the slakers and the absorber recirculation tank to maintain closed-loop operation. Scrubber outlet gas is reheated to 79°C (175°F) by indirect steam heat before entering the stack.

Specific Process Premises:

- The flue gas is assumed to be cooled from 149° to 53°C (300° to 127°F) in the presaturator at an L/G ratio of 0.5 ℓ/m³ (4 gal/10³ aft³).

- The absorber is a mobile-bed type with a flue gas superficial velocity of 3.8 m/sec (12.5 ft/sec) and a pressure drop of 2.14 kPa (8.6 inches H_2O), including the mist eliminator. An L/G ratio of 7.4 ℓ/m³ (55 gal/10³ aft³) is used.

• Stoichiometry is 1.05 mols of CaO to 1.0 mol of SO_2 removed and 1.0 mol of CaO to 2.0 mols of HCl removed.

Energy Requirements: For base-case conditions, reheat of the cleaned gas from 53° to 79°C requires 42.1 x 10^3 kg/hr (92,740 lb/hr) of 243°C (470°F) steam at 3.55 x 10^3 kPa absolute pressure (500 psig), equivalent to about 17.56 x 10^6 kcal/hr.

The electrical power demand for the base-case lime-sludge process is about 7,448 kW or 1.5% of the rated output of a 500-MW power plant. For 6,000 hours of operation, the annual electrical energy consumption is 44.7 x 10^6 kWh.

The total equivalent energy consumption for the base case is approximately 36.41 x 10^6 kcal/hr or 3.2% of the input energy required for the 500-MW power unit.

By-Product Management: Electrostatic precipitators remove 99.2% of the fly ash from the flue gas and, therefore, only a small amount of fly ash is found in the FGD process sludge. (Fly ash emission from oil-fired units does not exceed the EPA particulate emission standard and fly ash collection facilities are not included in oil-fired power plant design.) Projected mass flow rates of by-product wastes for the base case are shown below.

Component	Kilograms per Hour	Pounds per Hour
$CaSO_3·\frac{1}{2}H_2O$	16,550	36,480
$CaSO_4·2H_2O$	5,670	12,500
$Ca(OH)_2$	451	995
$CaCl_2$	433	955
Mg	123	271
Fly ash	149	329
Inerts	102	225
Total	23,478	51,755

Based on a 30-year life for both the power unit and the FGD unit, the sludge disposal pond for the base case requires approximately 104 ha (256 acres). It is designed for an optimum depth of approximately 5.8 m (19 ft).

Double-Alkali Sludge Process

Process Description: The double-alkali process evaluated has been generalized from several concentrated-mode double-alkali processes in the United States. A two-tray tower absorber with presaturator and mist eliminator is used. The scrubbing liquor is a solution of sodium salts of which sodium sulfite (Na_2SO_3) is the major active component. The SO_2 reacts with the Na_2SO_3 to form sodium bisulfite ($NaHSO_3$). A bleedstream of absorber liquor is treated with slaked lime to precipitate calcium sulfur salts and regenerate the Na_2SO_3.

Pebble lime is slaked and then reacted with a bleedstream of absorber effluent in agitated tanks. The reaction product, predominately calcium sulfite, flows to a thickener where the slurry is concentrated to 40% solids. This stream is further dewatered using drum filters to produce a cake containing 55% solids. The filter is designed with two wash sections to minimize sodium loss. The filter cake is conveyed to a reslurry tank where it is mixed with pond return water to a 15% solids slurry. The slurry is pumped to an earthen-diked clay-lined pond one mile away where the solids in the slurry settle to form a sludge containing approximately 40% solids. Makeup soda ash is added to the regenerated scrubber liquor at the thickener overflow storage tank.

Specific Process Premises:

- The flue gas is cooled from 149° to 53°C (300° to 127°F) and saturated in the presaturator. The presaturator has an L/G ratio of 0.5 ℓ/m³ (4 gal/10³ aft³).

- A two-tray tower absorber with a superficial velocity of 2.1 m/sec (7 ft/sec), and a pressure drop, including the mist eliminator, of 1.25 kPa (5 inches H_2O) is used. An L/G ratio of 0.5 ℓ/m³ (4 gal/10³ aft³) is used for recycle liquor to the absorber and an L/G ratio of 0.4 ℓ/m³ (3 gal/10³ aft³) is used for the regenerated scrubbing liquor to the absorber.

- Stoichiometry is 1.0 mol of CaO to 1.0 mol of SO_2 removed and 1 mol of CaO to 2 mols of HCl removed.

- Oxidation of 10% of the SO_2 removed to sulfate is assumed. The remainder is assumed to be in sulfite form.

Energy Requirements: For base-case conditions, reheat of the cleaned gas from 53° to 79°C requires 42.2 x 10³ kg/hr (93,060 lb/hr) of 243°C (470°F) steam at 3.55 x 10³ kPa absolute pressure (500 psig), equivalent to about 17.62 x 10⁶ kcal/hr.

The electrical power demand for the base case is estimated to be about 3,981 kW or 0.8% of the rated output of a 500-MW power plant. For 6,000 hours of operation the annual electrical energy consumption is 23.9 x 10⁶ kWh.

The total equivalent energy consumption for the base case is approximately 28.61 x 10⁶ kcal/hr or 2.5% of the input energy required for the 500-MW power unit.

By-Product Management: ESP units remove 99.2% of the fly ash from the flue gas and, therefore, only a small amount of fly ash is found in the FGD process sludge. (Fly ash emission from oil-fired units does not exceed the EPA particulate emission standard and fly ash collection facilities are not included in oil-fired power plant design.) Projected mass flow rates of by-product wastes for the base case are shown below:

Component	Kilograms per Hour	Pounds per Hour
$CaSO_3 \cdot \frac{1}{2}H_2O$	18,266	40,270
$CaSO_4 \cdot 2H_2O$	2,158	4,758
$Ca(OH)_2$	209	461
Na_2SO_3	319	703
Na_2SO_4	148	326
NaCl	456	1,006
Mg	117	258
Fly ash	149	329
Inerts	98	215
Total	21,920	48,326

Based on a 30-year life for both the power unit and the FGD unit, the sludge disposal pond for the base case requires approximately 99 ha (245 acres). It is designed for an optimum depth of approximately 5.8 m (19 ft).

Seawater Process

Process Description: The seawater process uses seawater from the power plant condensers as the scrubbing agent. Because the amount of condenser seawater is limited, the 3.5% and 2.0% sulfur coals are not included in the seawater process evaluation. The 149°C (300°F) flue gas is cooled to 53°C (127°F) in a presaturator and scrubbed in a countercurrent packed tower absorber to remove SO_2, SO_3, HCl, CO_2, and some residual fly ash. Seawater at 26°C (79°F) is used in both the presaturator and absorber in a one-pass flow. The flue gas is cooled to 27°C (81°F) in the SO_2 absorber.

Presaturator and absorber effluent at a pH of approximately 3.0 (based on SO_2 content only) is treated with additional condenser seawater to increase the pH to 6, treated with sparged air to oxidize 75% of the sulfite to sulfate, and returned to the sea. The chemical oxygen demand of the waste is estimated to be 3.1 mg O_2 per liter or less. Reheat of the flue gas to 79°C (175°F) is included in the process. A case variation of reheat to 53°C (127°F) is also included.

Specific Process Premises:

- A packed-bed absorber with a presaturator and mist eliminator is used. Pressure drop in the unit is 1.05 kPa (4.2 inches H_2O) and the superficial velocity in the absorber is 1.8 m/sec (6 ft/sec).

- The presaturator L/G ratio is 0.5 ℓ/m^3 (4 gal/10^3 aft^3) and the absorber L/G ratio is 8.0 ℓ/m^3 (60 gal/10^3 aft^3).

- The total alkalinity of the seawater is assumed to be 2.4 meq/ℓ as $CaCO_3$.

- Oxidation of SO_x to SO_4 is assumed to be 75% in the oxidation tank.

- Only condenser seawater from the power plant is used for neutralization; no alkali or additional seawater is added.

Energy Requirements: For base-case conditions, reheat of the cleaned gas from 27° to 79°C requires 68.7 x 10^3 kg/hr (151,500 lb/hr) of 243°C (470°F) steam at 3.55 x 10^3 kPa absolute pressure (500 psig) equivalent to about 28.70 x 10^6 kcal/hr.

The electrical power demand for the seawater process, base case, is estimated to be about 7,012 kW or 1.4% of the rated capacity of a 500-MW power plant. For 6,000 hours of operation, the annual electrical energy consumption is 42.1 x 10^6 kWh.

The total equivalent energy consumption for the base case is approximately 47.79 x 10^6 kcal/hr or 4.2% of the input energy required for the 500-MW power unit.

By-Product Management: ESP units remove 99.2% of the fly ash from the flue gas and, therefore, only a small amount of fly ash is found in the effluent which is pumped offshore for disposal. (Fly ash emission from oil-fired units does not exceed the EPA particulate emission standard and fly ash collection facilities are not included in oil-fired power plant design.) Projected mass flow rates of by-product wastes for the base case are shown below.

Component	Kilograms per Hour	Pounds per Hour
Total effluent	72,124,300	158,864,200
$SO_4^=$	4,040	8,900
$SO_3^=$	1,110	2,450
HCl	470	1,050
Fly ash	50	120

Approximate chemical oxygen demand (COD) for this effluent is estimated at 3.1 mg O_2 per liter.

Lime Gypsum (Saarberg-Hölter) Process

Process Description: The Saarberg-Hölter process removes SO_2 using a clear alkaline scrubbing solution in a Rotopart, a patented modular-design absorber. The flue gas enters the Rotopart vessel at 149°C (300°F) and is adiabatically cooled to 53°C (127°F) by contact with clear scrubbing solution. The flue gas and the scrubbing solution are cocurrently contacted in vertical scrubbing tubes of the Rotopart vessel. The scrubbing tubes have no packing but are equipped with injection nozzles and special shedding rings to promote liquid-gas contact. SO_2, SO_3, HCl, and some residual fly ash are removed from the gas stream. Gas and washing fluid are centrifugally separated in the separator section of the Rotopart vessel. No further demisting is required.

SO_2 in the flue gas reacts with $CaCl_2$ and $Ca(OH)_2$ in the scrubbing solution to form soluble $Ca(HSO_3)_2$ and HCl. The system uses formic acid as a buffering agent to maintain a pH of about 4.5. The acidic absorber effluent flows by gravity to an oxidizer vessel where air blown through the liquid oxidizes $Ca(HSO_3)_2$ to $CaSO_4 \cdot 2H_2O$ and H_2SO_4. $Ca(OH)_2$ is added at the oxidizer to neutralize the liquid and to replenish calcium in the solution. Makeup formic acid is also added at the oxidizer. Oxidizer effluent is pumped to a thickener where the suspension of the $CaSO_4 \cdot 2H_2O$ crystals is thickened to a 15% solids slurry. Slurry from the thickener is filtered to produce an 80% solids cake. The filter cake is conveyed to a storage area. Filtrate is recycled to the thickener and thickener overflow at a pH of 10.5 is returned to the absorber for use as scrubbing liquid.

Specific Process Premises:

- The Rotopart has a superficial velocity of 12.2 m/sec (40 ft/sec) in the scrubber tubes and a pressure drop through the unit of 2.44 kPa (9.8 inches H_2O). The L/G ratio is proprietary.
- The stoichiometry is 1.01 mols of CaO to 1.00 mol of SO_2 removed and 1.0 mol of CaO to 2.0 mols of HCl removed.
- Complete oxidation of SO_2 removed to sulfate in the form of gypsum is assumed.

Energy Requirements: For base-case conditions, reheat of the cleaned gas from 53° to 79°C requires 42.4 x 10³ kg/hr (93,460 lb/hr) of 243°C (470°F) steam at 3,550 kPa absolute pressure (500 psig), equivalent to about 17.7 x 10⁶ kcal/hr.

The electrical power demand for the base-case lime gypsum process is about 9,701 kW or 1.9% of the rated output of a 500-MW power plant. For 6,000 hours of operation, the annual electrical energy consumption is 58.2 x 10⁶ kWh.

The total equivalent energy consumption for the base case is approximately 39.70 x 10⁶ kcal/hr or 3.7% of the input energy required for the 500-MW power unit.

By-Product Management: Electrostatic precipitators remove 99.2% of the fly ash from the flue gas and, therefore, only a small amount of fly ash is found in the FGD process by-product. (Fly ash emission from oil-fired units does not exceed the EPA particulate emission standard and fly ash collection facilities are not included in oil-fired power plant design.) Projected mass flow rates of by-product for the base case are shown below.

Component	Kilograms per Hour	Pounds per Hour
$CaSO_4 \cdot 2H_2O$	27,640	60,880
$CaSO_3 \cdot \frac{1}{2}H_2O$	82	180
$CaCl_2$	440	968
$Mg(OH)_2$	295	650
Fly ash	149	329
Inerts	102	225
Total	28,708	63,232

The process is evaluated on the basis of 30-day storage of gypsum by-product. A 0.4 ha (1 acre) storage area has been provided for the base case and all fuel variations.

Jet-Bubbling Limestone (Chiyoda Thoroughbred 121) Process

Process Description: The Chiyoda Thoroughbred 121 process is a forced-oxidation limestone-scrubbing process which produces gypsum. It was developed from Chiyoda's dilute sulfuric acid process, the Thoroughbred 101.

Absorption, oxidation, and crystallization are accomplished in the same reactor vessel. The flue gas is cooled in a presaturator chamber and fed to the agitated jet-bubbling reactor. Air and a limestone slurry of 15% solids are introduced to the reactor where SO_2 is absorbed from the flue gas, oxidized to SO_4, and reacted with the limestone to form gypsum (calcium sulfate dihydrate). A bleedstream containing the gypsum crystals is pumped to a thickener. The thickener underflow containing 40% solids is filtered to approximately 80% solids. The filter cake is conveyed to a storage area and the filtrate is returned to the system for use in the wet ball mills.

Specific Process Premises:

- The reactor for absorption, oxidation, and neutralization is a Chiyoda jet-bubbling reactor equipped with a presaturator and mist eliminator. Pressure drop is 3.94 kPa (15.8 inches H_2O). Complete oxidation of absorber SO_2 to gypsum is assumed.

- A stoichiometry of 1.0 mol of $CaCO_3$ to 1.0 mol SO_2 removed and 1.0 mol of $CaCO_3$ to 2.0 mols of HCl removed is used.

Energy Requirements: For base-case conditions, reheat of the cleaned gas from 53° to 79°C requires 43.5 x 10³ kg/hr (95,900 lb/hr) of 243°C (470°F) steam at 3.55 x 10³ kPa absolute pressure (500 psig) equivalent to about 18.16 x 10⁶ kcal/hr.

The electrical power demand for the Chiyoda Thoroughbred 121 process, base case, is estimated to be about 8,090 kW or 1.6% of the rated production of a 500-MW power plant. For 6,000 hours of operation, the annual electrical energy consumption is 48.5 x 10⁶ kWh.

The total equivalent energy consumption for the base case is approximately 38.53 x 10⁶ kcal/hr or 3.4% of the input energy required for the 500-MW power unit.

By-Product Management: ESP units remove 99.2% of the fly ash from the flue gas and, therefore, only a small amount of fly ash is found in the FGD process by-product. (Fly ash emission from oil-fired units does not exceed the

EPA particulate emission standard and fly ash collection facilities are not included in oil-fired power plant design.) Projected mass flow rates of by-product for the base case are shown below.

Component	Kilograms per Hour	Pounds per Hour
$CaSO_4 \cdot 2H_2O$	27,751	61,180
$CaCl_2$	456	1,005
Mg	27	60
Fly ash	149	329
Inerts	884	1,948
Total	29,267	64,522

The process is evaluated on the basis of 30-day storage of saleable gypsum by-product. A 0.4 ha (1 acre) storage area has been provided for base case and all fuel variation estimates.

Magnesium Oxide Process

Process Description: The magnesium oxide process is an absorbent-regenerating process which produces sulfuric acid. The flue gas is scrubbed with an MgO slurry, producing $MgSO_3$ and $MgSO_4$, which are dried and calcined to regenerate SO_2 and MgO. The SO_2 is processed to sulfuric acid and the MgO is returned to the system. A spray grid absorber with a chloride scrubber and mist eliminator is used. Makeup and regenerated MgO are slurried into a bleedstream of liquor from the absorber and recycled to the absorber. Humidification losses are added as a fresh water upstream wash for the mist eliminator.

Effluent from the absorber, containing approximately 15% solids, is pumped to two parallel centrifuges to separate the solids from the liquor. The centrate is returned to the absorber system. The centrifuge cake is dried in an oil-fired rotary dryer. The dryer off-gas is cleaned in a cyclone and fabric dust collector. A portion of the gas is recycled to the dryer combustion chamber for temperature control and the remainder is routed to the absorbers.

The discharge from the dryer and filters is transferred to an oil-fired fluid-bed calciner which contains a single calcination bed operating at 871°C (1600°F). The calciner off-gas, containing SO_2, is partially cleaned in a cyclone, cooled to about 371°C (700°F) in a waste heat boiler, and fed to a fabric filter for final cleaning before entering the sulfuric acid unit. The MgO collected in the cyclone and bag filter is returned to the absorber feed preparation system.

A complete 390 metric ton/day contact sulfuric acid plant is provided for production of 98% acid from the SO_2 gas. The sulfuric acid is stored in 30-day capacity tanks. Tail gas from the acid plant is recycled to the absorber.

Specific Process Premises:

- The flue gas is assumed cooled from 149°C (300°F) to 53°C (127°F) and saturated in the presaturator chloride scrubber. The presaturator chloride scrubber L/G ratio is 0.5 ℓ/m³ (4 gal/10³ aft³).

- A mobile-bed absorber with a superficial velocity of 3.8 m/sec (12.5 ft/sec) and a pressure drop, including the mist eliminator, of 1.99 kPa (8 inches H_2O). The absorber L/G ratio is 3 ℓ/m³ (20 gal/10³ aft³).

- The stoichiometry is 1.05 mols of MgO to 1.0 mol of SO_2 removed.

Energy Requirements: For base-case conditions, reheat of the cleaned gas from 53° to 79°C requires 43.4 x 10³ kg/hr (95,700 lb/hr) of 243°C (470°F) steam at 3.55 x 10³ kPa absolute pressure (500 psig) equivalent to about 18.12 x 10⁶ kcal/hr.

The electrical power demand for the base-case magnesium oxide process is about 9,109 kW or 1.8% of the rated output of a 500-MW power plant. For 6,000 hours' operation, annual electrical energy consumed is 54.65 x 10⁶ kWh.

Fuel oil provides energy for the dryer and calciner. The total fuel oil consumption of approximately 3,690 ℓ/hr (975 gal/hr) is equivalent to 33.78 x 10⁶ kcal/hr.

Waste heat is recovered in the acid production area and in the calcining area to produce 5.76 x 10³ kg/hr (12,700 lb/hr) of 186°C (367°F) steam at 1.14 x 10³ kPa absolute pressure (150 psig). This steam is equivalent to 3.6 x 10⁶ kcal/hr and is considered a heat credit in the determination of total equivalent energy consumption.

The total equivalent energy consumption for the base case is approximately 70.97 Mkcal/hr or 6.26% of input energy required for the 500-MW power unit.

By-Product Management: ESP units remove 99.2% of the fly ash from the flue gas and, therefore, only a small amount of fly ash is found in the chloride scrubber effluent. (Fly ash emission from oil-fired units does not exceed the EPA particulate emission standard and fly ash collection facilities are not included for these units.) Projected mass flow rates of the by-products for the base case are shown below. HCl, SO_3, and ash are collected in the chloride scrubber and disposed of in the ash pond.

Component	Kilograms per Hour	Pounds per Hour
Product acid		
98% H_2SO_4	16,300	35,900
Chloride purge		
HCl	290	630
SO_3	40	80
Ash	150	330

The process is evaluated on the basis of 30-day storage of 98% sulfuric acid by-product. Two 3,250 m³ (858,000 gal) carbon steel tanks have been provided for the base-case design.

Sodium Sulfite (Wellman-Lord) Process

Process Description: The Wellman-Lord process is a sodium sulfite-based scrubbing process that produces SO_2. The SO_2 can be liquefied, or processed to sulfuric acid or elemental sulfur. The system evaluated is Wellman-Lord scrubbing combined with a sulfuric acid plant. An additional case variation is included that uses an Allied Chemical methane reduction unit for sulfur production. The flue gas from the common plenum enters a chloride scrubber where the gas is cooled and the chlorides are removed. From this scrubber the flue gas passes counter-currently to a recirculating sodium sulfite solution in a three-stage valve tray absorber where the SO_2 reacts with the sodium sulfite to form sodium bisulfite and a very small amount of sodium sulfate. Liquor is individually recirculated in each of the three stages to maintain efficient mass transfer. Each absorber has a chevron-type entrainment separator to control entrainment carryover in the gas stream.

A portion of the scrubber effluent is processed to remove Na_2SO_4 by evaporation and selective crystallization in a steam-heated, forced-circulation evaporator serving all four scrubber trains. The clear overflow, enriched in $NaHSO_3$, is returned to the regeneration area. The bottoms, consisting of a slurry enriched in Na_2SO_4 crystals, are centrifuged to produce a solid containing about two-thirds Na_2SO_4 and one-third Na_2SO_3. The centrate is returned to the regeneration area; the solids are dried in a steam-heated dryer and conveyed to a storage silo for sale or discard. There is a potential market available in the paper industry for this material.

The regeneration system consists of two trains of double-effect, forced-circulation evaporators. Scrubber effluent, combined with liquid from the sulfate removal process, is heated and 60% is pumped to the first-effect evaporators and 40% to the second-effect evaporators. The first effect is steam heated; the second effect is heated by combined first-effect vapor and sulfate crystallizer vapor. Some $Na_2S_2O_3$ formed in the first-effect evaporator is removed by a purge stream. Evaporator and stripper overhead vapor, containing H_2O and SO_2, is dried and compressed. The concentrated SO_2 stream is routed to a processing plant. SO_2-bearing condensate from the second-effect evaporator heater, the condensers, and the compressor is steam-stripped and combined with evaporator bottoms. The resulting mixture, enriched in Na_2SO_3, is returned to the absorber system.

The concentrated SO_2 stream is converted to H_2SO_4 in a single-contact, single-absorption acid plant. The tail gas containing unreacted SO_2 is returned to the scrubber. The Allied Chemical methane reduction process is used for a sulfur production case variation. In this process, about 60% of the SO_2 from the Wellman-Lord plant is reduced directly to elemental sulfur by reaction with methane in a primary reduction reactor. Gas leaving the primary reactor, containing H_2S and SO_2 in a molar ratio of 2:1, is routed to a catalytic Claus converter where total conversion of SO_2 to S is increased to about 95%. Sulfur, condensed from the gas stream after the primary and secondary reactors, is pumped to storage. Noncondensed gases from the last sulfur condenser are pumped to an incinerator to oxidize remaining sulfur compounds back to SO_2. After incineration the gases are recycled back to the SO_2 absorbers.

Specific Process Premises:

- The flue gas is cooled from 149°C (300°F) to 54°C (130°F) in the chloride scrubber at an L/G ratio of 1.3 ℓ/m^3 (10 gal/10^3 aft^3).

- A three-stage valve tray absorber with two chimney trays and a superficial velocity of 3 m/sec (10 ft/sec), an L/G ratio of 0.4 ℓ/m^3 (3 gal/10^3 aft^3), and a pressure drop, including the mist eliminator of 2.9 kPa (11.6 inches H_2O).

- Stoichiometry is 2.0 mols of Na_2CO_3 to 1.0 mol of Na removed in the sulfate purge stream.

Energy Requirements: For base-case conditions, reheat of the cleaned gas requires 36.2 x 10^3 kg/hr (79,700 lb/hr) of 243°C (470°F) steam at 3.55 x 10^3 kPa absolute pressure (500 psig), equivalent to about 15.1 x 10^6 kcal/hr. In the purge area, the sulfate dryer uses 0.7 x 10^3 kg/hr (1,600 lb/hr) of 243°C (470°F) saturated steam and the sulfate crystallizer uses 10.3 x 10^3 kg/hr (22,800 lb/hr) of 121°C (250°F) steam at 2.10 x 10^3 kPa (15 psig) absolute pressure. Total consumption in the purge area is equivalent to 6.1 x 10^6 kcal/hr. Steam consumption for the first-effect evaporators and the stripper in the regeneration area is

70.4 kg/hr (155,100 lb/hr) of 121°C (250°F) saturated steam, equivalent to 40.5 x 10^6 kcal/hr. The acid plant produces 3.6 x 10^3 kg/hr (7,900 lb/hr) of 121°C (250°F) saturated steam, equivalent to 1.98 x 10^6 kcal/hr. This is taken into account as a heat credit for the process.

The electrical power demand for the base case is about 11,300 kW or 2.3% of the rated capacity of a 500-MW power plant. For 6,000 hours of operation, the annual electrical energy consumption is 68.1 x 10^6 kWh.

The total equivalent energy consumption for the base case is approximately 92.35 x 10^6 kcal/hr or 8.1% of the input energy required for the 500-MW power unit.

By-Product Management: Electrostatic precipitators remove 99.2% of the fly ash in the flue gas. Most of the remainder is removed in the chloride scrubber; therefore, it is assumed that the by-products contain little, if any, fly ash. (Fly ash emission from oil-fired units does not exceed the EPA particulate emission standard and fly ash collection facilities are not included in oil-fired power plant design.) Projected mass flow rates of by-product streams for the base case are shown below.

Component	Kilograms per Hour	Pounds per Hour
Sulfuric acid		
98% H_2SO_4	15,100	33,300
Sodium sulfate		
Na_2SO_4	680	1,500
Na_2SO_3	450	1,000
Other Na salts	50	100
Total	1,180	2,600
Chloride purge		
HCl	290	630
SO_3	40	80
Ash	150	330

The by-product sulfuric acid is stored in a tank of 30-day capacity until sold. For the purposes of this study the sodium sulfate is assumed to be sold.

Carbon Adsorption Process

Process Description: The Bergbau-Forschung/Foster Wheeler process uses dry carbon adsorption followed by thermal regeneration to reactivate the adsorbent. A four-train adsorption system, each with a two-stage adsorber, is used. The adsorber stages are separate but adjacent vessels, each containing louvered moving beds of activated char.

Water is mixed with the flue gas to obtain an adsorber inlet temperature of 121°C (250°F). The gas then moves in horizontal crossflow through a vertical char bed where SO_2, SO_3, H_2O, and O_2 are adsorbed. Although some NO_x removal has been reported, the process is not designed for NO_x removal. The char bed also filters out some residual fly ash in the flue gas, which, leaving the first stage, is divided into upper and lower flow streams. The upper flows to the first-stage adsorber ID fan and then to the stack plenum, the lower, to the second-stage adsorber and a separate ID fan. Because of exothermic reactions the gases are exhausted about 14°C above inlet temperature, eliminating need for reheat.

The char moves by gravity in the adsorber beds. The velocity ranges from 0.3 to 0.91 m/hr (1 to 3 ft/hr). Char leaving the adsorbers is saturated with H_2SO_4 formed by reaction between SO_2, O_2 and H_2O. The saturated char is screened to remove fly ash and char fines and transported to the regeneration area. Saturated char and hot sand are gravity-fed into the regenerator vessel which operates with a reducing atmosphere. Sand enters the regenerator at 815°C (1500°F) and char enters at the flue gas temperature. The saturated char is heated to 650°C (1200°F) to liberate the adsorbed gases and return the char to its activated state. Some char reacts with oxygen in the gas and is chemically consumed. Sand and regenerated char leave the regenerator at 650°C (1200°F) and are separated by screens. The char is then routed to the char cooling area and the sand is heated and returned to the regenerator.

In the char cooling area, the char is cooled in two stages. First-stage cooling by indirect heat exchange with cleaned flue gas lowers char temperature to 200°C. The temperature is reduced to 120°C in the second stage by direct-contact water spray. The steam produced is unsuitable for use by the steam plant. It can, however, be used to preheat boiler feedwater. Makeup char is added to replace losses. The char is then transported back to the top of the adsorbers. The sand is simultaneously heated and pneumatically elevated to the regenerator by combustion gases. The heater gas may be routed either to the stack or to the adsorbers, depending on whether or not it contains SO_2.

SO_2 reduction is accomplished by the Foster Wheeler Resox process. The SO_2-rich gas is reacted with rice-sized anthracite coal to produce gaseous elemental sulfur. SO_2 and air are injected at the bottom of a Resox reactor containing downward-moving coal. The operating temperature is 800° to 850°C in the main reaction zone. In addition to sulfur vapor, some H_2S, COS, and CS_2 are formed in the reactor. The product gases, flowing out the top of the reactor at 350° to 400°C, are routed to a shell and tube exchanger to condense the sulfur product and produce steam. Recovered sulfur is then collected and pumped to storage. Noncondensed gases from the sulfur condenser are pumped to an incinerator to oxidize the remaining sulfur compounds back to SO_2. After incineration, the gases are cooled in a steam-producing waste heat boiler and recycled back to the SO_2 adsorbers.

Specific Process Premises:

- The adsorbers have superficial velocities of 7.6 m/sec (25 ft/ sec) in channels between the char beds and 0.3 m/sec (1 ft/ sec) approaching the char beds. Pressure drop is 0.5 kPa (2.0 inches H_2O) in the first stage and 0.3 kPa (1 inch H_2O) in the second stage.

- The saturated char SO_2 loading is 7.5 kg of sulfur to 100 kg of char.

Energy Requirements: For base-case conditions, steam consumption in the SO_2 reduction and sulfur storage area totals 1,300 kg/hr (2,860 lb/hr) of 148°C (298°F) steam at 450 kPa absolute pressure (50.3 psig) equivalent to about 0.66 x 10^6 kcal/hr. In addition to the steam consumed in this area, steam is produced in some areas. This, along with heat used to preheat boiler feedwater, is taken into account as a heat credit. The waste heat boiler produces 4,600 kg/hr (10,100 lb/hr) of 121°C (250°F) steam at 210 kPa absolute pressure (15.1 psig) equivalent to about 2.79 x 10^6 kcal/hr. The sulfur production unit produces 17,400 kg/hr (38,300 lb/hr) of 121°C (250°F) steam at 210 kPa absolute pressure (15.1 psig) equivalent to about 10.58 x 10^6 kcal/hr. The waste heat from the char cooler is used to preheat boiler feedwater from 38°C (100°F) to 88°C (190°F). This is equivalent to about 1.45 x 10^6 kcal/hr.

The electrical power demand for the base-case carbon adsorption process is about 2,650 kW or 0.53% of the rated capacity of a 500-MW power plant. For 6,000 hours of operation the annual electrical energy consumption is 15.92 x 10^6 kWh.

Fuel oil is used as a source of energy at an annual rate of 15.48 x 10^6 liters (4.09 x 10^6 gallons), equivalent to about 22.54 x 10^6 kcal/hr, to heat sand. For SO_2 reduction the annual consumption is 3.12 x 10^6 liters (0.82 x 10^6 gallons). The total fuel oil consumption for energy is equivalent to 27.10 x 10^6 kcal/hr. There is no reheat steam required for this process.

The total equivalent energy consumption for the base case is approximately 19.02 x 10^6 kcal/hr or 1.68% of the input energy required for the 500-MW power unit.

By-Product Management: ESP units remove 99.2% of the fly ash from the flue gas; some additional fly ash is removed with the char fines. The char fines and Resox waste may be burned in the boiler. These two materials have been assigned a monetary value based on their heating values and are treated as saleable by-products in the determination of net annual revenue requirements. (Fly ash emission from oil-fired units does not exceed the EPA particulate emission standard and fly ash collection facilities are not included in oil-fired power plant design.) Projected mass flow rates of by-product streams are shown below.

Component	Kilograms per Hour	Pounds per Hour
Sulfur	4,900	10,900
Char fines	1,660	3,660
Fly ash	150	330
Resox waste	2,600	5,700
Total	4,410	9,690

RESULTS

Process and economic evaluation results for the nine FGD processes evaluated are summarized in Tables 12.1 through 12.10. Figures 12.1 and 12.2 show the economic results geographically, illustrating the accuracy ranges projected for each process. Capital costs are in mid-1979 U.S. dollars. Revenue requirements are in mid-1980 U.S. dollars.

Because the evaluations are based on generalized premises and limited actual cost data are available for some of the processes, no attempt is made to compare the projected costs with costs for actual site-specific installations.

Table 12.1: Limestone Sludge Process Material, Energy, and Cost Summary

Fuel Percent Sulfur	Coal 0.8	Coal 1.4	Coal 2.0	Coal 3.5	Lignite 0.5	Oil 2.5
Raw materials, kg/hr						
Limestone	6,000	8,900	14,000	25,500	4,300	12,000
End product						
Sludge						
m³/hr	13.3	19.77	31.35	56.93	9.70	26.73
ft³/hr	470.0	698.0	1,107.1	2,010.4	342.4	943.9
kg/hr	17,561	26,079	41,364	75,112	12,792	35,265
lb/hr	38,715	57,493	91,190	165,590	28,200	77,744
wt % solids	40	40	40	40	40	40
Energy						
Reheat, kcal/hr	18.06 M	17.45 M	17.53 M	17.62 M	18.52 M	14.67 M
Electricity, kcal/hr	16.98 M	16.66 M	17.14 M	18.13 M	17.27 M	14.40 M
Total						
kcal/hr	35.04 M	34.11 M	34.67 M	35.75 M	35.79 M	29.07 M
Btu/hr	139.2 M	135.6 M	137.5 M	141.9 M	141.9 M	115.5 M
% input	3.27	3.18	3.23	3.33	3.34	2.71
Land						
Process, ha	3.7	3.7	3.7	3.7	3.7	3.7
Waste disposal, ha	38.6	51.8	74.6	123.0	29.8	65.8
Total						
ha	42.3	55.5	78.3	126.7	33.5	69.5
acres	104.5	137.1	193.5	313.1	82.8	171.7
Costs*, $						
Capital investment	39,995,000	41,634,000	45,516,000	53,083,000	39,036,000	40,643,000
Low range	34,112,000	35,526,000	38,863,000	45,376,000	33,285,000	34,705,000
High range	47,838,000	49,778,000	54,387,000	63,358,000	46,704,000	48,559,000
Annual revenue requirements	11,151,700	11,539,200	12,574,200	14,638,000	10,945,400	11,385,400
Low range	9,933,800	10,279,000	11,206,400	13,064,500	9,752,100	10,168,000
High range	12,775,200	13,219,500	14,397,200	16,736,100	12,536,500	13,008,300

*Accuracy range: +20 to -15%.

Source: EPA-600/7-80-142

Table 12.2: Limestone Sludge Process Size Variations, Material, Energy, and Cost Summary

	200	500	700	1,000
MW				
Fuel	Coal	Coal	Coal	Coal
Percent sulfur	3.5	3.5	3.5	3.5
Raw materials, kg/hr				
Limestone	10,400	25,500	35,300	49,300
End product				
Sludge				
m³/hr	23.28	56.93	78.82	110.07
ft³/hr	822.0	2,010.4	2,783.1	3,886.7
kg/hr	30,713	75,112	103,983	145,216
lb/hr	67,710	165,590	229,240	320,140
wt % solids	40	40	40	40
Energy				
Reheat, kcal/hr	7.21 M	17.62 M	24.39 M	34.07 M
Electricity, kcal/hr	7.72 M	18.13 M	24.74 M	33.68 M
Total				
kcal/hr	14.93 M	35.75 M	49.13 M	67.75 M
Btu/hr	59.2 M	141.79 M	194.95 M	268.83 M
% input	3.39	3.33	3.30	3.26
Land				
Process, ha	3.1	3.7	3.8	4.2
Waste disposal, ha	61.0	123.0	159.0	209.0
Total				
ha	64.1	126.7	162.8	213.2
acres	158.3	313.1	402.3	526.9
Costs*, $				
Capital investment	26,302,000	53,083,000	70,967,000	93,941,000
Low range	22,481,000	45,376,000	60,657,000	80,296,000
High range	31,397,000	63,358,000	84,713,000	112,135,000
Annual revenue requirements	7,029,100	14,638,000	19,734,300	26,384,700
Low range	6,260,400	13,064,500	17,619,200	23,573,800
High range	8,054,000	16,736,100	22,554,300	30,132,600

*Accuracy range: +20 to -15%.

Source: EPA-600/7-80-142

Table 12.3: Lime Sludge Process Material, Energy, and Cost Summary

Fuel Percent Sulfur	Coal 0.8	Coal 1.4	Coal 2.0	Coal 3.5	Lignite 0.5	Oil 2.5
Raw materials, kg/hr						
Lime	2,400	3,500	5,600	10,200	1,700	4,800
End product						
Sludge						
m³/hr	10.22	15.17	24.07	43.70	7.44	20.52
ft³/hr	360.3	535.0	848.6	154.1	262.4	723.4
kg/hr	13,394	19,891	31,549	57,290	9,756	26,898
lb/hr	29,529	43,851	69,553	126,300	21,509	59,298
wt % solids	40	40	40	40	40	40
Energy						
Reheat, kcal/hr	18.00 M	17.39 M	17.47 M	17.56 M	18.46 M	14.62 M
Electricity, kcal/hr	16.82 M	16.34 M	16.54 M	16.89 M	17.21 M	13.88 M
Total						
kcal/hr	34.82 M	33.73 M	34.01 M	34.45 M	35.67 M	28.50 M
Btu/hr	138.2 M	133.9 M	135.0 M	136.7 M	141.5 M	113.1 M
% input	3.25	3.14	3.17	3.21	3.33	2.51
Land						
Process, ha	2.5	2.5	2.5	2.5	2.5	2.5
Waste disposal, ha	32.5	44.8	63.5	104.0	25.5	54.9
Total						
ha	35.0	47.3	66.0	106.5	28.0	57.4
acres	86.5	116.9	163.1	262.4	69.2	141.8
Costs*, $						
Capital investment	37,795,000	38,912,000	41,952,000	47,743,000	37,165,000	37,286,000
Low range	32,244,000	33,216,000	35,841,000	40,853,000	31,695,000	31,855,000
High range	45,196,000	46,507,000	50,099,000	56,932,000	44,458,000	44,526,000
Annual revenue requirements	11,064,200	11,498,200	12,644,700	14,972,100	10,836,300	11,387,700
Low range	9,912,000	10,320,100	11,386,000	13,562,700	9,698,200	10,271,300
High range	12,600,400	13,069,200	14,323,000	16,851,500	12,353,800	12,876,100

*Accuracy range: +20 to −15%.

Source: EPA-600/7-80-142

Table 12.4: Double-Alkali Sludge Process Material, Energy, and Cost Summary

Fuel Percent Sulfur	Coal 0.8	Coal 1.4	Coal 2.0	Coal 3.5	Lignite 0.5	Oil 2.5
Raw materials, kg/hr						
Lime	2,277	3,381	5,363	9,737	1,659	4,572
Soda ash	186	277	439	797	136	374
End product						
Sludge						
m³/hr	6.98	10.37	16.45	29.87	5.09	14.03
ft³/hr	246.6	366.2	580.9	1,054.9	179.6	495.3
kg/hr	9,075	13,476	21,375	38,815	6,610	18,223
lb/hr	20,006	29,710	47,123	85,570	14,573	40,175
wt % solids	55	55	55	55	55	55
Energy						
Reheat, kcal/hr	18.06 M	17.45 M	17.53 M	17.62 M	18.52 M	14.67 M
Electricity, kcal/hr	8.08 M	7.99 M	8.33 M	9.03 M	8.18 M	7.02 M
Total						
kcal/hr	26.14 M	25.44 M	25.86 M	26.65 M	26.70 M	21.69 M
Btu/hr	103.7 M	101 M	102.6 M	105.8 M	106 M	86.1 M
% input	2.48	2.41	2.45	2.52	2.54	1.91
Land						
Process, ha	4.9	4.9	4.9	4.9	4.9	4.9
Waste disposal, ha	33.0	44.0	62.0	99.0	31.1	48.0
Total						
ha	37.9	48.9	66.9	103.9	36.0	52.9
acres	150.4	194.0	265.5	412.3	89.1	209.9
Costs*, $						
Capital investment	40,537,000	42,179,000	46,045,000	53,231,000	39,595,000	40,659,000
Low range	32,581,000	33,926,000	37,075,000	43,004,000	31,817,000	32,733,000
High range	52,472,000	54,559,000	59,500,000	68,662,000	51,261,000	52,546,000
Annual revenue requirements	11,173,000	11,793,600	13,080,000	16,010,600	10,828,000	11,879,500
Low range	9,528,500	10,091,700	11,258,600	14,359,800	9,217,700	10,251,200
High range	13,639,700	14,346,600	15,811,900	19,169,400	13,234,300	14,321,700

*Accuracy range: +30 to -20%.

Source: EPA-600/7-80-142

Table 12.5: Seawater Process Material, Energy, and Cost Summary

Fuel Percent sulfur	Coal 1.4	Coal 0.8	Lignite 0.5	Oil 2.5	Low Reheat Coal 1.4
End product					
Seawater					
ℓ/hr	71,385,400	51,526,200	35,692,700	83,214,800	71,385,400
kg/hr	72,124,300	52,041,400	36,062,200	84,071,900	72,124,300
lb/hr	158,864,200	114,731,700	79,432,100	185,346,900	158,864,200
Energy					
Reheat, kcal/hr	28.70 M	29.69 M	30.46 M	24.13 M	14.46 M
Electricity, kcal/hr	15.91 M	14.89 M	13.88 M	15.47 M	15.91 M
Total					
kcal/hr	44.61 M	44.58 M	44.34 M	39.60 M	30.37 M
Btu/hr	177.0 M	176.9 M	175.9 M	157.1 M	120.5 M
% input	4.22	4.22	4.21	3.49	2.82
Land					
Process					
ha	2.4	2.4	2.4	2.4	2.4
acres	6	6	6	6	6
Costs*, $					
Capital investment	30,048,000	29,590,000	29,068,000	27,937,000	28,582,000
Low range	21,157,000	20,838,000	20,471,000	19,691,000	20,106,000
High range	44,868,000	44,177,000	43,397,000	41,682,000	42,707,000
Annual revenue requirements	8,707,100	8,667,100	8,516,800	8,575,600	7,837,700
Low range	6,972,000	6,960,500	6,840,700	6,968,800	6,185,700
High range	11,597,700	11,511,800	11,310,300	11,253,400	10,590,900

*Accuracy range: +50 to −30%.

Source: EPA-600/7-80-142

Table 12.6: Lime Gypsum (Saarberg-Hölter) Process Material, Energy, and Cost Summary

Fuel Percent Sulfur	Coal 0.8	Coal 1.4	Coal 2.0	Coal 3.5	Lignite 0.5	Oil 2.5
Raw materials, kg/hr						
Lime	2,255	3,344	5,297	9,625	1,634	4,525
Formic acid	1.8	2.7	4.3	7.9	1.3	3.7
Flocculant	0.18	0.26	0.42	0.76	0.13	0.36
Nalco	0.02	0.03	0.04	0.08	0.01	0.04
End product						
Gypsum						
m^3/hr	6.06	9.00	14.27	25.92	4.41	12.17
ft^3/hr	214.1	317.6	503.9	915.3	155.7	429.8
kg/hr	8,263	12,260	19,445	35,320	6,008	16,586
lb/hr	18,200	27,000	42,830	77,800	13,230	36,530
wt % solids	80	80	80	80	80	80
Energy						
Reheat, kcal/hr	18.14 M	17.53 M	17.61 M	17.70 M	18.61 M	14.74 M
Electricity, kcal/hr	12.59 M	13.76 M	16.34 M	22.00 M	12.13 M	13.78 M
Total						
kcal/hr	30.73 M	31.29 M	33.95 M	39.70 M	30.74 M	28.52 M
Btu/hr	121.94 M	124.16 M	134.71 M	157.53 M	121.98 M	113.17 M
% input	2.89	2.93	3.17	3.67	2.89	2.51
Land						
Process, ha	4.46	4.46	4.46	4.46	4.46	4.46
By-product storage, ha	0.4	0.4	0.4	0.4	0.4	0.4
Total						
ha	4.86	4.86	4.86	4.86	4.86	4.86
acres	12	12	12	12	12	12
Costs*, $						
Capital investment	35,481,000	36,472,000	39,173,000	44,024,000	34,905,000	34,802,000
Low range	24,988,000	25,701,000	27,631,000	31,114,000	24,574,000	24,557,000
High range	52,970,000	54,424,000	58,407,000	65,541,000	52,124,000	51,878,000
Annual revenue requirements	9,825,900	10,303,700	11,443,300	13,706,300	9,571,100	10,347,600
Low range	7,791,500	8,214,400	9,201,300	11,192,900	7,568,900	8,362,400
High range	13,216,300	13,785,800	15,179,800	17,895,300	12,908,000	13,656,200

*Accuracy range: +50 to –30%.

Source: EPA-600/7-80-142

Table 12.7: Jet-Bubbling Limestone (Chiyoda Thoroughbred 121) Process Material, Energy, and Cost Summary

Fuel Percent Sulfur	Coal 0.8	Coal 1.4	Coal 2.0	Coal 3.5	Lignite 0.5	Oil 2.5
Raw materials, kg/hr						
Limestone	4,067	6,048	9,586	17,400	2,964	8,180
End product						
Gypsum						
m³/hr	6.19	9.18	14.57	26.45	4.50	12.42
ft³/hr	218.42	324.30	514.42	934.14	159.06	438.54
kg/hr	8,421	12,503	19,834	36,015	6,132	16,908
lb/hr	18,565	27,565	43,725	79,400	13,520	37,275
wt % solids	80	80	80	80	80	80
Energy						
Reheat, kcal/hr	18.61 M	17.98 M	18.07 M	18.16 M	19.09 M	15.12 M
Electricity, kcal/hr	17.32 M	16.97 M	17.42 M	18.35 M	16.63 M	14.64 M
Total						
kcal/hr	35.93 M	34.95 M	35.49 M	36.51 M	35.72 M	29.76 M
Btu/hr	142.57 M	138.68 M	140.82 M	144.87 M	141.74 M	118.09 M
% input	3.35	3.26	3.31	3.40	3.34	2.62
Land						
Process, ha	4.46	4.46	4.46	4.46	4.46	4.46
By-product storage, ha	0.4	0.4	0.4	0.4	0.4	0.4
Total						
ha	4.86	4.86	4.86	4.86	4.86	4.86
acres	12	12	12	12	12	12
Costs*, $						
Capital investment	42,002,000	42,007,000	43,819,000	47,017,000	42,095,000	38,532,000
Low range	29,538,000	29,545,000	30,826,000	33,091,000	29,602,000	27,118,000
High range	62,774,000	62,776,000	65,474,000	70,227,000	62,917,000	57,556,000
Annual revenue requirements	10,948,500	10,919,500	11,360,700	12,160,100	10,998,700	10,219,200
Low range	8,524,700	8,496,000	8,831,900	9,446,000	8,568,900	8,003,400
High range	14,988,200	14,958,600	15,575,600	16,683,500	15,048,200	13,912,300

*Accuracy range: +50 to -30%

Source: EPA-600/7-80-142

Table 12.8: Magnesium Oxide Process Material, Energy, and Cost Summary

Fuel Percent Sulfur	Coal 0.8	Coal 1.4	Coal 2.0	Coal 3.5	Lignite 0.5	Oil 2.5
Raw materials						
MgO, kg/hr	51	76	119	218	38	103
Catalyst, ℓ/hr	0.1	0.1	0.2	0.3	0.1	0.1
Agricultural limestone, kg/hr	431	832	1,255	420	398	0
End product						
Sulfuric acid						
ℓ/hr	2,041	3,026	4,800	8,717	1,485	4,090
kg/hr	3,735	5,538	8,785	15,951	2,717	7,484
lb/hr	8,233	12,210	19,367	35,167	5,990	16,500
wt % H_2SO_4	100	100	100	100	100	100
Energy						
Reheat, kcal/hr	18.57 M	17.95 M	18.03 M	18.12 M	19.05 M	15.09 M
Electricity, kcal/hr	16.38 M	16.57 M	17.87 M	20.66 M	16.38 M	12.94 M
Oil, kcal/hr	7.90 M	11.73 M	18.60 M	33.78 M	5.75 M	15.86 M
Heat credit, kcal/hr	(0.84 M)	(1.25 M)	(1.98 M)	(3.60 M)	(0.61 M)	(1.69 M)
Total						
kcal/hr	42.01 M	45.00 M	52.52 M	68.96 M	40.57 M	42.20 M
Btu/hr	166.7 M	178.6 M	208.4 M	273.6 M	161.0 M	167.4 M
% input	3.89	4.14	4.81	6.26	3.58	3.72
Land						
Process						
ha	4.86	4.86	4.86	4.86	4.86	4.86
acres	12	12	12	12	12	12
Costs*, $						
Capital investment	48,926,000	52,043,000	58,108,000	68,434,000	47,115,000	44,591,000
Low range	39,252,000	41,757,000	46,632,000	54,934,000	37,798,000	35,799,000
High range	63,436,000	67,472,000	75,324,000	88,685,000	61,090,000	57,779,000
Annual revenue requirements	12,949,500	13,651,200	15,114,300	17,546,900	12,573,900	11,984,600
Low range	10,999,500	11,576,400	12,759,400	14,809,500	10,697,100	10,208,300
High range	15,874,400	16,763,000	18,592,700	21,652,700	15,389,100	14,646,600

* Accuracy range: +30 to –20%.

Source: EPA-600/7-80-142

Table 12.9: Sodium Sulfite (Wellman-Lord) Process Material, Energy, and Cost Summary

Fuel Percent sulfur	Coal 0.8	Coal 1.4	Coal 2.0	Coal 3.5	Lignite 0.5	Oil 2.5	Sulfur Production Coal 3.5
Raw materials							
Sodium carbonate, kg/hr	218	323	512	930	159	437	930
Catalyst, ℓ/hr	0.07	0.10	0.16	0.29	0.05	0.14	—
Agricultural limestone, kg/hr	431	832	1,254	420	398	—	420
Filter aid, kg/hr	2	4	6	11	2	5	11
Natural gas, m³/hr	—	—	—	—	—	—	2,311
End product							
100% sulfuric acid							
ℓ/hr	1,887	2,803	4,446	8,073	1,375	3,790	—
kg/hr	3,453	5,129	8,136	14,774	2,516	6,935	—
lb/hr	7,613	11,307	17,937	32,570	5,546	15,290	—
Sodium sulfate							
kg/hr	277	410	650	1,181	201	555	1,181
lb/hr	610	903	1,433	2,603	443	1,223	2,603
Sulfur							
kg/hr	—	—	—	—	—	—	4,802
lb/hr	—	—	—	—	—	—	10,857
Energy							
Reheat, kcal/hr	15.47 M	14.95 M	15.02 M	15.10 M	15.87 M	12.57 M	15.10 M
Process steam, kcal/hr	10.90 M	16.19 M	25.68 M	46.64 M	7.94 M	21.90 M	47.20 M
Electricity, kcal/hr	17.99 M	18.68 M	20.90 M	25.73 M	17.72 M	15.44 M	25.59 M
Natural gas, kcal/hr	—	—	—	—	—	—	20.56 M
Heat credit, kcal/hr	(0.46 M)	(0.69 M)	(1.09 M)	(1.98 M)	(0.34 M)	(0.93 M)	(2.79 M)
Total – kcal/hr	43.90 M	49.13 M	60.51 M	85.49 M	41.19 M	48.98 M	105.66 M
Btu/hr	174.2 M	194.9 M	240.1 M	339.2 M	163.4 M	194.4 M	419.3 M
% input	4.13	4.64	5.73	8.14	3.78	4.53	9.93
Land							
Process – ha	3.2	3.2	3.2	3.2	3.2	3.2	3.2
acres	8	8	8	8	8	8	8
Costs*, $							
Capital investment	46,836,000	50,307,000	56,939,000	68,722,000	44,837,000	44,215,000	71,342,000
Low range	37,581,000	40,372,000	45,706,000	55,187,000	35,975,000	35,504,000	57,343,000
High range	60,719,000	65,210,000	73,789,000	89,025,000	58,131,000	57,280,000	92,341,000
Annual revenue requirements	12,218,300	13,081,000	14,802,200	17,886,400	11,754,700	11,801,900	21,015,700
Low range	10,434,500	11,164,000	12,629,900	15,257,400	10,048,100	10,124,600	18,296,500
High range	14,893,900	15,956,500	18,060,200	21,829,800	14,314,400	14,317,700	24,094,700

* Accuracy range: +30 to -20%.

Source: EPA-600/7-80-142

Table 12.10: Carbon Adsorption Process Material, Energy, and Cost Summary

Fuel	Coal	Coal	Coal	Coal	Lignite	Oil
Percent Sulfur	0.8	1.4	2.0	3.5	0.5	2.5
Raw materials						
Sand, kg/hr	106	151	257	454	76	212
Char, kg/hr	544	816	1,300	2,359	408	1,104
Anthracite coal, kg/hr	1,179	1,754	1,255	5,035	862	2,359
End product						
Sulfur						
lb/hr	2,533	3,800	6,000	10,900	1,867	5,133
kg/hr	1,149	1,724	2,752	4,944	847	2,328
Char fines and Resox waste						
lb/hr	2,267	3,400	5,366	9,733	1,633	4,567
kg/hr	1,028	1,542	2,434	4,415	741	2,071
Energy						
Electricity, kcal/hr	3.86 M	4.08 M	3.83 M	6.01 M	3.74 M	3.99 M
Process steam, kcal/hr	0.15 M	0.23 M	0.36 M	0.66 M	0.11 M	0.31 M
Fuel oil, kcal/hr	6.33 M	9.41 M	14.92 M	27.10 M	4.61 M	12.72 M
Heat credit, kcal/hr	(3.46 M)	(5.14 M)	(8.16 M)	(14.82 M)	(2.52 M)	(6.96 M)
Total						
kcal/hr	6.88 M	8.58 M	10.95 M	18.95 M	5.94 M	10.06 M
Btu/hr	27.3 M	34.0 M	43.4 M	75.2 M	23.6 M	39.9 M
% input	0.61	0.76	0.97	1.68	0.52	0.89
Land						
Process						
ha	4.86	4.86	4.86	4.86	4.86	4.86
acres	12	12	12	12	12	12
Costs*, $						
Capital investment	51,195,000	54,220,000	60,834,000	73,511,000	49,485,000	53,730,000
Low range	33,109,000	39,533,000	44,350,000	53,617,000	36,153,000	39,104,000
High range	69,978,000	78,698,000	88,308,000	106,666,000	71,704,000	78,105,000
Annual revenue requirements	13,899,100	15,803,600	19,982,700	28,489,400	12,780,100	17,542,000
Low range	10,747,500	12,844,600	16,658,600	24,473,000	10,094,500	14,600,600
High range	18,174,700	20,735,200	25,522,900	35,183,500	17,256,100	22,445,100

*Accuracy range: +50 to -30%.

Source: EPA-600/7-80-142

Figure 12.1: Base-Case Unit Investment Range for Alternate Processes

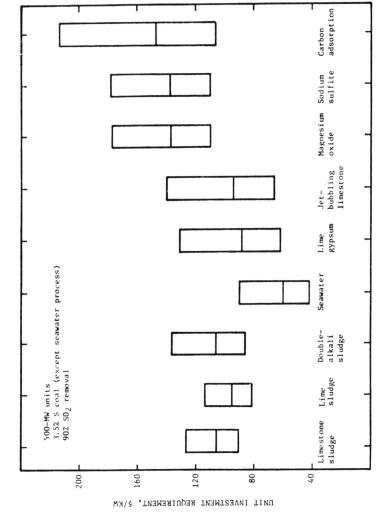

Source: EPA-600/7-80-142

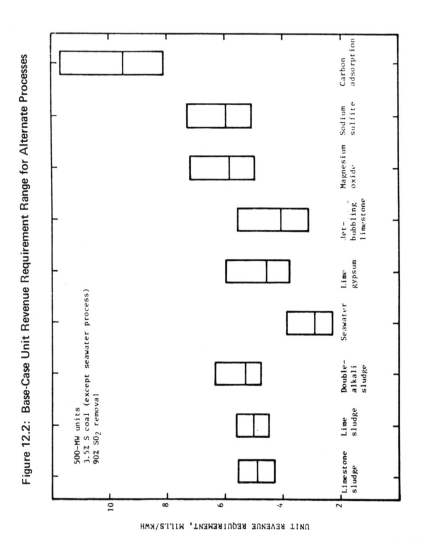

Figure 12.2: Base-Case Unit Revenue Requirement Range for Alternate Processes

Source: EPA-600/7-80-142

SOURCES UTILIZED

BuMines IC 8806
> *Citrate Process Demonstration Plant* by W.I. Nissen of Salt Lake City Metallurgy Research Center, Bureau of Mines and R.S. Madenburg of Morrison-Knudsen Co. for the Bureau of Mines, 1979.

BuMines RI 8374
> *Citrate-Process Pilot-Plant Operation at the Bunker Hill Company* by L. Crocker, D.A. Martin and W.I. Nissen of Salt Lake City Metallurgy Research Center, Bureau of Mines, 1979.

EPA-600/7-79-167a
> *Proceedings: Symposium on Flue Gas Desulfurization, Las Vegas, Nevada, March 1979,* Vol I, compiled by F.A. Ayer of Research Triangle Institute for the U.S. Environmental Protection Agency (referred to as Symposium—Vol 1).

EPA-600/7-79-167b
> *Proceedings: Symposium on Flue Gas Desulfurization, Las Vegas, Nevada, March 1979,* Vol II, compiled by F.A. Ayer of Research Triangle Institute for the U.S. Environmental Protection Agency (referred to as Symposium—Vol 2).

EPA-600/7-79-177
> *Definitive SO_x Control Process Evaluations: Limestone, Double Alkali, and Citrate FGD Processes* by S.V. Tomlinson, F.M. Kennedy, F.A. Sudhoff and R.L. Torstrick of TVA, Office of Power Emission Control Development Projects for the U.S. Environmental Protection Agency, August 1979.

EPA-600/7-79-224
> *Adipic Acid Degradation Mechanism in Aqueous FGD Systems* by F.B. Meserole, D.L. Lewis, A.W. Nichols and G. Rochelle of Radian Corporation for the U.S. Environmental Protection Agency, September 1979.

EPA-600/7-80-030
> *Survey of Dry SO_2 Control Systems* by G.M. Blythe, J.C. Dickerman and M.E. Kelly of Radian Corporation for the U.S. Environmental Protection Agency, February 1980.

EPRI CS-1381

Comparative Economics of Advanced Regenerable Flue Gas Desulfuriza-tion Processes prepared by M.R. Beychok for Electric Power Research Institute, March 1980.

EPRI CS-1428

Economic and Design Factors for Flue Gas Desulfurization Technology prepared by Bechtel National, Inc. for Electric Power Research Institute, April 1980.

NTIS PB-295 001

Flue Gas Desulfurization Pilot Study—Phase I—Survey of Major Installa-tions—Summary of Survey Reports on Flue Gas Desulfurization Processes prepared by the Flue Gas Desulfurization Study Group of the NATO Committee on the Challenges of Modern Society, F. Princiotta of the U.S. Environmental Protection Agency, Chairman, January 1979.

The following eight reports are appendices to the above NATO-CCMS study.

NTIS PB-295 002

Appendix 95-A—Limestone/Sludge Flue Gas Desulfurization Process prepared by F. Princiotta of the U.S. Environmental Protection Agency and R.W. Gerstle and E. Schindler of PEDCo Environmental, January 1979.

NTIS PB-295 003

Appendix 95-B—Lime/Sludge Flue Gas Desulfurization Process prepared by N. Haug of Umweltbundesamt and G. Oelert and G. Weisser of Batelle-Institut e.V., January 1979.

NTIS PB-295 005

Appendix 95-D—Sea Water Scrubbing Flue Gas Desulfurization Process prepared by R.I. Hagen and H. Kolderup of Foundation of Scientific and Industrial Research, Norway, January 1979.

NTIS PB-295 009

Appendix 95-H—Flue Gas Desulfurization by Scrubbing with Dilute Sulfuric Acid prepared by F. Princiotta of the U.S. Environmental Protec-tion Agency and R.W. Gerstle and E. Schindler of PEDCo Environmental, January 1979.

NTIS PB-295 010

Appendix 95-I—Magnesium Oxide Flue Gas Desulfurization Process prepared by F. Princiotta of the U.S. Environmental Protection Agency and R.W. Gerstle and E. Schindler of PEDCo Environmental, Jan. 1979.

NTIS PB-295 011

Appendix 95-J—Sodium Sulfite Scrubbing Flue Gas Desulfurization Proc-ess prepared by F. Princiotta of the U.S. Environmental Protection Agency, R.W. Gerstle and E. Schindler of PEDCo Environmental, Jan. 1979.

NTIS PB-295 012

Appendix 95-K—Carbon-Adsorption Flue Gas Desulfurization Process prepared by N. Haug of Umweltbundesamt and G. Oelert and G. Weisser of Battelle-Institut e.V., January 1979.

NTIS PB-295 013

Appendix 95-L—Copper Oxide Flue Gas Desulfurization Process prepared by F. Princiotta of the U.S. Environmental Protection Agency and R.W. Gerstle and E. Schindler of PEDCo Environmental, January 1979.